Springer Proceedings in Earth and Environmental Sciences

T0172097

Series Editor

Natalia S. Bezaeva, The Moscow Area, Russia

The series Springer Proceedings in Earth and Environmental Sciences publishes proceedings from scholarly meetings and workshops on all topics related to Environmental and Earth Sciences and related sciences. This series constitutes a comprehensive up-to-date source of reference on a field or subfield of relevance in Earth and Environmental Sciences. In addition to an overall evaluation of the interest, scientific quality, and timeliness of each proposal at the hands of the publisher, individual contributions are all refereed to the high quality standards of leading journals in the field. Thus, this series provides the research community with well-edited, authoritative reports on developments in the most exciting areas of environmental sciences, earth sciences and related fields.

More information about this series at http://www.springer.com/series/16067

Alojz Kopáčik · Peter Kyrinovič ·
Ján Erdélyi · Rinaldo Paar ·
Ante Marendić

Editors

Contributions to International Conferences on Engineering Surveying

8th INGEO International Conference
on Engineering Surveying and
4th SIG Symposium on Engineering Geodesy

INGEO&SIG2020, Dubrovnik, Croatia

 Springer

Editors
Alojz Kopáčik
Department of Surveying
Slovak University of Technology
in Bratislava
Bratislava, Slovakia

Ján Erdélyi
Department of Surveying
Slovak University of Technology
in Bratislava
Bratislava, Slovakia

Ante Marendić
Faculty of Geodesy
University of Zagreb
Zagreb, Croatia

Peter Kyrinovič
Department of Surveying
Slovak University of Technology
in Bratislava
Bratislava, Slovakia

Rinaldo Paar
Faculty of Geodesy
University of Zagreb
Zagreb, Croatia

ISSN 2524-342X   ISSN 2524-3438 (electronic)
Springer Proceedings in Earth and Environmental Sciences
ISBN 978-3-030-51955-1   ISBN 978-3-030-51953-7 (eBook)
https://doi.org/10.1007/978-3-030-51953-7

This Springer imprint is published by the registered company Springer Nature Switzerland AG
The registered company address is: Gewerbestrasse 11, 6330 Cham, Switzerland

Organization

Scientific Committee Chair

Alojz Kopáčik, Slovakia

Scientific Committee

Branko Božić, Serbia
Jiří Bureš, Czech Republic
Alessandro Capra, Italy
Ján Erdélyi, Slovakia
Vassilis Gikas, Greece
Maria Joao Henriques, Portugal
Boštjan Kovačič, Slovenia
Peter Kyrinovič, Slovakia
Ante Marendić, Croatia
Hans Neuner, Austria
Rinaldo Paar, Croatia
Miodrag Roić, Croatia
Martin Štroner, Czech Republic
Thomas Wunderlich, Germany

Organizing Committee Chair

Rinaldo Paar, Croatia

Organizing Committee

Almin Đapo, Croatia
Ján Erdélyi, Slovakia
Zdravko Kapović, Croatia
Peter Kyrinovič, Slovakia
Ante Marendić, Croatia
Marko Pavasović, Croatia
Mirko Živković, Croatia

Preface

The conference brings together experts, researchers, and users from the field of engineering surveying, monitoring of engineering structures, laser scanning, data processing, including instrument calibration and testing. Participants are coming from universities, governmental organizations, and engineering companies to discuss recent scientific and technical advancements and to study new technology and applications. The conference discussion focuses on current topics, mainly to the engineering of large investments, BIM, quality assurance, deformation measurement, as well as cultural heritage documentation.

INGEO2020 is the 8th event in the series of engineering surveying conferences organized by the Department of Surveying at the Slovak University of Technology in Bratislava. **SIG2020** is the 4th Symposium on Engineering Geodesy organized by the Croatian Geodetic Society and with the support of the University of Zagreb Faculty of Geodesy.

The conference is supported by FIG Commission 6—Engineering Surveying and their Working Groups. The conference program includes 32 regular papers and 11 posters arranged in 8 technical sessions and 1 poster session. From all submissions were 36 sent to a peer review process, and finally, 26 papers were selected for publishing in Springer series Contributions to International Conferences on Engineering Surveying, which belongs to the series of Proceedings in Earth and Environmental Sciences.

The joint conference **INGEO&SIG2020** was planned to be held in the Hotel Palace, Dubrovnik, Croatia, in spring months. Dubrovnik has been selected as the location of the joint conference since the Dubrovnik area is very interesting for the field of engineering geodesy. Two large and exciting projects have been and are performing in the Dubrovnik at the moment; the monitoring project of the vertical displacements of the Old City of Dubrovnik for the last 35 years—since Dubrovnik is located on a seismically active area and the construction of Pelješac Bridge which is currently the largest infrastructure project in Croatia (bridge will be 2404 m long, 55 m above sea level, with pilots 130 m in sea depth).

Due to the actual situation of the COVID-19 virus spreading around the world, but especially in Croatia, the organizers decided to hold the conference on-line. The new actual date of the meeting was set to October 22–23, 2020.

We would like to express our deepest gratitude to all authors of papers for their effort related to the preparation of papers and reviewers for the constructive remarks and discussion during the review process.

Bratislava, Slovakia

Alojz Kopáčik
On behalf of the Scientific Committee
of **INGEO&SIG2020**

Contents

BIM and Engineering of Large Investments

Misalignment—Can 3D BIM Overrule Professional Setting-out According to Plane and Height?

Thomas Wunderlich$^{(\boxtimes)}$

Technische Universität München, D 80290 Munich, Germany
th.wunderlich@tum.de

Abstract. Building Information Modeling (BIM) is a superb initiative to improve planning, construction and operation of structures and hence to avoid multiple databases and design errors by cooperative and standardized construction planning using Industry Foundation Classes (IFC). This holds for structural engineering with the main dimension of buildings in height. In contrast, long infrastructure projects of civil engineering will encounter considerable problems as the Cartesian 3D coordinate system and a scale fixed to 1 associated with BIM must collide with the necessary considerations for map projection and height of our well-founded geodetic representation of the real world, resulting in misalignment. In particular, in tunneling, where we have to observe tightest tolerances, the problem currently is left to the site surveyor, which should be aware of the difficulty and how to cope with it. This investigation will expose the affair along theoretical derivations as well as practical examples and present the state-of-the-art of how the BIM community is facing the problem and tries to overcome it, once initiated by the author, who will also give his latest considerations and advices. The current practice of ignoring the incompatibility of the two approaches for infrastructure projects with considerable extension in length and trying to get along with subdividing into smaller sections or introducing covert interim co-ordinate systems for setting-out will not only corrupt the splendid general concept of BIM but the more place the surveyor in danger of sliding into a legal dispute.

Keywords: BIM · Infrastructure projects · Setting-out · Coordinate system incompatibility

1 The BIM Process

1.1 Digital, Collaborative, Consistent, Up-to-date and Multidimensional

Building Information Modeling (BIM) is a process that involves creating and using a consistent and continuously updated 3D model to inform and communicate construction project decisions. Design, visualization, simulation, and collaboration enabled by BIM solutions provide greater clarity for all stakeholders across the project lifecycle from planning to building and operation [1]. Extra benefit is expected from extending the 3D model to 4D (schedule), 5D (budget), 6D (facility management), 7D (sustainability) to 8D (safety). Still, the core and foundation will always be the 3D geometry.

A. Kopáčik et al. (Eds.): *Contributions to International Conferences on Engineering Surveying*, SPEES, pp. 3–12, 2021. https://doi.org/10.1007/978-3-030-51953-7_1

1.2 BIM Geometry and Associated Coordinate System

In contrast to CAD systems, which as a rule rest upon line and surface models, BIM is aiming at volume models. The dominating realization is based on Constructive Solid Geometry (CSG) with construction rules for single solids, Boolean operators to merge, difference or intersect several basic objects and profile based generation procedures by extrusion and rotation. Moreover, the implicitly defined building component objects are usually linked with a parametric description (translations, rotations) to place the object in a desired spatial attitude at a fixed point in the coordinate frame [2].

In the planning phase, this coordinate frame is a 3D Cartesian System with scale 1, origin (0,0,0), an arbitrary orientation and limited dimensions. If it has to be used also for setting-out in the building phase, it is evident that pure shifts and rotations will not be sufficient to make the building model consistent with the geodetic projection system, be it the national or a specific one tailored to the project.

All geodata are referenced to a geodetic datum and have to account for the earth's curvature, vertical topography and projection distortions along mapping to the plane. This essential and traditional reduction of 3D space to 2.5 dimensions of plane projection coordinates attributed with a height of not only geometric but the more physical meaning must collide with an artificial Cartesian 3D flat earth world with scale 1.

2 BIM and Geodesy and Geoinformatics

2.1 General Considerations

BIM brings along enormous opportunities for our profession, not only for engineering surveyors, but also for experts in geoinformatics and photogrammetry. The process will create new geodetic tasks and can intensify the cooperation with civil engineers and architects in a prosperous way. New tasks on site are e.g. construction progress monitoring involving a lot of scanning and imaging by mobile mapping systems or advanced as-built surveying for building-in-stock projects. Some experienced professionals with early training in BIM could even become BIM coordinators. In science, the most urgent development concerns the integration of BIM and GIS data structures [3] to enable planning BIM in digital terrain and city models—a key task of geoinformatics. For engineering geodesy, the main missions are the introduction of deformation monitoring analyses and visualizations to BIM and, most of all, solving the inherent misalignment problem of the BIM coordinate system with respect to setting-out in the field [4].

As a matter of fact, the collision of the concepts only becomes apparent, if the construction project extends considerably in length as e.g. railway lines and base tunnels.

2.2 BIM Coordinate System

Before BIM adopted the concept, object oriented construction was long practiced in mechanical engineering (ME), where it proved extraordinarily useful for automation of production, in particular with parts prefabricated and delivered by other parties [5]. Defining and standardizing components by CAD to be manufactured by CAM and assembled according to a virtual flow chart became the backbone of successful industrial production, e.g. in vehicle production. The enormous advantages of designing parts and simulating assembly lines in a distributed and location-independent manner, but on a common and consistent model, to end up with a competitive product, proved convincing. However, the first crucial difference of ME to building is that ME mainly generates big series of identical products whereas architects and civil engineers primarily create unica. This provides a decisive gap in profitability and the late determination to start a complete digital and collaborative method like BIM may be charged to this. The second distinction originates in scale: while ME usually develops final products of rather limited dimension and dedicated to be transportable or mobile, there was no reason to use a coordinate system other than rectangular and fixed to scale 1.

When we look at vital traditions like the creative freehand sketches and model making of architects and the well-established planning procedures of civil engineers [6], it becomes readily understandable that the main driver to still go for a common and consistent digital model and workflow only was to overcome existing shortcomings, mainly collisions due to different plans of crafts and all too frequent missing of specific updates in other parties plans, repeatedly leading to delays and economic losses.

Tried CAD tools were present and used everywhere, eager sections of scientists oriented to computational engineering were ready for action and finally political emphasis appeared on the scene—so why not jump into the game? And so it started with BIM!

An incredible ongoing development happens since in various tasks to be solved and the present state is admirable. Practical use set in either on voluntary or legally base, at first in structural engineering. There, any deviations of the 3D model of the real situation could not be noticed at all—and it would have taken long until in course of a large infrastructure project of civil engineering the problem would have turned up, surely too late. That is the reason why the author, having become very early aware, started action. If a surveyor should set out a Cartesian 3D model of large extension in length, he would inescapable run into problems as the model would not fit into reality.

Any CAD kernel of a BIM still follows the ME path of being rectangular and scaled to 1, which must collide with the geodetic coordinate systems and frames. Moreover, all CADs are restricted in their numerical coordinate values to an extent not corresponding to large infrastructure extensions and only allow insufficient placing and orientation with respect to geodetic coordinate systems as mentioned in Sect. 1.2.

The manuals of prominent CAD programs used in BIM offer to define a base point, i.e. a surveyed project point with "real, site or world coordinates", to which the corresponding point of the BIM model is shifted, presuming that the xy-plane of the model is "horizontal". Following, the model can be rotated around this point to "geographic" North. The limitations found for maximum coordinate values are within ±5 to 16 km. Concerning treatment of the earth's curvature, nowhere reliable information was found.

2.3 Geodetic Coordinate Systems

It does not take much for geodesists to explain their most common coordinate systems and the relation to a specific geodetic datum. We will only give two approaches here.

Global Coordinate System

The most important global coordinate system at present is the World Geodetic System 1984 (WGS84) defined and used by GPS. Its geocentric coordinates XYZ refer to the origin in the earth's mass center, a Z-axis corresponding to the mean direction of the earth's rotation axis at a specific epoch and the XY-plane perpendicular to it defining the mean equatorial plane; the X-axis belongs to the mean meridian plane of Greenwich and the Y-axis points to the East. Fundamental relations allow conversion of the geocentric coordinates to ellipsoidal coordinates with respect to a defined mean earth ellipsoid. From the WGS84 the transit to a national or regional datum can be managed by a set of seven fixed parameters known from national survey by a spatial similarity transformation. For special project systems the parameters will be computed from an overdetermined (>3) set of identical points with coordinates in both systems. In most cases, subsequently a conformal map projection to gain plane coordinates and a height conversion taking into account the spacing in between ellipsoid and geoid follow.

Plane and Height Coordinate System

Splitting up the three dimension in 2D plane (horizontal) and 1D height (vertical) considers on the one hand that we are used to plane representations in the form of maps or plans and on the other hand that height must preserve its physical property of ensuring two points with the same altitude being inherent in the same level surface. Otherwise—and that could be the case for purely geometric heights of the type ellipsoidal—water could flow from one point to another with equal height. Geodesists know about the necessity to use two different reference surfaces, an ellipsoid for mapping and the geoid for establishing a reasonable height system. Because of the ellipsoidal shape of the earth, mapping affords a conformal projection, which must accept distance distortions increasing with the second power of the distance from the middle meridian. Therefore, in contrast to a geocentric coordinate system with all three axes scaled to 1, plane and height systems must have different scales for xy and height. All national surveys for mapping and cadastral purposes followed this approach and the control points established during the various surveying campaigns in dense networks therefore all bear projection coordinates and heights, thus serving as initial points for many project surveys. Their complement are reference stations for GPS (and other GNSS), which allow surveying operations based on satellite navigation. In geodesy, there are no collisions between the coordinate systems of different type; one can be rigorously transformed to the other using well-established procedures and algorithms.

As long as construction projects, in particular those of long infrastructure lines, rely on the same principles and complete their plans separated to plane and height, no problem arises at all when it comes to setting-out in the real world. Even the hybrid uses of satellite supported and terrestrial surveying strategies will not introduce any confusion.

2.4   The Problem of Possible Misalignment for BIM Infrastructure Projects

To appreciate the problem and its significance, it proves advisable to recall the difference between "position" and "location" as it once was highlighted when Location Based Systems came up. A position is a numerical description of a point with respect to a specific coordinate system in Euclidian space. A location pins down the position to the real world with respect to natural or man-made features and in absence of those to an established geodetic frame of known points. It is the surveyor's duty and competence to take a position from a virtual model and to indicate it in the real world, on ground or on a certain object—this responsible process is called "setting-out" [7].

The usual way of providing the link from the virtual model to a real site is to set up either a total station, orient it and iterate a prism to coincide with precalculated coordinates within a regulated tolerance, or to do the same by means of a GPS rover using predetermined transformation values from the global to the local system. The fundamental precondition is, that the virtual model and the geodetic frame are built up according to the same principles. Suppose a model based on projection coordinates and heights a.s.l. (we dispense here with the corrections for geoid height); we would just have to inversely apply the projection correction and raise the straightened distance to the station height to fit smoothly into the geodetic frame.

In case of BIM the situation is different: the model holds scale 1 and refers to a flat ground plane with height zero [8]. If we would now set out the model at a site distant from the projection system's central meridian and on an altitude other than m.s.l., it cannot fit into existing infrastructure, e.g. a projected railway line will not connect exactly to rail tracks at both ends. The misalignment will increase with distance from central meridian, with project height and with the length extension of the line. Imagining a tunnel bored from both sides, the engineering geodesist responsible for guidance of the TBMs and observing tight limits of breakthrough error could run into terrible problems. Even in case of being conscious of the problem and trying to compensate the impacts by applying any improvements by himself, the surveyor could experience a legal dispute due to disregarding the model proportions of the contract. This situation must end [9]!

3 Exemplary Dimensions of Possible Misalignment

A very thorough and extensive investigation in the theoretical problem can be found in [10], also containing tables and nomographs to quantify the influence of neglected corrections and reviewing the impact in the context of construction precision.

Here we give preference to show practical examples of infrastructure projects following conventional planning to present average divergences in case of BIM application. The first three come from personal experience, the following three are owed to [11] and the last one was taken from [12]. For most of the examples the planning was done referring to the respective national projection and height system; only the projects Koralm and Brenner use special project coordinate or height systems tailored to the project.

3.1 Second Suburban Railway Line, Munich (Under Construction)

The project of roughly 10 km extends East-West in a distance of coarsely 30 km of the central meridian and in an average height of 500 m. The inverse corrections to be applied to distances in the plane and height model amount for the Gauß-Krüger (GK) projection −13 mm/km and for the mean height +78 mm/km resulting to a combined scale factor of +65 mm/km. i.e. a misalignment of 65 cm on the 10 km length.

3.2 High-Speed Railway Line Köln-Rhein/Main, Section C (Completed)

The project of roughly 45 km extends SE-NW in a distance of coarsely 50 km of the central meridian and in an average height of 300 m. The inverse corrections to be applied to distances in the plane and height model amount for the Gauß-Krüger (GK) projection −0.8 mm/km and for the mean height +47.1 mm/km leaving accumulated a scale factor of only +16,3 mm/km. i.e. a misalignment of 73.4 cm on the 45 km length.

3.3 Schmitten Road Tunnel Zell Am See (Completed)

This 5.1 km long, N-S tunnel runs some 40.5 km away from the central meridian on an average height of 767 m. The corrections are −20 mm/km for GK and +120.4 mm/km for height, in total +100.2 mm/km corresponding to a difference in 51.2 cm for the tunnel.

3.4 Semmering Railway Base Tunnel (Under Construction)

The height correction itself for an average height of 556 m amounts to +87.3 mm/km resulting in a difference of 2.4 m for the total track length of approximately 28 km.

3.5 Short Road Tunnel in Vorarlberg (Completed)

Due to the considerable project height of 1700 m the corresponding scale factor reaches +266 mm/km causing a difference of 19 cm on the 700 m tunnel length.

3.6 Koralm Railway Base Tunnel (Under Construction)

The scale factor +59 mm/km related to a mean project height of 376 m would cause 1.95 m of length difference on the 33 km project length. Therefore, in the Koralm Tunnel project a specific projection system was agreed upon as moreover the project crossed the boundary of two national GK meridian strips. Nevertheless, the central meridian chosen in the middle of the project area was not able to make the projection corrections completely to vanish as the 33 km tunnel extends E-W.

3.7 Brenner Railway Base Tunnel (Under Construction)

The theoretical factor for the 55 km tunnel due to mean height would have come to + 121 mm/km for the average height of 770 m; the ones due to projection would have provoked a large total tunnel length difference, if the Italian UTM or the Austrian GK were used. That is the reason why the planners came to the agreement to use a project specific projection system with a middle meridian selected in a way to have the complete N-S project within a strip of ±10 km thus reducing the corrections to a negligible extent (8 ppm). This is a common way to handle transnational projects in an optimum way. To moreover choose a mean reference height for the height system, additionally minimized the amount of the height corrections, which however had to be applied.

The outstanding clever geodetic strategy made the Brenner Base Tunnel project to a splendid case study to investigate the possibilities of BIM in parallel to the conventional planning and construction, also concerning 3D alignment instead of separated design for track and gradient [13]. The other paramount example is represented by the London Crossrail project, in fact realized using BIM. There, natural benefit was taken from the London Survey Grid next to the zero meridian with a scale factor of 1 and from building close to seal level.

To tailor geodetic reference systems and frames to an infrastructure project, is a common geodetic strategy; current examples are explained very instructively in [14]. However, we must never disregard that such systems are a closed world of their own and anything, from present infrastructure, topography, cadastral boundaries and already existing control points must be properly transformed into this artificial cosmos!

4   Endeavors of the BIM Community to Overcome the Problem

To make BIM succeed in practice an enormous combined effort of universities, software developers, engineering enterprises, construction consortia and clients proves vital. Of highest value are international ambitions to standardize items and strategies like the buildingSMART initiative, with whom the Leonhard Obermeyer Center (LOC), the TUM Center for Digital Methods for the Built Environment, collaborates tightly.

buildingSMART started to develop a series of consensus-based standards, above all, the Industry Foundation Classes (IFC), an industry-specific data model schema. Any qualified person can contribute and foster improvements [15]. Invited by the LOC, the author was able to give a first presentation on the problem of colliding coordinate systems and its impact in course of a meeting with important experts at the TUM in 2016. The unconscious challenge was seriously discussed (starting with the belief, that the site surveyor should cope with the problem) and was finally spread. A first reaction came from Australia, where a wise surveyor, Lee Gregory, aware of the importance, initiated a world-wide questionnaire of geodetic datums and customs of synchronizing BIM with them. It resulted rather prompt in a first guideline of the IFC Infra Overall Architecture, in which a section "Geodetic Reference Systems" was incorporated to be cautiously considered by BIM users.

Yet, the majority of users rather blinded out the nasty matter, deliberately underestimating its impact. Hence, the author had to insist on attention by giving continued and ever more detailed presentations at conferences, seminars and workshops of the geodetic and the civil engineering community. At the same time scientists of the LOC, at the HTW Dresden and engineers responsible for prodigious infrastructure projects eagerly tried to find and elaborate workable solutions. These attempts will be discussed in the final section, but it will turn out that a rigorous solution in terms of engineering geodesy still has to be waited for.

5   Approaching Harmonization of the Colliding Systems

While for construction projects of limited extension approximated methods are acceptable [16], e.g. by using the IFCmapconversion entity, this is definitely not the case for long infrastructure projects for the main reason that the cited entity allows for a scale factor common to all three axes which would distort heights if used to compensate GK reduction. The entity moreover does not handle a map projection. For our problem we have three issued approaches to mitigate the inconsistencies and two in progress.

5.1   Dividing the Project into Sections

This represents a simple and straightforward strategy, reducing the impact of missing corrections. On the first glance it seems reasonable; e.g. long tunnels usually are subdivided by intermediate attacks which would support the concept to match. On the other hand, there will never be separated reference networks for setting-out the different sections; it will always be an all-embracing network for the sake of homogeneity. And there has to be taken a close watch on the sections' interfaces to ensure alignment. Nevertheless, for special situations the concept could be helpful, e.g. for metro projects, where the station construction could be planned in BIM and the connecting line sections conventionally according to plane and height as demonstrated in Doha [17].

5.2   Defining a Special Project Coordinate System

Transferring the familiar concept of choosing an average project height and an optimum projection system, repeatedly applied for conventional planning, to BIM, could as well reduce the interfering model misalignments as would be underpinned by the examples of the Koralm [14] and the Brenner Base [12] tunnels. Yet it is not a general solution; e.g. for the Semmering Base Tunnel, running in the first half N-S and in the second half E-W, no helpful coordinate system with respect to projection distortion could be determined. In addition, we must not forget that we are then working in an "island system" and have to transform any other data, in particular present connections, to it.

5.3   Insertion of an Intermediate Local Surveying Coordinate System

This strategy, sometimes applied in conventional project surveying in special cases, was taken up again [19], but rather aiming at BIM-GIS integration and building-in-stock. The idea is, to establish survey control points to provide a spatial link between the BIM and an intermediate local surveying coordinate system with short coordinates and scale 1. The physically existing points bear two sets of coordinates to enable in one direction the link to BIM and in the other direction to the national coordinate frame. It affords high attention, not to confuse coordinates and legal aspects are not resolved.

5.4   Options to Interpret Model Geometry

In a paper to appear soon [10], the authors investigate three different options and which one to choose in structural engineering or civil engineering. A key finding is that implicit geometry representations in BIM have to be made explicit to enable transformation of individual points, which then can be properly georeferenced and thus transformed back during setting-out. Based on a hypothetical railway line project, they also show that the impact of the models' collision goes far beyond length differences, but also prevents smooth cross-track and tangential transition into fixed tracks at the end points planned in BIM. Moreover, the hazard of misinterpretation is demonstrated well.

5.5 Looking for a Rigorous Transformation

The last idea came up during a discussion with a leading engineer of the DB Netz AG [20], who claimed with reference to the WGS84 that there is not the least problem in geodesy to transform a Cartesian coordinate system to any other one, so why in BIM? Indeed, if we imagined the project completed, it would be a simple procedure to determine its coordinates in WGS84 and transform it. The hypothesis is that an appropriate 6-parameter similarity transformation (leaving the 7th parameter scale 1 unchanged) to shift and rotate the surveyed model can be found so that it would have the straight connection of the endpoints in the ground plane of a height enabling all project points to have positive values and in an attitude of a tangent plane to the ellipsoid in the middle of the infrastructure model. Following this idea, we could try the same for the virtual project, supported by surveyed end points. The transformed system of Cartesian coordinates could brilliantly serve for planning in the present BIM systems with short coordinates and would be close enough to a local horizontal system. Transforming the project back by means of the inverse parameters delivered the project 1:1 in geocentric WGS84 coordinates to be further transformed to familiar geodetic plane and height!

6 Conclusions

The answer to the crucial question posed in the paper's heading is definitely "no", unless the responsible parties do not want to risk considerable economic losses due to misalignment! The good news is that the BIM community already found ways to mitigate the impact. There is no bad news just that we will have to be patient a little until a rigorous solution will arise! Now, as in an increasing number of countries BIM became mandatory for public projects, it is of much more importance to enforce that at the beginning of any project a negotiated determination of coordinate systems involved must take place and will become a compulsory part of the contract.

References

1. All Info about BIM - Blog, https://8dbim.weebly.com/, last accessed 2020/01/27.
2. Blankenbach, J., Clemen, C.: BIM-Methode zur Modellierung von Bauwerken. In: DVW e.V., Runder Tisch GIS e.V. (eds.) Leitfaden Geodäsie und BIM., version 2.0, pp. 20–32. Bühl, München (2019).
3. Borrmann, A., König, M., Koch, C., Beetz, J.: Building Information Modelling. VDI-Buch, Springer Vieweg, Wiesbaden (2015).

4. Wunderlich, T., Blankenbach, J.: Building Information Modeling & Absteckung. In: Vermessung aktuell - BIM. https://mediatum.ub.tum.de/doc/1435852/1435852.pdf, Innsbruck (2018).
5. Eastman, C., Teichholz, P., Sacks, R., Liston, K.: BIM Handbook. 2nd edn. Wiley, Hoboken (2011).
6. Günthner, W., Borrmann, A.: Digitale Baustelle – innovativer Planen, effizienter Ausführen. Springer, Berlin Heidelberg (2011).
7. You in charge of Revit coordinates – we need talk, https://www.linkedin.com/pulse/you-charge-revit-coordinates-we-need-talk-part-1-james-worrell/, last accessed 2020/02/02.
8. Pancera, M.: SwissFEL Neubau Freier Elektronenlaser - Erfahrungsbericht. In: Vectorworks Anwendertag 2016, IttenBrechbühl, Zürich (2016).
9. Becker, R., Clemen, C., Wunderlich, T.: BIM in der Ingenieurvermessung. In: DVW e.V., Runder Tisch GIS e.V. (eds.) Leitfaden Geodäsie und BIM., version 2.0, pp. 87–102. Bühl, München (2019).
10. Jaud, S., Donaubauer, A., Heunecke, O., Borrmann, A.: Georeferencing in the context of Building Information Modelling: A Thorough Analysis. Automation in Construction (in press) (2020).
11. Chmelina, K.: Personal Communication. Geodata, Vienna (2017).
12. Markic, S., Borrmann, A., Windischer, G., Glatzl, R., Hofmann, M., Bergmeister, K.: Requirements for geo-locating transnational infrastructure BIM models. In: Peila, D., Viggiani, G., Celestino, T. (eds.) TUNNELS AND UNDERGROUND CITIES, WTC 2019, Naples. Taylor & Francis, Abingdon (2019).
13. Windischer, G., Hofmann, M., Glatzl, R., Bergmeister, K.: Modellierung von Tunnelbauwerken in BIM-Systemen unter Berücksichtigung besonderer Referenzsysteme für den länderübergreifenden Lage- und Höhenbezug. In: Hanke, K., Weinold, T. (eds.) 20. IGW OBERGURGL 2019, Innsbruck. Wichmann, Berlin (2020).
14. Macheiner, K.: Drei große Eisenbahn-Tunnelprojekte in Österreich – ein Vergleich ausgewählter Aspekte aus der Sicht der ingenieurgeodätischen Praxis. VGI 103(4), 221–234 (2015).
15. Liebich, T.: IFC für Infrastruktur. In: 15. buildingSMART BIM-Anwendertag, Mainz (2017).
16. Wunderlich, T., Wasmeier, P., Wagner, A., Barth, W., Wiedemann, W., Raffl, L., Preuß, G., Fuchs, K, Reith, C.: Ingenieurgeodäsie. In: Zilch, K. et al. (eds.) Handbuch für Bauingenieure. Edition 3, pp. 1–61. Springer Vieweg, Wiesbaden (2020).
17. Jungwirth, J., Scholz, M., Deinhard, R., Schneider, M.: BIM in der Verkehrsinfrastruktur. Special 2015 - BIM, Ernst & Sohn, Berlin (2015).
18. Jaud, S.: Do BIM models intrinsically possess geodetic distortions or not? In: Wunderlich, T. (ed.) INGENIEURVERMESSUNG 20, Wichmann, Berlin (2020).
19. Kaden, R., Clemen, C.: Applying Geodetic Coordinate Reference Systems in Building Information Modeling (BIM). In: FIG Working Week 2017, Helsinki (2017).
20. Reifenhäuser, M.: Personal Communication. DB Netz AG, Frankfurt (2017)

Building Information Modelling of Industrial Plants

Ján Erdélyi(✉) ⓘ, Peter Kyrinovič, Alojz Kopáčik ⓘ, and Richard Honti ⓘ

Faculty of Civil Engineering, Slovak University of Technology in Bratislava, Radlinského 11, 810 05 Bratislava, Slovakia
jan.erdelyi@stuba.sk

Abstract. The information technology input for the building life cycle is more and more intensive, starting with design, through realization and including the use and the operation of the building. The progress in this field is represented by Building In-formation Modelling (BIM). BIM is in its base object-oriented 3D model of the building, which includes information about the structure, elements and their parameters. This paper deals with the possibilities of BIM model creation of industrial plants for documentation and management. The basic principles, of BIM, the possible content of three main parts of a BIM model of industrial plants (the graphical data, the non-graphical data and the documents) are discussed. Part of the paper is devoted to the role of the surveyors in the process of building information modelling of industrial plants. In addition to the above mentioned, a case study is part of the paper. The case study is aimed at the proposal of information model for documentation and management of pipeline bridges in the chemical company Duslo, a.s. (Slovakia). The measurement procedure of the geometry of a pipeline bridge, the additional information collection and the model creation in AutoCAD Plant 3D (used by the mentioned company) are briefly described.

Keywords: Building information modelling · Pipeline bridge · Industrial plant · Terrestrial laser scanning · As-built documentation

1 Introduction

Building information modelling (BIM) is in their base an object-oriented 3D model of the building, which includes information about the structure, elements and their parameters. Includes the development of the virtual data model, which describes not only the architecture, but simulate the construction and the operation of the new or renewed object, also. The aim of BIM is not only the creation of the building model, but the creation of correct, responsible, changeable and complete information about the building, including the access to this information for all stakeholders [1, 2]. BIM is based on interoperability and parametric behavior of its objects, change of one parameter generates an aggregated set of changes in all the parameters of a model object [3].

A. Kopáčik et al. (Eds.): *Contributions to International Conferences on Engineering Surveying*, SPEES, pp. 13–26, 2021. https://doi.org/10.1007/978-3-030-51953-7_2

The BIM model consists of three parts in terms of data (Fig. 1). It can be characterized as a combination of graphic and non-graphic data (information) and documents relating to the building project, provided that everything is stored and managed in a Common Data Environment (CDE). The detail of the BIM model can be defined as the Level of Development. According to the standards *ISO 19650 Organization and digitization of information about buildings and civil engineering works, including building information modelling (BIM)—Information management using building information modelling (Part 1: Concepts and principles, Part 2: Delivery phase of the assets)* LOD includes the level of detail of the graphical data—Level of Geometry (LOG) and the Level of Information (LOI), and thus describes the specification of the information model in both graphical and non-graphical terms.

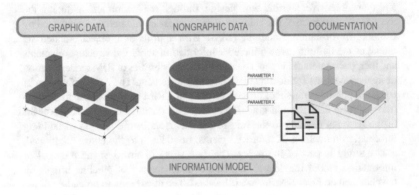

Fig. 1 Building information model

The 3D model or its graphic (geometric) representation (in native or exchange format) provides visual orientation, location and context, also defines relationships between spaces and other elements in the model. The 3D model is an information container for the non-graphical data (parameters) and potentially provides link to more information in other formats and locations. The non-graphical data is directly linked (associated) to graphical model, eventually are referenced to a specific building element. BIM allows assign physical or functional information to each building elements, so called parameters. The advantage of such data arrangement is easy search and extraction of data, which contributes to efficient access to information.

Individual LOD levels are defined from 100 to 500 as follows [4]:

- LOD 100—an element may be graphically represented in a model by a symbol or other generic representation that does not meet the requirements for LOD 200.
- LOD 200—the element can be generally graphically represented, and it is possible to assign basic geometric properties to it (approximate size, shape, location, orientation), and it is also possible to attach additional non-graphic information.
- LOD 300—the element is graphically represented in the model as a separate system, object, or device, defined by quantity, size, shape, location, and orientation. Additional non-graphic information may be added.

- LOD 350—an element is graphically represented in the model as a separate system, object, or device defined by quantity, size, shape, location, orientation, and also relationship to other elements. Additional non-graphic information may be added.
- LOD 400—elements in the model are graphically represented as a separate system, object, or device, defined by quantity, size, shape, location, orientation. They are processed in high detail including manufacturing documentation and installation information (assembly). Additional non-graphic information may be added.
- LOD 500—model elements are defined by the actual dimensions, shape, location, quantity and orientation. The information is verified in situ. Additional non-graphic information may be added.

The effectiveness of cooperation between the stakeholders of construction process can be significantly increased by sharing information from the BIM model. For efficient object-oriented interoperability, software used has to allow and ensure reliable data exchange. BIM itself is based on open cooperation between different professions within the whole building's lifecycle, including different software using data exchange formats (CIS/2, BACnet, cityGML, IFC, etc.). Widely used data exchange format is the Industry Foundation Classes (IFC), continually developed by international non-profit organization buildingSMART (formerly International Alliance for Interoperability). IFC is defined by the standard *ISO 16739-1 Industry Foundation Classes (IFC) for data sharing in the construction and facility management industries—Part 1: Data schema.*

2 Building Information Model of Industrial Plants

The first step in BIM implementation is the identification of the requirements from the users who will manage the information model created. The document defining how the project will be implemented, managed and controlled in relation to the declared requirements is the so-called BIM Execution Plan (BEP).

A BIM model of an industrial plant (object) is different from BIM model of a building. In addition to classical buildings (with all parts: architecture, structure, MEP and other building services), in an industrial plant there is machinery, production lines, conveyors and pipelines, each of which forms a separate unit and must be separately managed by the facility management (FM) of the plant. The plants are often spread over large areas of several tens of hectares or more, but the separate units can be connected to each other. When modeling large objects, the users often encounter software restrictions in terms of the dimensions of the objects modelled and inconsistencies between plane and height coordinate systems (orientation of coordinate axis, cartographic distortion, etc.). In the next chapters the above-mentioned topics are discussed with special regard to pipeline bridges, since the same rules apply to the BIM model of industrial buildings as to conventional buildings.

2.1 Buildings, Halls, Machinery

The creation of BIM models of buildings and halls in industrial plants are based on the principles of BIM of civil buildings. The main parts of the model are the architectural elements, the structural elements, electrical elements, plumbing elements (sanitary) and other building services (HVAC, backup power supply, etc.).

The individual elements of the model (foundations, walls, curtain walls, columns, floors, ceilings, roofs objects, stairs and elevators, openings, doors, windows, etc.) have to be modelled by volume models using available object libraries (predefined objects) or by user defined (modified) models arranged to the well-known hierarchy Category-Family-Type-Instance. To the graphical (3D) model some non-graphical information (parameters) can be added, which defines quantity, size, shape, location, orientation, relationship to other elements and some physical properties. To a 3D model of a wall is then linked information about size, shape, location, thickness of the layers (directly derived from the graphical model) and parameters defining e.g. the material composition, thermal properties, etc. Not all the properties and records of a building should be stored in the form of non-graphic data. Examples are manuals, specifications, or officially signed documents, such as contracts and certificates in static formats (PDF, JPG, etc.) [5].

The detail of the model should be carefully considered. Indeed, if the model is too generalized, it may not be sufficient for the purposes behind which the information model is developed. Conversely, if the model is very detailed, its creation will be more time-consuming and costly. In the case of creation of an information model of an existing building e.g. for FM, the LOD 300 (or LOD 350) can ensure sufficient detail.

Machinery (as part of a production line) in industrial plants creates separate category of objects. To create their models is recommended in close cooperation with their manufacturers. In the case of digitization of such objects, the same rules must be followed as for buildings. In addition to other parameters, space associations must be defined also (relationships to production lines, building, halls, spaces, etc).

2.2   Pipeline Bridges, Pipeline Routes

Pipeline bridges are structures built specifically to support pipes where adequate structure is not available. They are necessary for arranging often a large number of pipelines running through a plant and joining one equipment to another. In addition to the piping, cable ducts supporting electrical power cables are also often placed on them. The task of pipelines is to transport gaseous and liquid substances within the industrial plant. Especially in the chemical (petrochemical) industry, the number of these pipelines can be several hundred in the plant site. Therefore, they are a key part of manufacturing processes in the industrial plants.

Pipeline bridges usually consist of 3 main parts: the foundations, the supports (pillars) and the bridge sections in which the pipes are placed (Fig. 2). For the operators, it is important to have information about the pipeline routes, their characteristics, from where the route goes, who is the purchaser of the transported substance (which operating unit within the plant), the condition of the pipes as well as other facilities.

In the case of creation of an information model of an existing pipeline bridge, the data collection process is crucial. In most cases it is not necessary to model each element with the smallest details, because the graphic representation can be extended using images of the structure's part with complicated details (atypical valves, venting devices, blowdown devices, etc.). The LOD 300 (or LOD 350) can ensure sufficient detail of the geometry for the FM.

Fig. 2 Photo of a pipeline bridge

The information model of a pipeline bridge should consist of two main components, bridge structures and pipeline routes located on the bridges. The Fig. 3 shows the pipeline bridge BIM model's structure.

The bridge structure consists of four main parts: foundations, supports, bridge sections and accessories. The size, shape, location, orientation, and also relationship to other elements of each of them is defined by the graphic data (3D model) and can be derived from the model. Information that should be recorded in the BIM model (as attributes) about the bridge structures are: ID of the bridge, type (pipeline, cable, combination of both), length, slope, number of bridge sections, number of supports, number of floors (storage levels), ID of pipeline routes placed on them, material of the structure (steel, concrete, fibre reinforced polymer, other), type of the bridge structure (truss structure, simple beam), start and end of the bridge (according to existing objects and roads), date of last coating, presence of a walkway, operator, operating unit within the plant and the date of the last revision.

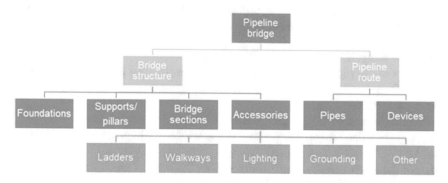

Fig. 3 Pipeline bridge BIM model's structure

The structure of the bridge sections can be made of steel, fiber reinforced polymer, reinforced concrete or other materials and very often consist of structural beams. In addition to the attributes directly derived from the graphical model (size, shape, location, orientation, etc.) it is recommended to defined for the parts of the pipeline bridges the:

- ID of the modelled part—footing, support (pillar), bridge section, structure element (structural beam),
- ID of the bridge to which it belongs,
- ID of the support/section to which the structure element belongs,
- material,
- depth of foundation, type of foundation, grounding (non-grounded, rod, strip),
- type of the structure,
- length of the bridge section,
- load capacity (designed value or value defined by an expertise)
- physical condition (OK, damaged, critical).

Integral parts of the pipe bridges are also accessories such as ladders, walkways, lighting, grounding, etc. For all the accessories should be defined attributes defining the belong of the element (ID of the bridge), the date of last revision, material, type of lighting, type of grounding, etc.

Pipe routes are composed of individual pipes, straight sections, branches, elbows and other fittings. Part of the pipeline routes are also devices as valves, venting devices, drainage devices, blowdown devices, gauges, etc. Important information for pipelines is their route, from supplier to purchaser. The supplier of the substance, which is transported by the pipeline, is the operation unit where the substance is loaded. The pipeline route usually starts there. Purchasers are the operation units that consume the substance transported by the pipeline for the purpose of producing their products.

Since pipeline route can run over several bridges, the user of the information model should be able to query on the base of the selected pipe bridge, but also on the base of the selected of pipeline route. For the management of the pipelines is needed the information about the substance transported, nominal diameter (internal diameter), nominal pressure (maximum pressure), operating pressure and temperature of the substance transported also. In terms of facility management, information about the operator and the pipeline supplier unit is needed also. The information about the pipeline routes defined in the BIM model as attributes should be:

- ID of the route, ID of the supplier, ID of the purchaser,
- diameter of the pipe, material of the pipe, thickness of the pipe wall, length of the route,
- insulation, heating (yes, no),
- substance, pressure (hPa), capacity (m^3),
- ID of the pipeline bridge, storage level of the bridge in which the pipeline is lying,
- operator,
- date of the last revision,
- existence of classified technical equipment.

Part of the documentation should be photographs. Some parts of the pipeline bridges and routes are difficult to model due to their complexity. In the case of devices with complex geometry it is more effective to create photographs of these objects in the field (Fig. 4) and then insert them into the information model as documentation (inserting link to the image). The images should be of uniform format (*.jpg, *.tiff, etc.) their resolution should be at least 300 dpi and the image size at least 1920 pixels x 1080 pixels (or panoramic) TrueColor color.

Fig. 4 Photograph of complex valves (right) and their representation in the BIM model (left)

2.3 Selected Limitations of BIM Platforms

Currently a number of software platforms support BIM concepts and principles [6]. The most widespread modelling tools are Allplan, ArchiCAD, Autodesk Revit, Bentley Architecture (in alphabetical order). These software are mainly used to create building structures and architecture and have some basic tools for bill of quantities preparation, budgeting and collision detection. In terms of professions (statics, MEP, HVAC, etc.), several computational and graphic solutions are also available, e.g. in the field of statics Dlubal, Scia, Sofistik, Tekla Structures (in alphabetical order), etc. and in the field of building services e.g. ArchiCAD MEP, AX 3000, DDS CAD, Autodesk Revit MEP, etc.

The software have their limitations in terms of dimensions of work plane. All geometry for the model, including geometry from an import or a link, should reside within the limits of the modelling work plane [7]. The work plane is interpreted as a flat surface, without taking into the account the Earth's surface curvature.

The coordinate system used in BIM software is a right-handed Cartesian system with a scale factor 1 and with limited numerical coordinate values. A geodetic coordinate system often has different orientation of coordinate axis (e.g. left handed system), it respects the Earth's curvature, the points are projected into a common level surface, and their projection from the Earth's surface to the plane of a map causes cartographic (projection) distortion (of lengths, angles in dependency on the projection used). Nevertheless buildingSMART has defined for current versions of IFC *IfcMapConversion* entity, it only allows to convert the local origin of the local engineering coordinate system to its place within a map (easting, northing, orthogonal height) and to rotate the x-axis of the local engineering coordinate system within the horizontal (easting/westing) plane of the map. It does not handle the projection of a map from the geodetic coordinate reference system [8]. These facts cause misalignment between the BIM (CAD) coordinate system and the geodetic coordinate system.

The BIM model should be created in a local coordinate system, defined for the information model. If needed, parts of the model can be transformed to the coordinate system in which the basic plant map is created (using Helmert transformation without the scale factor). In that case especially large models should be divided into smaller parts, to minimize the distortion caused by the cartographic projection. Another approach is creating a BIM model for each unit (building, pipeline bridge) separately (dividing the model to several parts/sections) [9]. The resulting model can then be displayed and managed using CDE software, e.g. Allplan BIM+, Bentley Navigator, Autodesk BIM 360, Dalux, Trimble Connect, etc.

3   Pipeline Bridge Information Model—A Case Study

In the followings the concept of creation of BIM model of pipeline bridges and routes in the chemical company Duslo Saľa, a.s. (Slovakia) is described. For the purpose of case study, a 184 m long, 9 m high and 5 m wide pipeline bridge (M1) was chosen.

The first step was the existing data and documentation collection about the pipeline bridge. In the first place, the data were obtained from the central archive of the company, where project documentations of investment constructions and maintenance are archived by the Department of Investments of the Plant. However, most of the data had to be obtained from operators of the pipeline routes. The project documentation of the reconstruction of the superstructure of the bridge from 1996, project documentation and as-build documentation of some pipeline routes placed on the bridge were provided.

Despite the availability of the above-mentioned documentation the data were collected in the field, because some of the document were out of date and inaccurate. Some of the pipeline routes were not drawn at all, some structural beams have different size than those specified in the documents and some bridge fields were of different shapes and dimensions also.

3.1   Measurements

To make the geometric data collection (measurement, in situ verification) more effective, the methods of close-range photogrammetry and terrestrial laser scanning can be used in addition to conventional surveying methods. For the purpose of case study, terrestrial laser scanning was used.

The whole structure was scanned from 20 stations of the instrument (Trimble TX5), with the parameters of the scanning 3 mm/10 m with 2 × repetition of the measured distances (Fig. 5). The accuracy of the position of discrete measured point by the instrument used was better than 3 mm in any cases. The reference points (used for transformation

Fig. 5 Scanning of the pipeline bridge (left) ang the resulting point cloud (right)

of individual clouds) were signalized by B&W checkerboard targets. Their coordinates were determined in a local coordinate system.

In addition to laser scanning photographs of the devices and equipment with complex geometry were taken These were used as documents linked to the model, so that every small detail doesn't had to be modeled.

3.2 Information Modelling

The creation of the information model itself consists of modeling objects either based on regression analysis or manually using the adjusted point cloud (transformed, filtered, etc.) and defining the properties of these objects in software for 3D modeling and BIM (Fig. 6). For the needs of the company Duslo Šaľa, a.s. the AutoCAD Plant 3D software was used for the proposal of the pipelines' BIM model. The software is used and therefore well-known by the employees of the plant, so the use of it ensures a very important part of the BIM implementation—proficient users.

Fig. 6 Modelling of pipes using adjusted point cloud

The disadvantage of the AutoCAD Plant 3D is the limited exchange of information using exchange formats, it enables using only its native formats. Therefore, to manage the model, it is necessary to use CDE software, allowing access to information for selected users or third parties. However, the model can be updated only by using the AutoCAD Plant 3D. The software allows to define attributes to individual building elements such as foundation blocks, beams, ladders, pipes, pipeline device, but does not allows to define attributes of the objects that are created by connection of several elements such as a pipe bridge, supports and bridge fields composed of multiple beams. Therefore, the Pipeline bridge BIM model's structure described in the chapter 2.2 has to be modified.

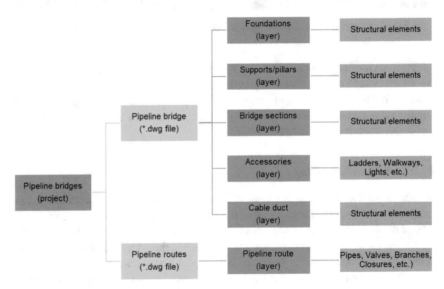

Fig. 7 Proposed information model structure

It is needed to set up a project—Pipeline Bridges (Fig. 7). In that project a *.dwg file must be created for each pipeline bridge separately and one * .dwg file for all pipelines. This creates a piping system, which will be continuous throughout the plant from the supplier to the purchaser and by loading the *.dwg file of the bridge required, its structural element can be displayed. The number of files for pipeline bridges will be equal the number of pipeline bridges in the plant while all pipeline routes will be stored in one file, but a separate layer should be created for each route. Within the pipeline bridge's file, it is suitable to divide its structural parts into layers as foundations, supports, bridge fields and accessories, even. cable ducts.

Modeling of the Pipeline Bridge. The first step was the modeling of individual bridge sections in the respective layer. These were modelled by structural beams (Fig. 8), while the adjusted point cloud represents the geometric base for this procedure. For each element were defined attributes, while a particular attention was paid to significant numbering of the parts. The second step was the modelling of the supports, then the foundations, cable ducts and the accessories (ladders, lights, walkways, etc.), All of these were modeled similar to the bridge sections using structural beams or predefined objects from the object library of the software used.

Fig. 8 Modeling using structural beams, the point cloud (left) and the resulting model (right)

In the attribute tables, the ID of the pipeline bridge (M1) was defined for each element. In addition to this a unique ID was defined in the form:

- Bridge section—BS_XX_L_NN
- Supports/pillars—S_XX_L_NN
- Foundations—F_XX_L_NN
- Cable ducts—M1_CNNN
- Other accessories—M1_NN

Where XX is the number of the bridge section, support or foundation. L defines the location (left, right, top, bottom or diagonal) from the point of view from the south end of the pipeline bridge to the north end. NN defines the number of the element modeled (beam, concrete block, ladder, etc.).

Modeling of the Pipeline Routes. The pipeline routes lying on the chosen bridge were modelled using the respective library of the software on the base of the adjusted point cloud. It is recommended to model each pipeline route with its all accessories (elements, devices) in a separate layer. Thus, the number of layers created (24) is equal to the number of pipelines located on the pipeline bridge M1. In the 3D model (Fig. 9), the outer perimeter of the pipe is modeled, i.e. the pipe together with the insulation. Therefore, it is not possible to determine from the graphical part of the model whether the 3D model of the pipeline is directly a pipeline, or it is a pipe insulation together with a pipeline. The actual pipe diameter and insulation thickness is then defined in the attribute table. A specific case are pipes that are heated by a steam pipe (e.g. nitrogen pipes). In those cases, the isolation layer containing the steam, the pipeline itself and the steam supply pipe were modelled separately.

Fig. 9 BIM model of the pipeline bridge M1

Complex devices such as valves, metering devices were generalized by a simple device, and a photograph of the device was attached to the model using a hyperlink as the Fig. 4 shows. The image is displayed by clicking on the hyperlink in the pop-up menu that opens by right-clicking on the device in the model.

The IDs of the pipeline routes have the form RRRR_SS_DDD_MMMM_II, where RRRR is the route's number, SS is the code of the substance transported, DDD is the inner diameter of the pipe, MMMM is the code of the material of the pipe and II is the code of the material of insulation. In the attribute tables (line list) the following information is recorded for each pipe:

- ID of the pipeline route,
- type and the standard—e.g. Pipe DIN 2448,
- outer diameter, inner diameter, length, wall thickness,
- material,
- substance transported,
- temperature,
- operating pressure maximum pressure,
- Insulation—YES/NO, thickness,
- capacity,
- input/output object, direction of flow,
- existence of classified technical equipment,
- existence of heating,
- date of last revision, ID of the revision,
- storage level of the bridge in which the pipeline is lying,
- operator.

LOD, Accuracy, Degree of Generalization. The accuracy of the graphical part of the model created depends on the accuracy of the measured points (point cloud) and the degree of generalization. Regular shapes are easy to model; however, the problem arises with irregular shapes of objects, with slightly twisted pipes, or slightly curved beams. It is not possible in terms of time and cost to model every detail, and therefore it is necessary to define the degree of generalization of the model. For the case study described it was defined at the level of 50 mm. That means that the beams or pipes, whose geometry deviates from the regular shape by a higher value, were modelled as separate element, and the model resulted respects the as-built geometry.

The LOD of the created BIM model can be defined as LOD 300, however the parts with complex geometry were generalized, and a photograph was attached to the model using a hyperlink. In those cases, the LOD is 200, so it is difficult to define the level of development of the whole model by a single LOD.

4 Conclusions

Building information modelling is in their base an object-oriented 3D model of the building, which includes information about the structure, elements and their parameters. The paper is aimed at the proposal of information model for documentation and management of industrial plants with special focus on pipeline bridges. The content and the structure of three main parts of a BIM model (the graphical data, the non-graphical data and the documents) are proposed. A BIM model of an industrial plant is different from BIM model of a building. In addition to classical buildings (with all parts: architecture, structure, MEP and other building services), in an industrial plant there is machinery, production lines, conveyors and pipelines, each of which forms a separate unit and must be separately managed by the facility management of the plant. A case study of information modeling of a pipeline bridge in a chemical company is also described. The proposal is based on the experience of the authors and do not represent the final form or the single option of the BIM model creation.

Acknowledgements. „This work was supported by the Slovak Research and Development Agency under the Contract no. APVV-18-0247".

References

1. Smith, D. K., Tardif, M.: Building Information Modeling, A Strategic Implementation Guide. John Wiley & Sons, New Jersey (2009).
2. Eastman, Ch., Teicholz, P., Sacks, R., Liston, K.: BIM Handbook. John Wiley & Sons, New Jersey (2008).
3. Smith, M.: BIM in Construction. https://www.thenbs.com/knowledge/what-is-building-information-modelling-bim.
4. Funtík, T.: Open Data Exchange in BIM. In: Eurostav. Vol. 22, No. 3, pp. 60–61. Eurostav, Bratislava (2016).
5. Eastman, CH., Teicholz, P., Sacks, R., Liston, K.: BIM Handbook – A Guide to Building Information Modeling for Owners, Managers, Designers, Engineers, and Contractors. . John Wiley & Sons, New Jersey (2011).

6. Nawari, O.N., Kuenstle, M. Building Information Modleing – Framework fo Structural Design. CRC Press Taylor & Francis Group, London (2015).
7. Autodesk Knowledge Network. https://knowledge.autodesk.com/support/revit-products/troubleshooting/caas/CloudHelp/cloudhelp/2019/ENU/Revit-Troubleshooting/files/GUID-3F79BF5A-F051-49F3-951E-D3E86F51BECC-htm.html.
8. buildingSMART, IfcMapConversion https://standards.buildingsmart.org/IFC/RELEASE/IFC4/ADD2_TC1/HTML/schema/ifcrepresentationresource/lexical/ifcmapconversion.htm.
9. Kaden, R., Clemen, C.: Applying Geodetic Coordinate Reference Systems within Building Information Modeling (BIM). In: FIG Working Week 2017 Surveying the world of tomorrow - From digitalisation to augmented reality, FIG, Helsinki (2017).

Possibility of Use of BIM in Automated Geometry Check of Structures

Richard Honti[⊠] [iD], Ján Erdélyi[iD], Gabriela Bariczová, Alojz Kopáčik[iD],
and Peter Kyrinovič

Faculty of Civil Engineering, Department of Surveying, Slovak University of Technology in
Bratislava, Bratislava, Slovakia
richard.honti@stuba.sk

Abstract. An integral part of the construction works is the as-built documentation of the constructed structures and buildings and the check of the geometry of their parts. Moreover, in the nowadays era of automatization of processes, this procedure is often semi- or fully automated. For the data acquisition except for conventional surveying methods, terrestrial laser scanning (TLS) and other photogrammetric methods are used. The main task of geometry verification is the 3D model creation of the structure and the quantification of the deviations from the design. For automatization of 3D model creation, initial information about the position and orientation of structures is very useful. Building Information Modelling (BIM) represents significant progress in this field. BIM is an object-oriented 3D model of the building enabling the derivation of the geometric parameters of the structures in 3D space, identical to the one in which the measurements are made. The paper deals with the automated geometry check using data from TLS and BIM. Part of the paper is devoted to the derivation of information about the designed geometry from BIM exchange format Industry Foundation Classes (IFC). The automated wall (plane segment) estimation, the determination of deviations from the design and the as-built wall flatness quantification are described in a case study.

Keywords: Building information modeling · Automated geometry verification · Wall flatness · Terrestrial laser scanning · As-built documentation

1 Introduction

Today, terrestrial laser scanning is one of the main techniques for 3D data acquisition with high resolution. TLS allows non-contact documentation of the behavior of the monitored structure with all his structural elements. The accuracy of the 3D coordinates determined by current laser scanners reaches several millimeters. The precision can be increased using suitable data processing. The results of the measurement by TLS are the point clouds, that are often used as basic data for geometry generation for BIM models.

BIM is an intelligent 3D model-based process that enables to more efficiently plan, design, construct, and manage structures and buildings. BIM is in their base an object-oriented 3D model of the building, which includes information about the structure,

A. Kopáčik et al. (Eds.): *Contributions to International Conferences on Engineering Surveying*,
SPEES, pp. 27–37, 2021. https://doi.org/10.1007/978-3-030-51953-7_3

elements, and their parameters. Includes the development of the virtual data model, which describes not only the architecture but simulates the construction and the operation of the new or renewed object, also. The aim of BIM is not only the creation of the building model but the creation of correct, responsible, changeable and complete information about the building, including the access to this information for all stakeholders. BIM is based on the interoperability and parametric behavior of its objects, change of one parameter generates an aggregated set of changes in all the parameters of a model object [1, 2]. A BIM model consists of three parts in terms of data. It can be characterized as a combination of graphic and non-graphic data (information) and documents relating to the building project, provided that everything is stored and managed in a common data environment (CDE). In some definitions, metadata is included in addition to the three components as part of the information model. To make BIM complete, all this information for each product, material, and system that is designed in the project is needed. Not all the properties and records of a building should be stored in the form of non-graphic data. Examples are manuals, specifications, or officially signed documents, such as contracts and certificates, as they document the historical record of the project's progress, and so it is not the information about the building itself, which is usually supplied in static formats (PDF, JPG, etc.), but can also be referenced such as a video file with maintenance instructions. Documents should be well organized, labeled and stored in a way that allows access for those who need to use the information in time [3].

There are many ways in which data can be managed in BIM, and as a result, there are several different file formats that can be used. The most common non-proprietary file format is the Industry Foundation Classes (IFC), which aims at providing an open and neutral data format for storing and exchanging building information model data between heterogeneous software applications [4]. Within IFC, building components are stored as instances of objects that contain data about themselves. This data includes geometric descriptions (position relative to building, the geometry of object) and semantic ones (description, type, relation with other objects) [5].

Building industry interests for timely and accurate information on the progress of the construction project are increasing. This includes the measurement of the as-built geometry and comparison with the project plan (design). Manual visual observations and traditional geometry measurement can be time-consuming, costly expensive and ineffective. Automatization of the processes of the construction progress monitoring can contribute to increasing the efficiency and precision of this process. Laser scanners show the potential for supporting automatic construction progress tracking.

The goal of the paper is to show the possibility of the use of BIM and TLS point clouds in automated geometry check of structures. In general, a 3D CAD model would be sufficient for this issue, however, in nowadays trend of digitalization of the construction process, BIM models will be the main available product. With the method proposed, there is no need for creating another 3D model.

In this paper, a novel technique for the information derivation about the design geometry from the IFC exchange format is proposed. The derived information (IFC wall objects) with the scanned point clouds of the as-built structure is used for the determination of deviations from the design, and the as-built wall flatness quantification. For automation of the as-built wall (plane segment) estimation from the measured point cloud by TLS, an algorithm is proposed. The described methodology is presented in a case study.

2 As-Built Documentation and Geometry Check of Structures

Nowadays, the use of BIM in the design and construction phases of structures has dramatically increased, but in the operation and maintenance phase, the owners are often not utilizing their BIM models. The main reason can be that the accuracy and level of detail obtained from construction models do not always reflect as-built conditions needed by maintenance personnel. Moreover, multiple changes can occur after the construction, that is rarely documented. TLS represents a relatively fast and simple method for digitizing the 3D geometry of real-world objects with high detail, so it can be one of the possible solutions for this issue. With the technology of TLS, the as-built conditions of BIM models and changes from the design can be documented and verified [6].

Today, one of the main applications of TLS is the reconstruction of as-built 3D BIM models from the scanned 3D point clouds, the so-called Scan-to-BIM process. The other process is the Scan-versus-BIM (as an analogy to Scan-to-BIM), that compares the scanned point clouds of the built structures with the design BIM model. A few research teams have already demonstrated the potential value of Scan-versus-BIM processes for tracking progress and dimension quality control of structural works.

The Scan-versus-BIM process generally consists of several steps. The first step is to align the scanned point clouds in the same coordinate system as the BIM model. This step can be done in multiple approaches, e.g. using some survey points or based on overlaps of geometric shapes (e.g. planes, cylinders, etc.) recognized in point clouds and the BIM model, or using other automated or semi-automated registration techniques [7]. The next step can be the deviation computation, that can also be done in several ways. The most common way is to compute the minimum distance from each point of the scanned point cloud to its nearest surface in the BIM. Other methods compute the distance along the user-specified directions, like the X, Y, Z direction of the common coordinate system or the direction of the surface normal. In some cases, the point cloud is segmented into regions based on the objects recognized in the designed BIM model, i.e. for each scan, each point is matched with a 3D model object. Most often, the points are matched with the closest 3D objects from the BIM model with some tolerance using a metric combining two criteria, proximity and surface normal similarity. The last step is the deviation visualization using several visualization techniques (e.g. color maps, contour lines, etc.). Besides, a statistical analysis can be used to analyze the deviation patterns [8–11].

The current trend is the automatization of the steps of the Scan-versus-BIM process. One of the options for the automated geometry check of structures is to compare the selected geometric shapes (e.g. walls—plane segments) from the designed BIM model with the as-built geometry gathered from TLS point clouds.

3 Algorithm Proposal for Automated Geometry Check

The algorithm proposed compares the design geometry of a BIM model (in IFC format) with the as-built geometry from the scanned point clouds. The process of the algorithm is shown in Fig. 1. The algorithm requires that the 3D point cloud and the BIM model are registered in the same coordinate system.

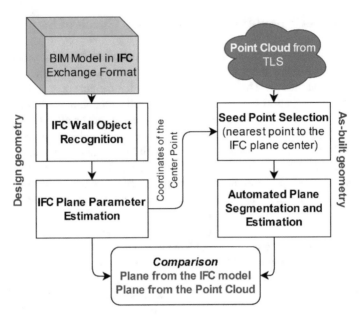

Fig. 1 Flowchart of the algorithm proposed

3.1 Derivation of Information from IFC

The Industry Foundation Class is a platform-neutral, open file format specification that is not controlled by a single vendor or group of vendors. It is an object-based file format with a data model, which is intended to describe building and construction industry data and exchange these data between AEC (Architecture, Engineering, and Construction) software application. The IFC model specification is open and available. It is registered by ISO and is an official international standard ISO 16739:2018 Industry Foundation Classes (IFC) for data sharing in the construction and facility management industries [4].

The architecture of the IFC data structure is defined by four definition layers, to which schemas are strictly assigned. The main layer is the domain-specific layer, that organizes definitions according to industry discipline (building control domain, structural elements domain, electrical domain, etc.). Consequently, it is hierarchically arranged by the shared element layer, which provides more specialized objects and relationships shared by multiple domains. Then hierarchically follows the core layer, that provides the basic structure, the fundamental relationships and the common concepts for all further specializations. The fourth layer is the resource definition layer, which consists of supporting data structures [4]. From the geodetic point of view, this layer is the most important, since it contains the coordinates of the objects within the model as well as other geometric parameters.

This work is focused on wall objects in IFC, defined as *IfcWallStandardCase*. An example of the definition of a wall object in IFC is shown in Fig. 2.

```
111  #134= IFCCARTESIANPOINT((52414.1410933094,25630.2336971545,0.));- reference point
112  #136= IFCDIRECTION((-0.537758036394573,-0.843099219720344,0.)); - unit normal vector
113  #138= IFCAXIS2PLACEMENT3D(#134,#19,#136);
114  #139= IFCLOCALPLACEMENT(#125,#138);
115  #141= IFCCARTESIANPOINT((3825.,-0.));
116  #143= IFCPOLYLINE((#9,#141));
117  #145= IFCSHAPEREPRESENTATION(#96,'Axis','Curve2D',(#143));
118  #148= IFCCARTESIANPOINT((1912.49999999999,1.39266376208980E-12));
119  #150= IFCAXIS2PLACEMENT2D(#148,#25);        wall length
120  #151= IFCRECTANGLEPROFILEDEF(.AREA.,$,#150,3024.99999999999,139.999999999999);
121  #152= IFCAXIS2PLACEMENT3D(#6,$,$);        wall height
122  #153= IFCEXTRUDEDAREASOLID(#151,#152,#19,2500.);
123  #154= IFCCOLOURRGB($,0.709803921568627,0.709803921568627,0.709803921568627);
124  #155= IFCSURFACESTYLERENDERING(#154,0.,$,$,$,$,IFCNORMALISEDRATIOMEASURE(0.5),IFCSPECULAREXI
125  #156= IFCSURFACESTYLE('Betonov\X2\00E9\X0\ Tv\X2\00E1\X0\rnice',.BOTH.,(#155));
126  #158= IFCPRESENTATIONSTYLEASSIGNMENT((#156));
127  #160= IFCSTYLEDITEM(#153,(#158),$);
128  #163= IFCSHAPEREPRESENTATION(#98,'Body','SweptSolid',(#153));
129  #166= IFCPRODUCTDEFINITIONSHAPE($,$,(#145,#163));                        wall width
130  #170= IFCWALLSTANDARDCASE('1dzFimzZbEyw6Xo3xbqS21',#41,'Basic Wall:Obecn\X2\00E9\X0\,140mm
```

Fig. 2 An example of wall object definition in IFC

In the first two rows, the reference (initial) point (*IfcCartesianPoint*) of the wall and the direction (*IfcDirection*) by the unit normal vector is defined. In row *IfcExtrudeAreaSolid* the wall height, in row *IfcRectangleProfileDef* the wall length is defined. The last row shown is the mentioned *IfcWallStandardCase*, where the wall IDs and the wall width is defined.

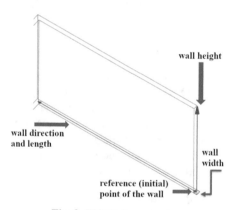

Fig. 3 IFC Wall parameters

The necessary parameters (Fig. 3) for the plane estimation of one wall object are as follows: the reference (initial) point; the direction; the length, the height and the width of the wall.

For the information derivation from IFC, an algorithm is proposed in Matlab® software, that automatically identifies the mentioned rows (Fig. 2) of the wall object and extracts the necessary data (Fig. 3) from these rows.

From the extracted wall data, the parameters needed for the plane estimation are calculated. First, the 3D coordinates of the four boundary points and the center point of the wall-side (plane) are calculated using the spatial polar method. Based on these wall points calculated, the plane coefficients are estimated. The equation of the plane can be described as follows:

$$a.X + b.Y + c.Z + d = 0, \tag{1}$$

where a, b and c are the parameters of the normal vector of the plane; X, Y, Z are the coordinates of a point lying on the plane; and d is the scalar product of the normal vector of the plane and the position vector of any point of the plane. To calculate the elements of the normal vector the cross product from the boundary points is used.

3.2 Point Cloud Plane Segmentation

After a plane is derived from the designed IFC model, the belonging plane is segmented from the point cloud. First, the algorithm selects the nearest point from the scanned point cloud to the center point of the estimated plane from IFC. This selected point is used as a seed point for plane segmentation from the point cloud. For this issue, an algorithm proposed in [12] is used.

The algorithm in [12] is a combination of elements of the modified RANSAC algorithm and the region-based segmentation method. The first plane is estimated from the 10 nearest neighbors to the selected seed point. The estimation of the plane is done by orthogonal regression, which minimizes the perpendicular distances to the plane. The solution is based on the general equation of a plane (Eq. 1). To calculate the elements of the normal vector, the Singular Value Decomposition (SVD) is used:

$$A = U.\Sigma.V^T, \tag{2}$$

where A is the design matrix with dimensions of nx3; n is the number of points used for the calculations. The column vectors of U are normalized eigenvectors of the matrix AA^T. The matrix Σ contains eigenvalues on the diagonals. The normal vector of the regression plane is the column vector of V corresponding to the smallest eigenvalue from AA^T. The plane is then defined by the elements of the normal vector a, b and c. Parameter d is calculated by fitting the elements of the normal vector and the coordinates of the centroid of the point cloud according to the formula

$$d = -(a.X_0 + b.Y_0 + c.Z_0). \tag{3}$$

The orthogonal distances of the points from the plane are calculated as follows:

$$d_p = \frac{|(a.X + b.Y + c.Z + d)|}{\sqrt{a^2 + b^2 + c^2}}. \tag{4}$$

The standard deviation of the regression plane, based on the orthogonal distances of the inlier points from the plane, is calculated as follows:

$$\sigma_\rho = \sqrt{\frac{d_{p,\rho}^T.d_{p,\rho}}{n-1}}, \tag{5}$$

This model is tested against the selected neighbors, while the points lying on this plane are identified. The plane parameters are re-estimated using all the identified points—inliers.

a)   b)   c)   d)

Fig. 4 Process of the algorithm for automated plane segmentation from point clouds. **a** measured point cloud; **b** 6,400 nearest points; **c** 102,400 nearest points; **d** inlier points for the selected plane

This process is done iteratively, with a gradual increase of the points tested (Fig. 4). The plane re-estimation is repeated until the plane size stops increasing, so there are no more inliers for the selected plane.

At this step, the plane parameters are derived from the IFC model and the belonging plane points from the point cloud are segmented. The next step of the algorithm is the IFC plane and the point cloud plane comparison, by calculating the orthogonal distance of the segmented points from the design model (IFC plane).

In Fig. 5 the orthogonal deviations are visualized between the IFC plane and the segmented plane from the point cloud. The positive values of these deviations are for the points "over" the IFC plane (closer to the position of scanning) and the negative values are the points "under" the IFC plane. Alternatively, the differences between the estimated parameters of the two planes can be compared.

In the last step of the algorithm, the wall flatness quantification is performed (Fig. 6), estimating the best-fit regression plane of the segmented points using orthogonal regression [12], and calculating the distances for each segmented point to this regression plane.

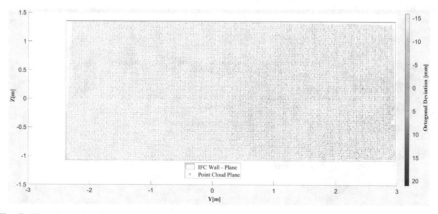

Fig. 5 Visualization of the deviations between the design model and the as-built geometry for one plane

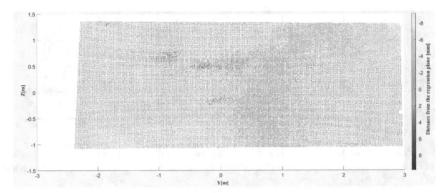

Fig. 6 Wall flatness quantification

The whole process (Fig. 1) is repeated semi-automatedly, while all the wall objects from the IFC are geometrically verified. The results are the spatial deviations between the planes of the design model (IFC) and the as-built geometry (from point clouds) in every point of the point cloud plane shown in a table with the visualization (Fig. 5) of the deviations for each plane, and the wall flatness quantification.

In case if the building tolerances are provided, a comparison of the deviations between the design and the as-built geometry with these tolerances can be executed, to determine if the tolerances are not exceeded.

4   Case Study of an Apartment

For experimental testing of the above-described proposed algorithm, a BIM model in IFC exchange format and a point cloud from TLS of an apartment (Fig. 7) was used. The apartment consists of 5 rooms, and from 25 walls (the inner side plane segment for each wall was considered).

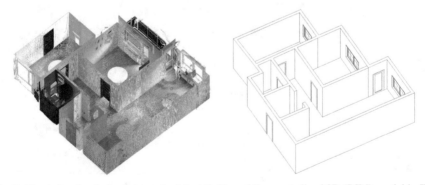

Fig. 7 The point cloud of a case-study object (left), and the generalized 3D (BIM) model in IFC of the case-study object (right)

The scanning was performed with the Trimble TX5 3D laser scanner. The scanning was executed with a resolution of 2 mm/10 m. The distance between the scanner and the scanned object was always below 5 m. The whole point cloud of the apartment consists of approx. 10 million points. With the mentioned scanner and considering the conditions (maximum distance of the scanned objects from the scanner, resolution and quality setting of the scanner, etc.) during the measurement, the accuracy of a single measured point was less than 3 mm in any cases. For practical application, it is necessary to include the transformation error (transformation of the point cloud into the model's coordinate system) applying error propagation law.

The deviations were calculated for every plane segment of the apartment. The maximum values (under and over the designed wall) of these deviations are shown in Tables 1 and 2.

Table 1 Maximum values of the deviations for *Room 1, and 2*

Room	Wall	Max. deviation [mm]	Max. deviation [mm]
1	1	14	−15
	2	12	−20
	3	21	−5
	4	14	−21
2	1	16	−13
	2	5	−24
	3	19	−14
	4	14	−20

In Table 1 the maximum values of the deviations between the two models are shown for *Room 1* and 2. The maximum deviation for *Room 1* reaches 21 mm for the points "over" and −21 mm for the points "under" the IFC plane. For *Room 2* the maximum values are 19 mm and −24 mm.

In Table 2 the maximum values for *Room 3, 4,* and 5 are shown. For *Room 3* the maximum values are 25 mm and −19 mm, for *Room 4,* 20 mm and −15 mm, and for *Room 5,* 25 mm and −21 mm.

The maximum values for wall flatness quantification (Fig. 6), calculated based on the maximum distance of the segmented points from the regression plane, reached from −9 mm to 9 mm in any cases of the tested object.

5 Conclusion and Future Work

The technology of TLS and BIM models offers to improve the reliability, efficiency, and completeness of geometry check of structures. This paper focused on geometry check using geometric shapes (plane segments) and proposed an algorithm that integrates TLS point clouds and BIM models in IFC exchange format to automate this process. First,

Table 2 Maximum values of the deviations for *Room 3, 4,* and *5*

Room	Wall	Max. deviation [mm]	Max. deviation [mm]
3	1	15	−15
	2	25	−8
	3	4	−11
	4	8	−15
	5	10	−19
4	1	9	−10
	2	18	−8
	3	15	−9
	4	20	−15
5	1	17	−17
	2	21	−20
	3	13	−10
	4	25	−8
	5	14	−20
	6	15	−21
	7	9	−19
	8	6	−8

the walls are recognized in the IFC model, and the parameters of the plane segments belonging to these walls are calculated. In the next step, from the scanned point cloud, the planes belonging to the planes in the IFC model are automatically segmented. Next, these planes from IFC and TLS are compared, and the orthogonal deviations are calculated between them. In the last step, the wall flatness quantification is performed. The algorithm was tested on a model of an apartment with 5 rooms, the results are shown in tables.

In future work for simplification, an application based on the proposed algorithm will be developed in Matlab® software. Furthermore, this procedure could be extended for geometry check based on other geometrical shapes, likes spheres, cylinders. To apply and test the proposed algorithm on point clouds with IFC models of complex scenes from several areas is also planned. Besides, it is planned to integrate the possibility of the registration of the design model with the as-built point cloud.

Acknowledgements. This work was supported by the Slovak Research and Development Agency under Contract no. APVV-18-0247.

References

1. Eastman, Ch., Teicholz, P., Sacks, R., Liston, K.: BIM Handbook. 1st ed. John Wiley & Sons, New Jersey (2008).
2. Smith, D. K., Tardif, M.: Building Information Modeling, A Strategic Implementation Guide. 1st ed. John Wiley & Sons, New Jersey (2009).
3. Erdélyi, J., Funtík, T.: The Current State of BIM Implementation in the Slovak Republic. Slovenský geodet a kartograf (Slovak), 22 (4), 5–10 (2017).
4. ISO. "ISO 16739-1:2018 Industry Foundation Classes (IFC) for data sharing in the construction and facility management industries — Part 1: Data schema" (2018).
5. Thomson, Ch., Boehm J.: Automatic Geometry Generation from Point Clouds for BIM. Remote Sensing, 7 (9), 11753–11775 (2015). https://doi.org/10.3390/rs70911753.
6. Giel, B., Issa R. R. A.: Using laser scanning to access the accuracy of as-built BIM. In: Proceedings of ASCE International Workshop on Congress on Computing in Civil Engineering, 665–672 (2011). https://doi.org/10.1061/41182(416)82.
7. Kim, Ch., Son, H., Kim, Ch.: Fully automated registration of 3D data to a 3D CAD model for project progress monitoring. Automation in Construction, 35, 587–594. (2013). https://doi.org/10.1016/j.autcon.2013.01.005.
8. Bosché, F. et al.: Automating surface flatness control using terrestrial laser scanning and building information models. Automation in Construction, 44, 212–226 (2014). https://doi.org/10.1016/j.autcon.2014.03.028.
9. Tang, P., Anil, E. B., Akinci, B., Huber, D.: Efficient and Effective Quality Assessment of As-Is Building Information Models and 3D Laser-Scanned Data. In: Proceedings of ASCE International Workshop on Congress on Computing in Civil Engineering, 486–493 (2011). https://doi.org/10.1061/41182(416)60.
10. Turkan, Y., Bosche, F., Haas, C. T., Haas, R.: Automated progress tracking using 4D schedule and 3D sensing technologies. Automation in Construction, 22, 414–421 (2012). https://doi.org/10.1016/j.autcon.2011.10.003.
11. Bosché, F. et al.: The value of integrating Scan-to-BIM and Scan-vs-BIM techniques for construction monitoring using laser scanning and BIM: The case of cylindrical MEP components. Automation in Construction, 49 (B), 201–213 (2015). https://doi.org/10.1016/j.autcon.2014.05.014.
12. Honti R., Erdélyi J., Kopáčik A.: Plane segmentation from point clouds. Pollack Periodica, 13 (2), 159–171 (2018). https://doi.org/10.1556/606.2018.13.2.16.

Quality Assurance

Empirical Evaluation of Terrestrial Laser Scanner Calibration Strategies: Manufacturer-Based, Target-Based and Keypoint-Based

Tomislav Medić$^{(\boxtimes)}$, Heiner Kuhlmann, and Christoph Holst

University of Bonn, Institute of Geodesy and Geoinformation, Bonn, Germany
{t.medic,c.holst}@igg.uni-bonn.de, heiner.kuhlmann@uni-bonn.de

Abstract. To assure accurate terrestrial laser scanner (TLS) point clouds, the instruments should be repeatedly calibrated in the manufacturer's facilities, either in regular time intervals or if they fail the examination on a test-field. This workflow has drawbacks for the end-users: there is no deeper understanding of the calibration procedure and it is time- and money-wise burdening. As an alternative, scientists developed several user-oriented calibration approaches, which are not well accepted in practice. We investigated the differences between the factory calibration and two user-oriented calibration approaches, as well as their interaction. The investigation focuses on the raw point clouds without factory calibration. Our results show that the user-oriented calibration can replace the manufacturer calibration to some degree. However, the user-oriented calibration on top of the manufacturers' calibration improves the measurement quality beyond the factory calibration. Hence, user-oriented calibration should be considered as an alternative for the re-calibration to deliver up-to-date calibration parameters.

Keywords: TLS · Point cloud · Accuracy · Quality assurance · Features · Systematic errors

1 Introduction

Terrestrial laser scanners (TLSs) are used for a variety of high accuracy demanding applications such as deformation monitoring, industrial quality inspection and reverse engineering [1]. To meet the requirements of such uses, the manufacturers put considerable efforts into the accurate assembly of all instruments' mechanical components. The remaining small mechanical misalignments are accounted for mathematically—by applying the calibration parameters (CPs), which are estimated within the specialized calibration facilities [2–4]. The TLS calibration is a laborious procedure and has a high impact on the final price of the instruments.

Additionally, the calibration parameters change over time due to various causes, such as wear and tear, suffered stress [5, 6]. Hence, it is necessary to repeatedly calibrate the instruments. Typically, the instruments are re-calibrated in the manufacturer's

A. Kopáčik et al. (Eds.): *Contributions to International Conferences on Engineering Surveying*,
SPEES, pp. 41–56, 2021. https://doi.org/10.1007/978-3-030-51953-7_4

facilities. This occurs either in regular time intervals (once every 1 or 2 years based on manufacturers' recommendations) or when the TLS fails the testing procedure recommended by DIN or ISO norms [7]. In each case, the re-calibration is a costly event for the end-users, typically requiring several weeks and several thousands of euros (personal correspondence with one of the manufacturers).

To address these difficulties, scientists have invested substantial efforts in the development of different TLS calibration approaches, primarily aiming at user-oriented re-calibration. Based on different calibration strategies, the existing approaches can be divided in plane-based [8], cylinder-based [9], paraboloid-based [10], target-based [5] and keypoint-based [11] calibration approaches. They are all realized as self-calibration approaches requiring no reference-values measured with instruments of higher accuracy, making it cost-efficient. The mentioned approaches can be subdivided into two categories: the ones aiming at the a priori calibration, before the measurements, and the ones aiming at the in situ calibration, during the measurements [12]. Both categories have certain advantages and disadvantages that need to be considered when selecting the calibration strategy for individual measurement tasks.

Despite the two passed decades of successful implementations and enhancements of the latter approaches, the user-oriented calibration is not well accepted in the commercial sector. Hence, more studies are necessary to demonstrate their value for the end-users. In this study, we investigate in detail two user-oriented self-calibration approaches, one aiming at the a priori calibration (target-based self-calibration or TBSC) and one aiming at the in situ calibration (keypoint-based self-calibration or KBSC). We compare these approaches with the success of the manufacturer's calibration (MBC) by analyzing the raw point cloud without the initial manufacturer's calibration. To summarize, the main questions addressed in this article are:

- Do we need the factory calibration of TLSs?
- Can user-oriented calibration approaches help us?
- How to choose between the keypoint-based and the target-based calibration or should we combine them?

Additionally, to achieve the best calibration results, we introduced several improvements for the existing keypoint-based calibration algorithm [11]. Based on the results, we discuss the role of the user-oriented calibration approaches in the commercial sector and their optimal integration in the usual workflows of the end-users.

2 Materials and Methods

2.1 Instrument

We investigated the Zoller + Fröhlich Imager 5016 panoramic TLS [13]. It is a high-end instrument with high measurement accuracy (1 mm + 10 ppm range accuracy and 14.4" angular accuracy), allowing two-face measurements, equipped with a dynamic dual-axis compensator, and facilitating a wide field-of-view of 360° × 320°. The instrument is comprehensively calibrated in the manufacturer's facilities upon the assembly [4]. The range measurement systematic errors are estimated on the comparator bench/track by comparing the TLS measurements with the reference (interferometer). The angular systematic errors are estimated using the target-based calibration on a dedicated calibration field. The calibration field consists of numerous black and white planar targets with accurately estimated reference coordinates. The TLS accuracy in the datasheet is calculated as the RMSE of the calibration adjustment residuals [4, 13].

2.2 Measurements

Our investigation is based on two datasets. The first dataset is acquired on the calibration field of the University of Bonn developed for the user-oriented TBSC of panoramic TLSs [14]. The calibration field consists of 14 BOTA8 black and white planar targets developed for high-accuracy [15]. The targets are placed on predefined locations that are sensitive towards detecting all mechanical misalignments relevant for high-end panoramic TLSs (10 in total, Table 1)—[16].

Table 1 Calibration parameters (CPs) relevant for the high-end TLSs

CP	Tilts /angular errors [″]	CP	Offsets/metric errors [mm]
x_4	Vertical index offset	x_{1n}	Horizontal beam offset
x_{5n}	Horizontal beam tilt	x_{1z}	Vertical beam offset
x_{5z}	Vertical beam tilt	x_2	Horizontal axis offset
x_6	Mirror tilt	x_3	Mirror offset
x_7	Horizontal axis error (tilt)	x_{10}	Rangefinder offset

Dataset 1: The first dataset comprises two two-face scans (overall four scans) made from two scanner stations. The scans were made with the scanning resolution of 1.6 mm at 10 meters and with the lowest "quality settings", i.e. without any onboard measurements averaging. The instrument was placed on a heavy-duty wooden tripod and secured using the stabilization. The scanning lasted approximately 1 h, and during the measurements, the atmospheric conditions were monitored and they remained stable (temperature ± 0.5 °C, pressure ± 1.5 hPa, humidity ± 2.5%). Before the calibration, the TLS was warmed-up for approximately 1 h. The whole calibration is described in detail in [14]. To fairly compare the user-oriented calibration approaches against the manufacturer's calibration, these four scans were de-calibrated by the manufacturer. To do so, they extracted the raw TLS measurements without applying the mathematical corrections for the systematic errors estimated during the factory calibration.

Dataset 2: The second dataset is a single two-face scan of a building facade (same instrument settings and stabilization). The measurements lasted approximately 30 min and they took place roughly 2 months after the acquisition of the 1st dataset. In this case, the manufacturer's calibration remained. Both datasets were already partially analyzed in [14], however with a different focus.

2.3 Calibration

Target-based: For the TBSC, the target centers of all four scans of Dataset 1 were estimated using the template-matching algorithm described in [15]. To obtain the CPs, these observations were processed in the TBSC adjustment described in [14]. As a result, we obtained a set of 10 CPs (Table 1). The TBSC was applied only on Dataset 1 because it is typically used for the a priori calibration and it requires an adequate distribution of targets, which was not achieved for Dataset 2.

Keypoint-based: We applied the KBSC on both datasets, where in the case of the 1st dataset, we used only 2 out of 4 scans (one two-face measurement). In short, the KBSC relies on representing the 3D point cloud as a spherical intensity image based on the scanner's polar coordinates. Then, within this spherical image the well-defined keypoints (typically corners) are automatically detected using the Förstner operator (image feature detection algorithm) [17]. Consequently, these keypoints are uniquely described using the BRISK feature description algorithm based on their surroundings. Using these descriptions, the corresponding keypoints between the first and the second scan of two-face measurements are searched for and matched together. Finally, each keypoint is given one range, horizontal and vertical angle measurement and they are all processed together in the two-face calibration adjustment. The whole procedure is described in [11]. As the KBSC adjustment relies on only one set of two-face measurements from a single station, it cannot be used to estimate all CPs from Table 1. Namely, in the original implementation, the CPs that are not estimable are: x_2, x_{1n}, x_{1z} and x_{10}. Herein, we made 5 improvements in the KBSC approach:

One: First, we limited the search and matching of corresponding keypoints by distance. Because the intensity image based on the scanner's polar coordinates is spherical, we used spherical distance between the potentially corresponding keypoints. We set the distance threshold of 10 times the scanning resolution. This step reduces the number of outliers and increases number of observations in the KBSC adjustment.

Two: As the spherical intensity image for each scan is huge in data size (typically more than 1 gigabyte), it cannot be processed as one single image or matrix. Hence, it is segmented in patches to reduce the computational costs. In the original code, in this process, some keypoints are lost on the borders of the patches. Therefore, we created an overlap between different image patches. After the keypoint detection and matching for all patches, we removed eventual duplicates within the same pixel of the intensity image. This modification increased the number of keypoints for nearly 6%.

Three: Each spherical intensity image patch has the size of $10° \times 10°$ (horizontal and vertical angles directions). As the 3rd improvement, we adjusted some of the Förstner operator parameters specifying the feature detection process so that they change with the vertical angle of the image patch. More specifically, we set the rule that the parameters should increase the values with the sinus of the vertical angle (Table 2). The vertical angle (θ) equals the lowest vertical angle of an image patch. This way the feature detection is optimized for the spherical nature of the intensity image as the patches close to poles (zenith and nadir) hold much less information than patches close to horizon. This change yields more correctly matched keypoints on higher and lower elevations, which were critical in the original implementation [11].

Table 2 Förstner operator parameters changes regarding the original implementation in [11]

Selected parameters for the Förstner operator	
Size of derivative filter (sigma)	$1.5/|\sin(\theta)|$ *
Size of integration filter	$5/|\sin(\theta)|$ *
Threshold for precision of points	$1/|\sin(\theta)|$ *

*θ represents the lowest vertical angle of the intensity image patch

Four: We improved the range measurements, which were one of the main bottlenecks of the original approach. In short, we estimated the range differences for the matched keypoints and 440 measurements in their proximity (440 pixels in the intensity image, 21×21 pixel windows around the keypoint). Then, we removed the outliers in the differences with the threshold of 3 times the scaled median absolute deviation (MAD) from the median ($3\sigma = 3* \times 1.4826 \times$ MAD) [18]. Finally, each keypoint is assigned the mean from the remaining range measurements. This allowed estimating two additional CPs regarding the original implementation (x_2 and x_{1n}).

Five: Finally, we created a covariance matrix based on the statistical analysis of a local neighborhood of each keypoint. The standard deviations in the horizontal and vertical angle direction are estimated with the Förstner operator [17]. The standard deviations for the ranges were estimated from the above-mentioned range differences as 1.41σ. The covariance matrix is formed as a diagonal matrix due to the unknown correlations. This allowed relaxing the precision threshold (Table 2), which defines the acceptable standard deviation of angular measurements and increased the overall number of observations.

Figure 1 shows the distribution of the detected and successfully matched keypoints in the instrument's field-of-view for the new (blue) and the original (red) implementation of the KBSC. Now, the number of keypoints is substantially larger (~18 times) and the distribution is much more uniform. In the original KBSC, the keypoints on the higher elevations were sparse. This impacts the calibration results, as the observations on higher elevations are mandatory for the accurate estimation of multiple CPs [16]. The benefit of all improvements will be furtherly pointed out in Sect. 3.

2.4　Method for Evaluating the Calibration Results

To compare different calibration approaches, we evaluated the impact of the estimated CPs on the point cloud accuracy for both datasets. The point cloud accuracy is analyzed by computing the differences between the first and the second scan of two-face measurements. The differences are computed using the M3C2 point cloud comparison algorithm, which computes the differences in the direction of the local surface normal and de-noises them based on the point cloud local statistics [19]. All external causes for the systematic

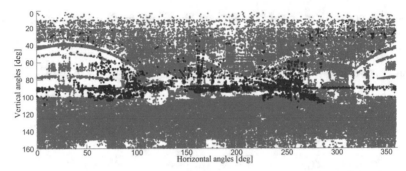

Fig. 1 Distribution of observations (keypoints) in the scanner's field-of-view for the original implementation (red) and the new implementation (blue) of the keypoint-based calibration

TLS measurement errors act identically on both scans of the two-face measurements due to: same measurement configuration, same atmospheric conditions in the short period, and the same properties of the measured object. Hence, the only point cloud differences originate from the instrument's systematic errors and measurement noise. Therefore, in the case of the perfectly calibrated TLS, they would be normally distributed with zero mean and standard deviation equal to or lower than the measurement noise (due to M3C2 point cloud de-noising).

Using this evaluation strategy based on two-face measurements has the limitation that the influence of some calibration parameters cannot be detected. In the case of the factory-calibrated high-end TLSs, the rangefinder offset (Table 1) is the only relevant parameter not influencing the difference of two-face measurements. Hence, its influence on the point cloud quality is not analyzed within this study.

3 Results: Evaluating the Calibration Strategies

3.1 Dataset 1—the Indoor Calibration Field

First, we analyzed the point cloud accuracy of the calibration field (Sect. 2.4) in four cases: point clouds without the factory calibration, with the manufacturer-based calibration (MBC), with target-based self-calibration (TBSC), and with keypoint-based self-calibration (KBSC). The results are separated on three following topics:

3.1.1 User- Vs. Manufacturer- Calibration (Main Calibration Parameters)

In the TBSC, we estimate all CPs from Table 1, while in the KBSC two parameters (x_{10} and x_{5z}) were omitted due to the limitation of the calibration approach (2.3). The results are presented in Fig. 2, where the green color represents the zero difference and it changes gradually until the defined threshold. The remaining colors depend on the direction of the point cloud differences (red—positive and blue—negative). We present the results twice for better visual analysis, with different thresholds: once at 3 cm (Fig. 2, top and middle) and once at 1 mm (bottom).

From the top part of Fig. 2, it is apparent that all calibration approaches notably improved the point cloud quality. Before any calibration, nearly the complete point cloud

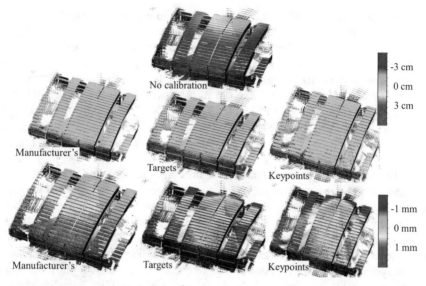

Fig. 2 M3C2 differences for 1st and 2nd scan of two-face measurements: factory, targets and keypoints calibration (green no difference, solid blue/red over thresholds ±3 cm or ± 1 mm)

had differences (due to the systematic errors) larger than 3 cm reaching the maximum of approximately 6 cm. When the point clouds are calibrated, with any of the three methods, all differences dropped notably below 3 cm. When we further analyze the corresponding histograms of the point cloud differences (Fig. 3), we can notice that the MBC improved the point cloud accuracy most. Namely, in the latter case, practically

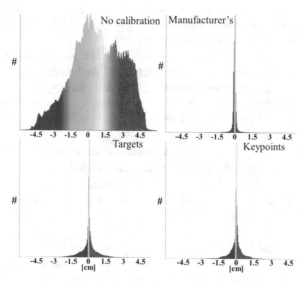

Fig. 3 Histograms of the point cloud differences (M3C2) presented in Fig. 2

all of the differences are within ± 3 mm, while in the case of the TBSC and KBSC a large portion of the values has the differences up to 15 mm.

These ± 3 mm in the differences of two point clouds mean that each point is maximally displaced for 1.5 mm in the comparison to its true position. This value is better than the values given in the manufacturer's specifications, where the maximal expected error (3σ) on e.g. 25 meters is 6.6 mm (25 meters is roughly the mean distance of all points represented in Figs. 2 and 3)—[4, 13]. Hence, our data confirms that applying the manufacturer's CPs, which were a priori estimated on a dedicated calibration field (approximately one month before the measurements), assures the measurement accuracy given in the instrument's specifications.

On the other hand, the user-oriented calibration approaches failed to reach the latter aim, which is not surprising. Namely, the manufacturers have in-depth knowledge of the instrument's assembly and all mechanical misalignments. Hence, their number of CPs is extensive. On the contrary, 10 CPs used in the user-based calibration (Table 1) is comprehensive only if some requirements for high-end TLSs are fulfilled. Either the instrument has to have angular encoders with multiple reading heads (eliminating multiple systematic errors—[2]) or the angular encoders need to be a priori calibrated, e.g. on a rotary table. For Dataset 1 this is not the case, as all CPs, including the ones relevant for the angular encoders, are removed.

3.1.2   User- vs. Manufacturer- Calibration (Calibration Parameters Extended)

We improved the user-based calibration by introducing further CPs, which brings us to the first big difference between the KBSC and the TBSC. Namely, the calibration field for TBSC is specifically designed to detect and estimate the a priori defined set of the CPs with pre-defined quality criteria [14]. Therefore, the number of observations is limited to a minimum to reduce the efforts for the assembly and maintenance of the calibration field, measurement acquisition and processing time. Hence, we cannot use this calibration approach to detect eventually missing CPs. We can only estimate the CPs that we are a priori expecting to exist.

On the contrary, the KBSC delivers a huge number of well-distributed observations covering the full instrument's field-of-view (depending on the surrounding). To compare, in the TBSC, the number of observations is 168, while in the KBSC, it is approximately 440 000. Hence, the KBSC is well suited to detect missing CPs. Figure 4 (top) presents the KBSC adjustment residuals of the horizontal angles, plotted regarding the field-of-view and colored according to the their magnitude and sign.

As can be seen, positive and negative residuals are clustered together, changing with the sine of the horizontal angles. This indicates the presence of a large systematic error in horizontal angle measurements due to the encoder eccentricity [20]. To prove this, we introduced the CPs for the encoder eccentricity and repeated the calibration (Fig. 4, bottom). We can observe that the magnitude of the residuals is lower and that the previously visible trend is removed. However, by removing the latter large trend, a previously undetectable trend appeared, which corresponds again to the sine of the horizontal angle, but with an increased frequency (repeating the patter every 120°).

We analyzed all measurement residuals and found overall 14 new significant CPs. Out of these 14, only 4 (including encoder eccentricity), had large values and a strong

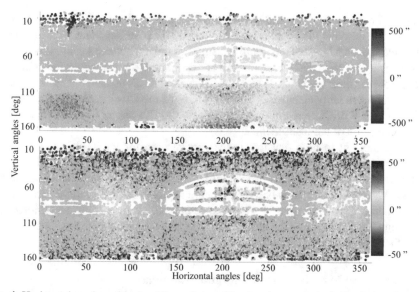

Fig. 4 Horizontal angle residuals of keypoints calibration in scanner's field-of-view for main calibration parameters (top), accounting for encoder eccentricity (bottom), mind colorbar scales

impact on the point clouds. We also estimated these 4 CPs with the TBSC. Figure 5 (top) presents the M3C2 differences of Dataset 1 for: MBC (equal to Fig. 2 bottom left), KBSC and TBSC (both user-based calibrations with the extended set of CPs). Additionally, Fig. 5 (bottom) presents the corresponding histograms (1 mm threshold for solid red/blue color).

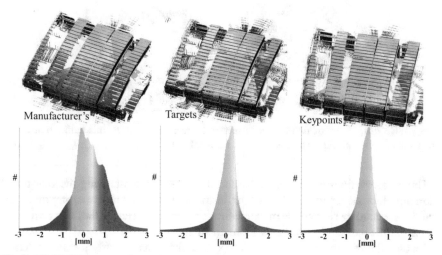

Fig. 5 M3C2 differences of two-face measurements: factory, targets and keypoints calibration with extended CPs set (top) and related histograms (bottom) - solid blue/red: ± 1 mm threshold

From the given data, we can observe that the point cloud accuracy after the user-oriented calibration is now comparable to the manufacturer's calibration. Moreover, it seems that it even surpasses the improvement of the manufacturer's calibration. This is not surprising, as the manufacturer's CPs are estimated a priori within the calibration field, before the measurements. Hence, the values of the CPs at the moment of capturing the Dataset 1 probably changed due to instability of the CPs [5]. On the contrary, the TBSC and the KBSC are realized in situ. Hence, the CPs are up-to-date, reflecting the current condition of the instrument.

It is important to note that this is the special case for the TBSC, as the in situ use is only possible within the calibration field. Typically, the TBSC is used for the a priori calibration and the CPs are used in the following measurements. Nevertheless, the results achieved with both user-based calibration approaches are almost the same, which makes a good cross-validation. This also points out that modeling only 4 additional CPs concerning Table 1 removes the majority of the remaining systematic errors in the case of not-factory-calibrated instrument.

3.1.3 User- and Manufacturer- Calibration Combined

The latter analysis demonstrates the possibility of the user-based calibration to substitute the manufacturer's calibration, if the functional model of the CPs is comprehensive. However, the user-based calibration is typically used to re-calibrate the already factory calibrated instrument. Hence, it is applied on top of the existing CPs to account for changes in the CP values. Figure 6 presents the M3C2 differences after the TBSC and the KBSC realized on top of the manufacturer-calibrated point clouds using 10 CPs from Table 1. Also, Fig. 7 presents the corresponding histograms.

Fig. 6 M3C2 differences of two-face measurements: targets and keypoints calibration of the factory-calibrated TLS, calibration parameters from Table 1—solid blue/red: ± 1 mm threshold

Three observations can be made based on our results. First, using the user-based calibration approaches to re-calibrate the instrument (for in situ cases) improves the point cloud accuracy beyond the manufacturer's calibration. Hence, they can be used to account for the internal instrument's geometrical change.

Second, the improvement achieved by both calibrations is nearly identical. Hence, estimating the CPs beyond certain accuracy and involving tens of thousands of observations does not bring noticeable benefits for the in situ calibration. This means that in situ calibration can be realized by processing only a portion of the whole instrument's

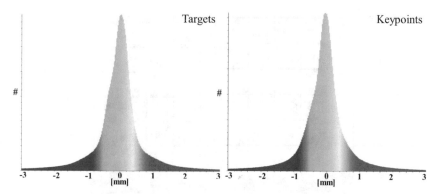

Fig. 7 Histograms of the point cloud differences presented in Fig. 6

field-of-view. This could lead to smaller computational requirements and allow nearly real-time and on-board calibration with the instrument's processing unit.

Third, we repeated this combined manufacturer's and user-based calibration with an extended set of CPs from Sect. 3.1.2 (+4 for the TBSC and +14 for the KBSC). The M3C2 differences and the histograms were nearly identical. Hence, there are no noticeable differences when using more CPs for the factory-calibrated TLS. Hence, this confirms that for the factory calibrated high-end TLSs, the angular encoder errors can be neglected. This serves as a validation of both the CPs in Table 1, as well as the validation of the comprehensiveness of the designed calibration field.

3.2 Dataset 2—A Building Façade

In Dataset 1, we compared the success of the user-oriented calibration with the manufacturer's calibration based on the de-calibrated point cloud. Although it gives important insights, it can hardly be considered as a typical TLS use case. Namely, there, we investigated the in situ quality improvement of both the TBSC and the KBSC. However, in the normal workflow, the TBSC on a dedicated calibration field would be made a priori, before some measurement task and these previously estimated CPs would be applied on later measurements. On the contrary, the KBSC is primarily developed for the in situ calibration and it is typically realized during the measurements. But, the surrounding geometry found during the measurements is not necessarily well suited for the comprehensive calibration.

Therefore, Dataset 2 represents a more realistic measurement setup. A building façade is measured from a single station from a distance of approximately 25 meters. The goal is to analyze the point cloud accuracy and to compare: the MBC, the a priori TBSC, the in situ KBSC, as well as the combination of the latter two. In this case, both user-oriented calibration approaches are applied to the previously factory-calibrated TLS (on top of the MBC). The M3C2 differences are presented in Fig. 8 and the related histograms are given in Fig. 9 (the red/blue threshold is ±1.5 mm).

From Fig. 8, it is apparent that some systematic errors exist in the point cloud when we rely only on the manufacturer's calibration. Again, these values are notably below the values defined in the specifications [13]. However, they are visually observable. If

Fig. 8 M3C2 differences of two-face measurements of a façade: factory calibration only and enhanced with targets or keypoints calibration - solid blue/red: over ± 1.5 mm threshold

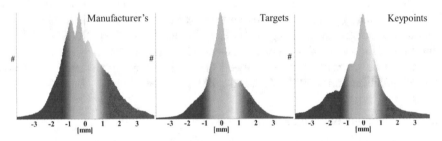

Fig. 9 Histograms of the point cloud differences presented in Fig. 8

we use the a priori estimated CPs using the TBSC, we can see that the point cloud accuracy increased and that the M3C2 differences are smaller and starting to resemble more to a normal distribution (Fig. 9). Hence, applying the CPs a priori estimated on our calibration field, roughly two months in advance, improves the point cloud quality beyond the manufacturer's calibration (already presented in [14]).

Further, we tried to estimate all 8 relevant two-face sensitive CPs (Table 1, except x_{5z} and x_{10}), using the KBSC and using only the point clouds of the building façade. If we analyze the histograms, the results achieved are comparable to the a priori TBSC, showing no further benefit of the in situ KBSC calibration with the up-to-date CPs. Furthermore, in Fig. 8 we can see the systematic trend, which increases with the vertical angle.

We analyzed the estimated CPs of the KBSC and noticed that some of them were almost perfectly correlated (6 out of 8 shared ≥ 98% correlations). Additionally, some offset parameters had unrealistically high values. The horizontal beam offset (x_{1n}) had a value of 4.1 mm, which is an order of magnitude higher than expected for the factory-calibrated TLS. These results showed that the KBSC partially failed due to the measurement configuration that was insufficient to correctly estimate all CPs. This insufficient configuration led to the extreme parameter correlations making some CPs inseparable their estimates biased.

From the literature, it is known that tilt and offset CPs (Table 1) are hardly separable if the sufficient variation in the distances for similar elevation angles cannot be achieved [16]. Therefore, to account for the latter problem, we used the offset CPs estimated on the calibration field as given values and applied the KBSC to estimate only the remaining tilt CPs. The results are presented in Fig. 10. The point cloud accuracy is noticeably better than in Figs. 8 and 9. This indicates that the adequate combination of the a priori and in situ calibration strategies can deliver the highest measurement accuracy.

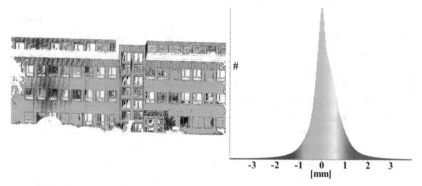

Fig. 10 M3C2 differences of two-face measurements combining factory, targets and keypoints calibration (top), corresponding histogram (bottom) - solid blue/red: ± 1.5 mm threshold

4 Discussion

The results presented in this study confirm that measurements of the TLS Z + F Imager 5016 are a few times more accurate than the values provided in the datasheet [13]. This is analyzed on two common measurement subjects, a façade, and an indoor hall. This points out that the further calibration is only relevant for the tasks demanding extraordinary measurement accuracy on the limit of the instrument's possibilities.

Directly comparing the in situ success of two user-oriented calibration approaches against the manufacturer's calibration indicates that the user-oriented calibration can achieve comparable results, when two-face measurements are analyzed. Hence, if necessary, it could substitute the costly manufacturer's calibration to some extent. However, each of the calibration approaches has certain advantages. The advantage of the manufacturer's calibration is that they have an in-depth knowledge of the instrument assembly that surpasses one of the end-users. Hence, the manufacturer's functional model of the CPs is more comprehensive and their calibration procedure is optimized for the estimation these CPs. However, this approach is made a priori in the manufacturer's facilities, upon the instrument's assembly. Hence, the CPs are likely to partially change with time. Therefore, the advantage of user oriented-calibration is that it is more up-to-date than the manufacturer's calibration. So it can account for the small geometrical changes that occurred in the instrument.

Hence, for achieving the highest accuracy, the initial manufacturer's factory calibration is necessary for the comprehensiveness, while the user-based calibration should be used to account for changes of the CPs. In this case, as our data suggests (Fig. 6), that focusing only on the most relevant set of the CPs (Table 1) is sufficient.

When comparing the target-based and keypoint-based user-oriented calibration approaches, some important advantages and disadvantages should be considered. The TBSC is more comprehensive and it is not limited to two-face sensitive CPs. Additionally, the calibration field can be specifically designed to accurately estimate all relevant CPs with pre-defined quality criteria, such as precision, correlations and reliability [14]. The main disadvantage of the TBSC is that it is intended for the a priori calibration, similarly to the manufacturer's calibration. Hence, it also suffers from the fact that the CPs partially change over time. The second disadvantage originates from the fact that when designing a calibration field we aim at a cost-efficient solution with minimal time and effort necessary for the calibration. Hence, if the functional model of the CPs for the specific instrument is not a priori well known, this approach can hardly be used to detect and model missing CPs.

The KBSC can be considered as a complementary for the TBSC. Namely, the main advantages of this approach are the possibility of the in situ use, delivering up-to-date CPs and that it has highly redundant observations, allowing the detection of eventually missing CPs. However, it is limited to two-face measurements, and it strongly depends on the geometry of the environment found during the measurements. Hence, it cannot guarantee comprehensive calibration with pre-defined quality criteria.

To summarize, the optimal workflow comprises the manufacturer's exhaustive calibration upon the instrument's assembly, a priori TBSC in regular time intervals as a cost-efficient alternative for the costly factory re-calibration. Finally, the in situ calibration should account for the remaining small changes for all CPs that can be estimated regarding the limitations of the given measurement geometry.

Finally, there is a possibility to use the combination of both user-based calibration approaches. In the case of the a priori calibration, this would lead to comprehensiveness due to target based calibration and improved calibration parameter precision due to redundant observations from keypoints. However, this increase in precision is doubtfully beneficial for the later CP use, as it is known that the CP stability is limited. A larger gain could be achieved with the in situ combination of both approaches, where the targets could supplement an eventually deficient measurement configuration, which would lead to more comprehensive and accurate calibration results.

5   Conclusion

In this study, we analyzed different strategies for the calibration of panoramic terrestrial laser scanners. In short, we compared: manufacturer's calibration, two user-oriented calibrations (target-based and keypoint-based) and their combinations. The special focus is placed on the point clouds depleted of the factory calibration. Our results indicate that, if necessary, both user-oriented calibrations can substitute the factory calibration, if the list of the calibration parameters is comprehensive. Furthermore, our data demonstrated that the most accurate point clouds are acquired when all three calibration approaches are appropriately combined, which successfully combines the main advantages of each individual approach. These findings can help the end-users to decide if they require additional instrument calibration for their measurement task. In the case they do, the demonstrated experiments can help to select the optimal strategy regarding the cost-efficiency and accuracy demands of the end-users.

Finally, we introduced several improvements to the original implementation of the keypoint-based calibration approach. This yielded approximately 18 times higher number of well-distributed observations in comparison to the original implementation leading to improved calibration results and the possibility to eventually detect eventually missing calibration parameters.

Acknowledgements. The authors would like to express gratitude to Zoller + Fröhlich GmbH for providing us the unique opportunity to investigate their flagship TLS Imager 5016.

References

1. Mukupa, W.; Roberts, G.W.; Hancock, C.M.; Al-Manasir, K. A review of the use of terrestrial laser scanning application for change detection and deformation monitoring of structures. *Surv. Rev. 2017, 49,* 99–116.
2. Walsh, G. Leica ScanStation P-Series – Details that matter. Leica ScanStation - White Paper, Online: http://blog.hexagongeosystems.com/(accessed on May 21, 2019).
3. Holst, C.; Neuner, H.; Wieser, A.; Wunderlich, T.; Kuhlmann, H. Calibration of Terrestrial Laser Scanners / Kalibrierung terrestrischer Laserscanner. *Allg. Vermessungs-Nachrichten 2016, 123,* 147–157.
4. Mettenleiter, M.; Härtl, F.; Kresser, S.; Fröhlich, C. *Laserscanning—Phasenbasierte Laser-messtechnik für die hochpräzise und schnelle dreidimensionale Umgebungserfassung,* Süd-deutscher Verlag, Munich, Germany, 2015.
5. Lichti, D.D. Error modelling, calibration and analysis of an AM-CW terrestrial laser scanner system. *ISPRS J. Photogramm. Remote Sens. 2007, 61,* 307–324.
6. Reshetyuk, Y. Self-calibration and direct georeferencing in terrestrial laser scanning, KTH Stockholm, 2009.
7. International Organization for Standardization (ISO) Optics and optical instruments – Field procedures for testing geodetic and surveying instruments – Part 9: Terrestrial laser scanners. *2018.*
8. Gielsdorf, F.; Rietdorf, A.; Gruendig, L. A Concept for the Calibration of Terrestrial Laser Scanners. *Proc. FIG Work. Week. Athens, Greec 2004,* 1–10.
9. Chan, T.O.; Lichti, D.D.; Belton, D. A rigorous cylinder-based self-calibration approach for terrestrial laser scanners. *ISPRS J. Photogramm. Remote Sens. 2015, 99,* 84–99.
10. Holst, C.; Kuhlmann, H. Aiming at self-calibration of terrestrial laser scanners using only one single object and one single scan. *J. Appl. Geod. 2014, 8,* 295–310.
11. Medić, T.; Kuhlmann, H.; Holst, C. Automatic in-situ self-calibration of a panoramic TLS from a single station using 2D keypoints. In *ISPRS Annals of the Photogrammetry, Remote Sensing and Spatial Information Sciences;* 2019.
12. Vosselman, G.; Maas, H.G. *Airborne and Terrestrial Laser Scanning;* Whittles Publishing, 2010; ISBN 9781439827987.

13. GmbH, Z. + F. Z + F Imager 5016 Preliminary Data Sheet 2016, 1–3.
14. Medić, T.; Kuhlmann, H.; Holst, C. Designing and evaluating a user-oriented calibration field for the target-based self-calibration of panoramic terrestrial laser scanners. *Remote Sens. 2020*, 12(1), 15.
15. Janßen, J.; Medić, T.; Kuhlmann, H.; Holst, C. Decreasing the uncertainty of the target centre estimation at terrestrial laser scanning by choosing the best algorithm and by improving the target design. *Remote Sens. 2019, 11 (7).*
16. Medić, T.; Kuhlmann, H.; Holst, C. Sensitivity Analysis and Minimal Measurement Geometry for the Target-Based Calibration of High-End Panoramic Terrestrial Laser Scanners. *Remote Sens. 2019, 11*, 1519.
17. Förstner, W.; Gülch, E. A Fast Operator for Detection and Precise Location of Distict Point, Corners and Centres of Circular Features. In *Proceedings of the ISPRS Conference on Fast Processing of Photogrammetric Data*; Interlaken, 1987, 281–305.
18. Förstner, W.; Wrobel, B.P. *Photogrammetric Computer Vision*; Springer International Publishing Switzerland, 2016; ISBN 9783319115498.
19. Lague, D.; Brodu, N.; Leroux, J. Accurate 3D comparison of complex topography with terrestrial laser scanner: Application to the Rangitikei canyon (N-Z). *ISPRS J. Photogramm. Remote Sens. 2013, 82*, 10–26.
20. Muralikrishnan, B.; Ferrucci, M.; Sawyer, D.; et al. Volumetric performance evaluation of a laser scanner based on geometric error model. *Precis. Eng. 2015, 40*, 139–150.

Determining Variance-Covariance Matrices for Terrestrial Laser Scans: A Case Study of the Arch Dam Kops

Gabriel Kerekes$^{(\boxtimes)}$ ⓘ and Volker Schwieger ⓘ

Institute of Engineering Geodesy – University of Stuttgart, Geschwister-Scholl-Str. 24D, Stuttgart, Germany
gabriel.kerekes@iigs.uni-stuttgart.de

Abstract. Deformation and displacement monitoring of arch dams is usually conducted by measurement methods that gained widespread acceptance. Mostly pointwise measurements acquired within different epochs serve as the basis for deformation analysis. The uncertainty information of these observations is generally established. When referring to surface-based deformation monitoring, terrestrial laser scanning (TLS) is intensively hyped up. Nevertheless, current knowledge about stochastic modelling of TLS observations is scarce and very often reduced to a diagonal variance-covariance matrix (VCM). This neglects the exiting correlations within the point clouds. Aiming to fill the gap, this paper shows one possibility of obtaining a fully populated variance-covariance matrix using the elementary error model (EEM). Previous publications on this topic described the application of EEM for the Leica HDS 7000 laser scanner and showed results on simulated data for laboratory objects. Within this paper instrumental errors of the same scanner are modeled differently, according to recent work. A real scan of the arch dam Kops in Austria highlights influences on the variances, covariances and correlations of points within the point cloud for two instrument error models. Findings show that the model choice does not bring a notable change in the error of positions, but influences the correlation coefficients up to a level of $\Delta\rho = 0.2$.

Keywords: Terrestrial laser scanning · Variance-covariance matrix · Elementary error model

1 Introduction

Surface-based deformation monitoring by Terrestrial Laser Scanning (TLS) measurements is gaining acceptance in dam monitoring. In the Dam Surveillance Guide, the International Commission on Large Dams (ICOLD) describes laser scanning as a recent development for deformation monitoring of a dam's surface [1].

As regards the geodetic community [2], efforts are made in adapting methods used for pointwise monitoring to achieve rigorous deformation analysis and significance tests [3]. One of the main issues is lack of knowledge regarding the modeling of the complete error budget also referred to as stochastic model [4].

A. Kopáčik et al. (Eds.): *Contributions to International Conferences on Engineering Surveying*, SPEES, pp. 57–68, 2021. https://doi.org/10.1007/978-3-030-51953-7_5

This is crucial for high precision tasks, like deformation monitoring without the need of signalizing discreet points on the objects or having access to them [5]. First examples of concrete dam TLS monitoring are given by Grimm-Pitzinger and Rudig [6], Alba et al. [7], Heine et al. [8] and Eling [9]. According to Neuner et al. [10], none of these methods are considered established in engineering geodesy, therefore more research is needed.

One method of defining a stochastic model is the Elementary Error Model (EEM). Previous publications on this topic (cf. [11, 12]) set the foundation for the EEM applied on TLS and results were shown on simulated datasets of small laboratory objects. A separation between the influencing error sources was not discussed.

The current paper is concentrated only on TLS instrumental error sources based on available knowledge about the panoramic laser scanners. Differences between two models are outlined. Examples are shown on a real scan of the dam Kops in Austria.

Continuing this paper, Sect. 2 shortly reviews the importance and possibilities of determining variance-covariance matrices (VCMs). In Sect. 3 the instrumental error models of the Leica HDS 7000 are discussed. A case study of the Arch Dam Kops and the EEM results are presented in Sect. 4. The focus is set on presenting the impact on the error of position and correlation coefficients for both models. Also the individual contribution of each instrumental error is studied. Section 5 concludes this contribution.

2 Determining Variance-Covariance Matrices in Geodetic Monitoring

In point-wise deformation monitoring each epoch leads to a set of parameters needed for the analysis. These include adjusted coordinates, VCMs and variance factors of the unit weight for each epoch [13]. An established straightforward analysis is the congruency test. The VCMs play an essential role, defining the stochastic model in the global congruency test and directly influencing the test statistic within the hypothesis test [14]. In surface-based approaches the stochastic model is possibly needed to estimate an epoch representative surface. One example is the estimation of freeform surfaces [15]. Harmening et al. [16], use the fully populated VCM to compute a weighting matrix for control points in NURBS.

One straight forward method of filling a VCM of the observations is using the instrument manufacturer accuracy information. This leads to a diagonal matrix where no correlations between the observations are assumed. Zhao et al. [17] proved that a simplified stochastic model used to estimate B-Spline surfaces of a simulated data set returns misleading conclusions of the congruency test, especially in the case of small deformations. A similar conclusion is drawn by Jurek et al. [18], in which neglected correlations of TLS observations falsify the estimation of plane parameters.

In the EEM, a stochastic model is defined by a so called "synthetic variance-covariance matrix" Σ_{ll}, which describes the impact of variances, covariances and correlations of point clouds. Multidimensional normal distributed observations are assumed within the model. The impacts are classified into instrumental, atmospheric and object related error sources. For details concerning the general EEM theory, the reader is advised to consult the abovementioned publications and Schwieger [19]. Other methods like

GUM [20] or Monte-Carlo simulations [21] are also used to define stochastic models, but this will not be further detailed in this contribution.

Without asserting that one stochastic model is better than the other, the possibility of deriving a VCM using the EEM is presented and the topic is restricted to instrumental errors.

3   The Elementary Error Model for Terrestrial Laser Scanners

3.1   General TLS Error Sources

Applying the EEM theory on laser scanners requires a classification of all error sources. To begin with, instruments are only realizations of an idealistic measurement system, therefore affected by physical manufacturing accuracies. All measurements, weather outdoor or indoor, are more or less affected by the environment through which the electromagnetic waves are travelling [22]. Further influencing sources are related to the measured object properties. Surface material, roughness and color all play an important role on the distance measurements. Scanning geometry, on the other side, is directly interconnected with the aforementioned object related errors [23]. Moreover, instrumental and environmental errors are also dependent on scanning geometry; therefore the risk of modelling the same errors twice is high. Other authors prefer classifying error sources in four categories (cf. [4, 22]). Further on, the focus is set only on instrumental elementary errors. All other error sources can be consulted in [11].

3.2   Instrumental Elementary Errors of the Leica HDS 7000 Scanner

Similar to previous publications [11], instrumental elementary errors for the HDS 7000 scanner are modeled as non-correlating and functional correlating errors. The angular measurement noise σ_λ, σ_θ and range noise σ_R are classified as non-correlating elementary errors. Opposite to how the term range noise is defined by the manufacturers as linearity RMS error [24], we adopt a constant value as the internal noise of the range measurement unit, therefore the variance-covariance matrix for the non-correlating term is structured diagonally as follows:

$$\sum_{\delta\delta,k} = diag\left(\sigma_\lambda^2\,\sigma_\theta^2\,\sigma_R^2\right). \tag{1}$$

As known, calibration models are proprietary information of the manufacturers and made available only in rare situations. Due to this, researchers are obliged to redefine working calibration models.

Further on, two functional models are presented in parallel. Both of them model instrumental errors of panoramic TLS. At the end of Sect. 3.3., a brief comparison with advantages and disadvantages of the models is given.

Since not all the calibration parameters (CPs) from one have a correspondent in the other, new CPs names are adopted. This gives a better overview of the comparison (Table 1). Original names are only given to be coherent with the model authors. The

Table 1. Overview of the instrumental elementary errors with standard deviations

Correlation type	Parameter model 1	Standard deviation	Parameter model 2	Standard deviation	Equivalence in total station model
Non-correlating errors	$\sigma_R = 0.5$ [mm]		$\sigma_R = 0.5$ [mm]		Identical
	$\sigma_\lambda, \sigma_\theta = 0.8$ [mgon]		$\sigma_\lambda, \sigma_\theta = 0.8$ [mgon]		
Functional correlating errors	Not modeled		a_1 [mm]	0.21	No equivalent
			a_2 [mm]	0.03	
	b [mm]	0.20	b [mm]	0.20	Identical
	c [ppm]	0.30	Not modeled		No equivalent
	d [mgon]	0.44	d [mgon]	0.44	Identical
	e [mgon]	0.48	e_1 [mgon]	0.48	Partially modeled
			e_2 [mgon]	0.48	
			e_3 [mgon]	0.48	
	f [mgon]	0.53	f [mgon]	0.53	Identical
	g [mgon]	0.13	Not modeled		No equivalent
	h [mm]	0.08	h_1 [mm]	0.08	Partially modeled
			h_2 [mm]	0.08	
			h_3 [mm]	0.08	

first model (*Model 1*) is based on a parallel between TLS and total stations [25]. For comprehension the new annotations are presented [25, 26]:

$$R_k = b + R \cdot c \tag{2}$$

$$\theta_k = \mathrm{acos}(\cos(e) \cdot \cos(d) \cdot \cos(\theta + \Delta\theta + f) - \sin(e) \cdot \sin(d)) \tag{3}$$

$$\lambda_k = \lambda + \Delta\lambda + \frac{h}{R_k \cdot \sin(\theta + \Delta\theta + f)} + \mathrm{atan}\left(\frac{\cos(e) \cdot \tan(d)}{\sin(\theta + \Delta\theta + f)} + \frac{\sin(e)}{\tan(\theta + \Delta\theta + (f))} \right) \tag{4}$$

$$\Delta\lambda = g \cdot \sin\lambda_z \cdot \cot\theta_z \tag{5}$$

$$\Delta\theta = g \cdot \cos\lambda_z \tag{6}$$

Here the index k defines the corrected observations, R_k range, λ_k horizontal angle and θ_k vertical angle. The other calibration parameters (CPs) are defined as follows with their original notation in brackets:

b (k_0) zero point error; c (m_0) scale error; d (c_0) collimation axis error; e (i_0) horizontal axis error; f (h_0) vertical index error; g (v_0) tumbling error; h (e_z) eccentricity of the collimation axis.

This model also considers the tumbling error which may occur when the scanner is rotating around the vertical axis [27]. The impact of this error can be approximated by Eqs. 5 and 6. Here λ_z represents the horizontal direction of the projection of the zenith angle into the horizontal plane and θ_z defines the angle between the vertical axis and the direction to the zenith. The notations have been chosen to maintain coherence as much as possible with their original authors.

A more recent functional model is established by Muralikrishnan et al. [28] where a set of 18 CPs are used to model the instrumental errors of panoramic laser scanners. Not all 18 parameters are determinable through typical calibration routines and some of them are not separable or sensitive to two-face measurements. For this reason, a simplified version of this model has been adapted by Medić et al. [29, 30], further referred to as *Model 2*. Here, 11 relevant CPs are used to possibly reduce the systematic instrumental errors (cf. [31]). They are mathematically modeled as part of the polar coordinates. Only the correction term is given here:

$$\Delta R = h_2 \Delta \sin \theta + b \tag{7}$$

$$\Delta \lambda = \frac{a_1}{R \cdot \tan \theta} + \frac{a_2}{R \cdot \sin \theta} + \frac{e_3}{\tan \theta} + \frac{2 \cdot d}{\sin \theta} + \frac{h_1}{R} \tag{8}$$

$$\Delta \theta = \frac{h_3 \cdot \cos \theta}{R} + f + e_1 \cdot \cos \theta - \frac{a_1 \cdot \sin \theta}{R} - e_2 \cdot \sin \theta \tag{9}$$

where the equivalent names a to h and original according to [28, 29] are:

$h_1 (x_{1n})$ horizontal beam offset; $a_1 (x_{1z})$ vertical beam offset; $h_2 (x_2)$ horizontal axis offset; $a_2 (x_3)$ mirror axis offset; $f (x_4)$ vertical index error; $e_1 (x_{5n})$ horizontal beam tilt; $e_2 (x_{5z})$ vertical beam tilt; $d (x_6)$ collimation axis error; $e_3 (x_{5z-7})$ combined vertical beam tilt and horizontal axis error (tilt); $b(x_{10})$ zero point error, $h_3 (x_{1n-2})$ combined effect of x_{1n} and x_2.

For detailed explanations of the CPs and their combined effects (e.g. e_3, h_3) the reader is advised to consult [30, 31]. Recently, it has been shown that some CPs can be determined with a certain variance level [32]. It is therefore interesting to use this in the EEM and consider the reported variances in the VCM of the functional correlating errors $\Sigma_{\xi\xi}$ (cf. [11]). Influencing matrices F contain the partial derivatives of Eqs. 2–4 (Model 1) and Eqs. 7–9 (Model 2) with respect to each CP. According to the matrix structure [11], this yields a fully populated matrix.

3.3 Classification Overview and Comparison of Instrumental Error Sources

In order to have a plausible comparison of the calibration models, the physical meaning [33] and value of the CPs must be same. This is in some cases not possible, because some of them do not exist in the calibration model (e.g. $a_2(x_3)$ mirror axis offset and c scale factor), or their equivalence is only partially defined. If CPs are only partially modeled, an average value is given to maintain comparability. For more details about how the equivalence is defined, the reader is advised to read [33].

Both models try to model the instrumental errors of panoramic TLS as good as possible considering that (a) some error sources may not be relevant in certain situations and (b) that the effect of some error sources can be neglected for individual observation types (R, λ, θ). For example, model 1 considers a scale factor which may be advantageous for long distances. There is however, no scale factor in model 2. Also the tumbling error is accounted for in model 1, which is neglected in model 2. On the other side, model 2 has parameters that are specific only for panorama scanners with 45° rotating mirrors (a_1, a_2), which makes it appropriate for modeling errors of this specific type of TLS, fact proven by the number of recent publications on the topic (cf. [29–31]), but these parameters are closely analyzed in Sect. 4.2.

4 Study Case—Arch Dam Kops

4.1 Arch Dam Kops

Between the 29th of July and of 2nd of August 2019, a first measurement campaign of the Kops dam took place in the Austrian Alps. The Kops water dam is a storage concrete dam built between 1962 and 1969. It is considered a hybrid type made out of a gravity dam and arch dam with artificial counterfort or abutment [34]. It retains a volume of almost 43 Mil. m³ of water, thus creating the 1 km² "Kopssee" lake [35]. Only measurements of the downstream (airside) arch dam are considered, since this can be interesting to model by means of B-Spline surfaces. For this reason, its dimensions are mentioned to give a general impression. The crown spans over 400 m, its height is 122 m from foundation to crest and has a crest width of 6 m.

Even though laser scanning is not used by the Vorarlberger Illwerke AG measurement team for monitoring the Kops dam, there have been several studies that involved scanning of its airside. Heine et al. [8] report differences between two scanning epochs in September 2008 and April 2009 of maximum 5 cm at the arch crest middle point. This corresponded to the optical plummet measurement of the Vorarlberger Illwerke AG team. The comparison was made directly between point clouds or to a triangulation mesh without considering statistical tests.

4.2 Synthetic Variance-Covariance Matrix of Kops Dam

Within this campaign, a part of the TLS measurements were conducted with the Leica HDS 7000. Phase-based scanners are known to reach better distance measurement accuracies [24] than for e.g. time-of-flight (TOF) scanners, fact that has been recently reconfirmed [36]. One exception to this rule may be TOF scanners that use waveform digitizing technology (e.g. Leica P50).

Nevertheless, the measurement site is chosen as close as possible to the dam, while covering most of its surface. The station point is situated approximately 50 m at the base of the dam's airside and the distances at the crest extremities reach 170 m. Along the crest, zenith angles vary between 29 and 61 gons. An important notice is that the scanner was not calibrated in situ, but values from literature (Sect. 3.3) are used to exemplary show the error impact in each case.

Before applying the EEM on the point clouds, only points on the dam are selected. Coordinates (X, Y, Z) are considered instead of observations (R, λ, θ), because the VCM will be used for estimating the geometric primitives of a B-Spline surface in the future (cf. [15]). Afterwards, a subsampling is necessary due to technical reasons. The reduction is argued by the fact that the EEM has been implemented in Matlab and computing VCMs with sizes of 21000×21000 is in the moment a limitation. The subsampling is achieved using a spatial sampling toolbox in Cloud Compare [36] and set to a minimum distance of 1.5 m between the points. This means that a random raster of neighboring points is selected based on the smallest neighbor distance of 1.5 m.

After determining the VCM for the reduced point cloud, the error of position is computed and visualized. Both EEM models consider only the instrumental error sources. This leads to mean errors of positions smaller than 5 mm (Fig. 1) which is in the same dimension order of a few mm as in [9]. This is expected, given the small variation level of CPs (few mgon and mm). It can be seen how both models follow the same linear pattern, constantly growing from the base to the dam's crest, reaching the maximum value where distances are up to 170 m. Differences in standard deviations between model 1 and 2 are nearly imperceptible on the whole point cloud, therefore a more in depth analysis follows.

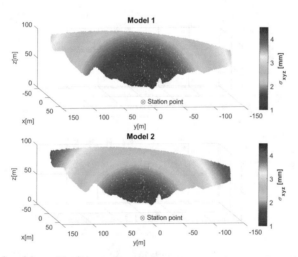

Fig. 1. Error of position with all instrumental elementary errors (Model 1 up, Model 2 down)

Having a fully populated VCM, it is interesting to analyze the existing spatial correlations between the coordinates in the point cloud. After computing the correlation matrix from the synthetic VCM, correlation coefficients can be analyzed. Six transversal sections and five vertical sections are analyzed as depicted in Fig. 2.

Corresponding spatial correlations are determined in each case from one point to all other points in the section. Note that the point distance is about ten times smaller than in the case of the whole point cloud, so a point spacing of approx. 15 cm is used in each section. In the vertical section, correlations between the first point of the section and all other points of the sections are presented. In the crest and base sections, the middle point is considered. Standard deviations of points are also presented along the respective

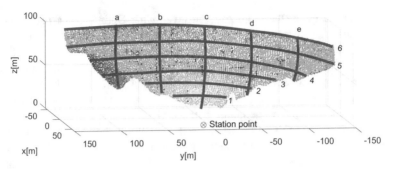

Fig. 2. Sections used for spatial correlation analysis

coordinate axis (e.g. only Z or only Y). Figure 3 shows how spatial correlations vary along the vertical section and do not differ too much for both models.

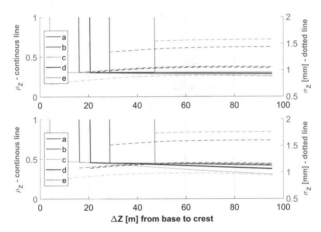

Fig. 3. Correlations and standard deviations for vertical sections (model 1 up, model 2 down)

Having a closer look at the vertical sections (Fig. 3) it can be observed how the trend for the correlation coefficients is almost the same for both models. Noteworthy is the fact that in model 1 (Fig. 3 up) all of them remain at a level of $\rho = 0.3$, whereas in model 2 (Fig. 3 down), the coefficients are higher having an offset of about 0.2. Likewise, the standard deviations are lower for model 1.

Proceeding to the horizontal sections (Fig. 4), a fist impression is that correlations coefficients vary along the length of the dam for both models. The same finding, that an offset in present between the correlation levels, is also valid here. Model 2 shows for all sections coefficients greater then $\rho = 0.5$, whereas model 1 computed for the same sections shows again lower coefficients with the same amount of 0.2. An interesting effect occurs along the crest (Fig. 4 up) at profile 6 for model 1, where a local minimum is reached. Therefore the individual effect of the instrumental errors helps in determining an explanation for this phenomenon. This difference between the models occurs as result of how the influencing parameters work in the EEM. As described in Table 1, parameters

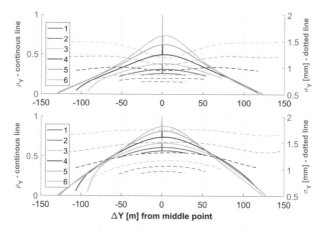

Fig. 4. Correlations and standard deviations transversal sections (model 1 up, model 2 down)

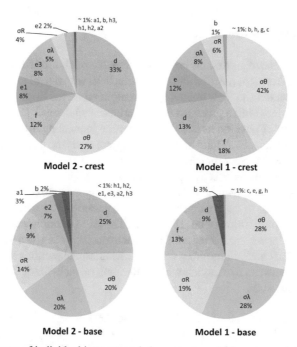

Fig. 5. Influence of individual instrumental elementary errors for a two points on profile c

$\sigma_R, \sigma_\lambda, \sigma_\theta, b, d, f$ have the same physical meaning, but only the modelling is different, meaning that the partial derivatives of Eqs. 2–9 are of different nature, therefore at the same level of variance (see d collimation axis error), the influence is up to 20% different. If the effect of d is closely examined (cf. [27], pp. 4, Fig. 2), it can be concluded that the local minimum is reached near the minimum distance in the horizontal plane. Regarding noise level, a big difference is notable in the influence of vertical angle noise where again a difference of 15% was found. In this case, model 1 (only functional-correlating errors) contributes 44% to the total error budget and the rest is covered by non-correlating elementary errors. Model 2 on the other side, has a contribution of 64%.

If the same situation is analyzed for a point near the scanner at approx. the collimation axis height (base of profile c), the results are as follows, in model 1, only 26% of the errors are due to the functional correlating error sources. Model 2 brings almost half of the contribution with 47%. Another noteworthy finding is, that the model 2 specific parameters (a_1, a_2), have a very small influence on the error of position at this level of variance; at a range of ca. 107 m less than 1% and at 48 m only a_1 reaches 3%. At this level of variance, one can conclude that some of the parameters can be neglected in both cases. All these differences are also responsible for the change in correlation level.

This shows that the individual contribution of each instrumental elementary error differs with regard to the relative position between scanner and observed object. This is why a further classification of scanning geometry as an extra error source is not seen as necessary by the authors.

5   Conclusion and Outlook

In this contribution the focus to show how two different models used for TLS instrumental errors influence the error of position and correlation level with regard to a real object. Two models were presented and their impact was exemplified on the Kops dam, highlighting the error of position within a given set of CP variances. Outcomes showed comparable values for the error of position as in [9], but the more interesting findings are related to the correlation coefficients. A total of 11 sections were analyzed and a general offset $\Delta\rho = 0.2$ between correlation coefficients of the same section was noticed. This is important in a future deformation analysis using B-Spline surfaces.

Furthermore, it was shown how the error budget is comprised of the individual error contribution. This lead to the conclusion that some parameters can be neglected in both models, having a contribution of ca. 1% or less. The EEM is improved by: (a) knowing which parameters have a small impact on the error of position at a given variance level, (b) knowing the spatial correlations along vertical and horizontal sections and (c) having the option to choose between two models when estimating surfaces for deformation analysis based on synthetic VCM.

Funding.   The authors cordially thank DFG (Deutsche Forschungsgemeinschaft) for funding the investigations of the IMKAD II Project under the sign SCHW 838/7-3.

We also want to express our gratitude to Illwerke vkv AG, especially to Dr.-Ing. Ralf Laufer for allowing and supporting the measurement campaign in 2019.

References

1. ICOLD (2018) Dam Surveillance Guide, Bulletin 158, pp. 107–108.
2. Heunecke, O.; Kuhlmann, H.; Welsch, W.; Eichhorn, A.; Neuner, H. (2013) Handbuch Ingenieurgeodäsie-Auswertung geodätischer Überwachungsmessungenm, Chapter 15, in Möser, M.; Müller, G.; Schlemmer, H. (Ed.) 2nd edition, Wichmann, Berlin, Germany.

3. Wunderlich, T.; Niemeier, W.; Wujanz, D.; Holst, C.; Neitzel, F.; Kuhlmann, H. (2016) Areal Deformation Analysis from TLS Point Clouds–the Challenge, Allgemeine Vermessungsnachrichten, Vol. 123, no. 11–12.

4. Kuhlmann, H.; Holst, C. (2018) Flächenhafte Abtastung mit Laserscanning-Messtechnik, flächenhafte Modellierung und aktuelle Entwicklungen im Bereich des terrestrischen Laserscanning, pp. 168–207. In: Ingenieurgeodäsie-Handbuch der Geodäsie, ed. W. Schwarz, Springer-Verlag, Berlin, Germany.

5. Gordon, S.J.; Lichti, D.D. (2007) Modeling Terrestrial Laser Scanner Data for Precise Structural Deformation Measurement, Journal of Surveying Engineering Vol. 133, No.22.

6. Grimm-Pitzinger, A.; Rudig, S. (2005) Laserscannerdaten für flächenhafte Deformationsanalysen, Proceedings of 13th Internationale Geodätischen Woche, Obergurgl, Austria.

7. Alba, M.; L. Fregonese, F. Prandi, M. Scaioni, P. Valgoi (2006) Structural Monitoring Of A Large Dam By Terrestrial Laser Scanning, IAPRS & SIS, 36(5), pp. 6–12.

8. Heine, E.; Reiner, H.; Weinold, T. (2009) Deformationsmessungen mit terrestrischen Laserscannern am Beispiel der Kops Staumauer, Chesi G., Weinold T. (Ed.): 15. Internationale Geodätische Woche Obergurgl, Herbert Wichmann Verlag.

9. Eling, D. (2009) Terrestrisches Laserscanning für die Bauwerksüberwachung, In: Wissenschaftliche Arbeiten der Fachrichtung Geodäsie und Geoinformatik der Leibniz Universität Hannover, Nr. 282.

10. Neuner, H.; Holst, C.; Kuhlmann, H. (2016) Overview on Current Modelling Strategies of Point Clouds for Deformation Analysis, Allgemeine Vermessungsnachrichten, Vol. 123, no. 11–12.

11. Kauker, S. & Schwieger, V. (2017) A synthetic covariance matrix for monitoring by terrestrial laser scanning. Journal of Applied Geodesy, 11(2), pp. 77–87.

12. Kauker, S.; Harmening, C.; Neuner, H.; Schwieger, V. (2017) Modellierung und Auswirkung von Korrelationen bei der Schätzung von Deformationsparametern beim terrestrischen Laserscanning. In: Lienhart, W. [Ed.], Proceedings of 18. Internationalen Ingenieurvermessungskurs, Graz, 2017, Wichmann Verlag, Berlin, pp. 321–336.

13. Ogundare, J. O. (2016) Precision Surveying: The Principles and Geoamtics Practice, John Wiley & Sons, Inc., New Jersey, pp. 300–301.

14. Niemeier, W. (2008) Ausgleichungsrechnung, 2nd edition, Walter de Gruyter, Berlin, Germany.

15. Harmening, C.; Neuner, H. (2015) Continous modelling of point clouds by means of freefrom surfaces, Österreichische Zeitschrift für Vermessung und Geoinformation (VGI),103.

16. Harmening, C.; Kauker, S.; Neuner, H-B.; Schwieger, V. (2016) Terrestrial Laserscanning-Modeling of Correlations and Point Clouds for Deformation Analysis, FIG Working Week 2–6 May 2016 Christchurch, New Zealand.

17. Zhao, X.; Kermarrec, G.; Kargoll, B.; Alkhatib, H.; Neumann, I. (2019) Influence of the simplified stochastic model of TLS measurements on geometry-based deformation analysis. J. of Applied Geodesy, 13(3), pp. 199–214.

18. Jurek, T.; Kuhlmann, H. & Holst, C. (2017) Impact of spatial correlations on the surface estimation based on terrestrial laser scanning. J. of Applied Geodesy, 11(3), pp. 143–155.

19. Schwieger, V. (1999) Ein Elementarfehlermodell für GPS Überwachungsmessungen, Schriftenreihe der Fachrichtung Vermessungswesen der Universität Hannover, Vol. 231.

20. Kirkup, L.; Frenkel, B. (2006) An Introduction to Uncertainty in Measurement: Using the GUM. Cambridge University Press.

21. Metropolis, N.; Ulam, S. (1949) The Monte Carlo Method. Journal of the Americal Statistical Association, Vol. 44, No. 247, pp. 335–341.

22. Soudarissanane, S. S. (2016) The Geometry of Terrestrial Laser Scanning-identification of errors, modeling and mitigation of scanning geometry, retrieved from: repository.tudelft.nl, last accessed on 20.01.2020.

23. Zámečníková, M.; Neuner, H.; Pegritz, S.; Sonnleitner, R. (2015) Investigation on the influence of the incidence angle on the reflectorless distance measurement of a terrestrial laser scanner. Österr. Z. Vermess. Geoinform 2015, 103, 208–218

24. Leica Geosystem AG (2011) Datasheet HDS 7000 Laser Scanner. Retrieved from http://w3.leica-geosystems.com/downloads123/hds/hds/HDS7000/brochures-datasheet/HDS 7000_DAT_en.pdf, last accessed on 20.01.2020.

25. Neitzel, F. (2006) Gemeinsame Bestimmung von Ziel-, Kippachsfehler und Exzentrizität der Zielachse am Beispiel des Zoller + FröhlichImager 5003, Luhmann/Müller(Ed.)– Photogrammetrie-Laserscanning-Optische 3D-Messtechnik, Proceedings oft he Oldenburger 3D-Tage 2006, Wichman, Berlin, Germany.

26. Strahlberg, C (1997) Eine vektorielle Darstellung des Einflusses von Ziel-und Kippachsfehler auf die Winkelmessung, Zeitschrift für Vermessungswesen, Vol. 5/1997, pp. 225–235.

27. Neitzel, F. (2006b) Untersuchung des Achssystems und des Taumelfehlers terrestrischer Laserscanner mit tachymetrischem Messprinzip. In: Proceedings of the 72. DVW-Seminar Terrestrial Laserscanning 2006 in Fulda, Volume 51, Wißner, Augsburg, pp. 15–34.

28. Muralikrishnan, B.; Ferrucci, M.; Sawyer, D.; Gerner, G.; Lee, V.; Blackburn, C.; Phillips, S.; Petrov, P.; Yakovlev, Y.; Astrelin, A.; Milligan, S.; Palmateer, J. (2015) Volumetric Performance Evaluation of a Laser Scanner Based on Geometric Error Model, Precise Engineering, Vol. 40, p. 139–150.

29. Medić, T.; Holst, C.; Kuhlmann, H. (2017) Towards System Calibration of Panoramic Laser Scanners from a Single Station, Sensors, Vol. 17, No. 5.

30. Medić, T.; Holst, C.; Kuhlmann, H. (2019) Sensitivity Analysis and Minimal Measurement Geometry for Target-Based Calibration of High-End Panoramic Terrestrial Laser Scanners Remote Sensing, Vol. 11, no. 1519.

31. Holst, C.; Medić, T.; Kuhlmann, H. (2018). Dealing with systematic laser scanner errors due to misalignment at area-based deformation analyses. J. of Applied Geodesy, 12(2), pp. 169–185.

32. Holst, C.; Medić, T.; Blome, M.; Kuhlmann, H. (2019) TLS-Kalibrierung: in-situ und/oder a priori?, 184th DVW-Seminar, Terrestrisches Laserscanning 2019, Fulda, Germany.

33. Chow, J. C. K.; Lichti, D. D.; Glennie, C.; Hartzell, P. (2013) Improvements to and Comparison of Static Terrestial LiDAR Self-Calibration Methods Sensors 2013, 13.

34. Ganser O. (1975) Staumauer Kops, Anlage der Drainagebohrungen, Auswirkung Dieser Massnahmen auf die Höhe des Bergwasserspiegels und die Grösse des Sohlenwasserdruckes. In: 12. Talsperrenkongreß in Mexiko 1976. Die Talsperren Österreichs, vol 22. Springer, Vienna, Austria.

35. Illwerke vkw AG (2020) Website: https://www.illwerkevkw.at/kopssee.htm, last accseed on 22.01.2020.

36. 36. Suchocki, C. (2020) Comparison of Time-of-Flight and Phase-Shift TLS Intensity Data for the Diagnostics Measurements of Buildings. In: Materials 2020, 13, pp. 353.

37. Girardeau-Montaut, D. (2019) Point processing with Cloud Compare, Presentation of Point Cloud Processing Workshop 2019 Stuttgart, retrieved from: pcp2019.ifp.uni-stuttgart.de/pre sentations/04-CloudCompare_PCP_2019_public.pdf, last accessed 24.01.2020.

Assessing the Temporal Stability of Terrestrial Laser Scanners During Long-Term Measurements

Jannik Janßen(✉), Heiner Kuhlmann, and Christoph Holst

Institute of Geodesy and Geoinformation, University of Bonn, Bonn, Germany
{j.janssen,c.holst}@igg.uni-bonn.de, heiner.kuhlmann@uni-bonn.de

Abstract. Due to improved technology terrestrial laser scanners (TLS) are increasingly used for tasks demanding high accuracy, such as deformation monitoring. Within this field, often long-term measurements are acquired, for which the temporal stability of the laser scanner's observations need to be assured or at least its magnitude and influence factors should be known. While these influence factors have been investigated for most of the geo-sensors taking part in long-term monitoring, it has not been investigated for TLS yet. In this study, we empirically reveal the stability of terrestrial laser scanner observations at long-term measurements. With these investigations, we can analyze the drifts, which occur in the polar observations during the warm-up phase and after it. It is shown that the drifts cause both rigid body movements and inner shape deformations of the point cloud. By re-stationing the scanner during long-term measurements, the drifts in the vertical angle in particular can be reduced by half.

Keywords: Laser scanner registration · Deformation monitoring · Sensor drift · Geo-monitoring · Free stationing

1 Introduction

By improved technology, calibration strategies, and registration approaches, terrestrial laser scanners (TLS) are increasingly used for tasks demanding high accuracy. One emerging field is TLS-based deformation monitoring. Within this field, often long-term measurements are acquired. This holds for applications in which a high point density is required (i.e. each measurement epoch is quite long) as well as for applications in which laser scans are repeated frequently (i.e. there are many short measurement epochs). In both cases, the temporal stability of the laser scanner's observations need to be assured or at least its magnitude and influence factors should be known. While this aspect has been investigated for most of the geo-sensors taking part at a long-term monitoring it has not been investigated for terrestrial laser scanners yet.

In Woźniak et al. [1] and Odziemczyk [2], for example, the temporal stability of total station measurements over several days is investigated. It is found that temperature

A. Kopáčik et al. (Eds.): *Contributions to International Conferences on Engineering Surveying*,
SPEES, pp. 69–84, 2021. https://doi.org/10.1007/978-3-030-51953-7_6

is a decisive factor for the stability of the measurements. Gucevic et al. [3] investigate the compensation error of levelling instruments at different temperatures and Elias et al. [4] investigate the influence of temperature on the calibration of measuring cameras. There are, however, barely any studies on the stability of terrestrial laser scanning measurements. Reshetuyuk [5] already investigated the influence of temperature on the calibration of laser scanners. He determined, for example, a range drift up to 3 mm during the warm-up phase of the scanner, but it is doubtful whether his results are still valid for newer scanners due to improved technologies.

The aim of this paper is to assess the stability of long-term measurements of TLS by means of empirical data. In general, we want to answer the following questions:

1. Which drifts can occur in the polar observations of the scanner during long-term measurements? Here, measurements over several days as well as the warm-up phase of the scanner are analyzed. In addition, the relations between drift and ambient air temperature and internal scanner temperature are investigated.
2. Can a simple rigid body movement of the scanner explain the previously determined drifts or do the drifts cause a shape deformation in the point clouds? While a rigid body movement of the scanner can be compensated only by translations and rotations of the point cloud, this is not the case for shape deformations. Shape deformations can, for example, be caused by temporal unstable calibration parameters and lead to scaling or shearing of the point clouds.
3. What is the consequence of the rigid body movement detected in relation to re-stationing of the laser scanner during long term measurements?

2   Instruments and Testing Procedures

For the investigations in this paper, repeated scans are recorded in a measurement laboratory at the University of Bonn. With each scan, ten BOTA8 targets [6] are measured with a Z + F Imager 5016. Seven targets are scanned in the first face and three targets in the second face of the scanner. All targets are fixed approximately in a distance of about 5 m to the scanner and at the height of the scanner. Figure 1 shows a sketch and a picture of the experimental setup. Figure 2 shows a panoramic intensity image from the scanner's point of view.

Two data sets of long-term measurements are captured:

- **Data set A**: The scanner is already warmed up completely (2 h warm-up scans). Every 20 min a full field of view scan is recorded with a scan resolution of 3.1 mm at 10 m. Due to the limit of the internal hard disk, 175 scans are captured, which results in a measuring time of approximately 59 h or 2.5 days.
- **Data set B**: The scanner has not warmed up yet. It is switched on and the consecutive scans are started immediately. 75 full field of view scans with a resolution of 3.1 mm at 10 m are recorded. In contrast to data set A, no pauses are performed between the individual scans. The scans are taken directly one after the other, resulting in long-term measurements of about 4:50 h. To record the warm-up phase more than once, the procedure is repeated on 5 days.

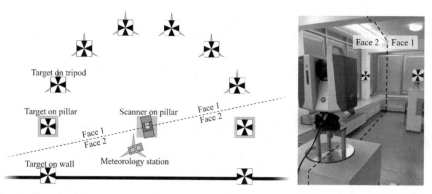

Fig. 1. Sketch of the experimental setup (left) and photo of the scanner and two targets (right)

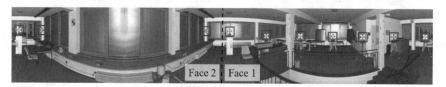

Fig. 2. Panoramic intensity image from the scanner's point of view with marked targets

For both data sets the ambient air temperature near the scanner is recorded in parallel to the scans using a high-precision meteorology station. Additionally, the internal temperature of the scanner is retrieved for each scan. This internal scanner temperature is a temperature issued by the manufacturer. It is not known where in the scanner this temperature is measured or for which parts of the scanner it is representative. Nevertheless, the recording of temperatures allows for a comparison between the ambient air temperature, the internal scanner temperature and any drifts.

During all scans conducted for the experiment, the compensator was switched on. The scanner is mounted on a concrete pillar as well as two targets. Two more targets are mounted in magnetic nests on the wall; the other six targets are mounted on heavy metal tripods. All tripods, pillars and nests are protected against direct sunlight by darkening the room with the help of sun blinds. To avoid short-term fluctuations of the temperature, we closed the laboratory to persons during scans.

The Z + F Imager 5016 scanner is specified by the manufacturer with an angular accuracy of 14.4" ($\hat{=}$0.35 mm at 5 m), a linearity error of less than 1 mm + 10 ppm and a distance noise between 0.2 mm and 0.3 mm to 10 m distance [7]. However, the empirical precision of the target centers is much higher: The target centers are estimated by means of an algorithm using image correlations [6]. For these estimated target centers standard deviations of less than 3" ($\hat{=}$0.1 mm at 5 m) for the angular precision and a standard deviation of less than 0.1 mm for the range precision are realistic values [6, 8].

It should also be noted that the assessments of the temporal stability in this paper do not refer to individual scan points. Only the polar coordinates of targets are considered. These coordinates of the targets have a higher precision than single points, so that even small drifts in the polar coordinates become visible. However, the findings on drifts should also be transferable to other modelled objects, such as planes or cylinders.

3 Analysis of the Long-Term Measurement

In Sect. 3.1, the long-term measurements with an already warmed-up scanner and scans over 2.5 days (data set A) are analyzed in detail. Section 3.2 analyzes the shorter 5 h measurements, which also include the warm-up phase of the scanner (data Set B). In Sect. 3.3, the target stability is verified by means of keypoints in the scans.

3.1 Analysis of Data Set A

To investigate the long-term stability of the scanner observations during the 2.5 days, the target coordinates of the first scan are subtracted from the target coordinates of all subsequent 174 scans. The resulting drifts for the polar observations of the targets related to the first scan (range r, horizontal direction h, vertical angle v) are visualized in Fig. 3.

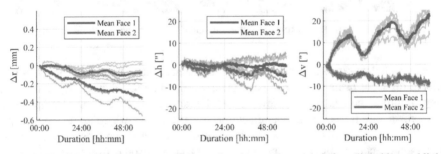

Fig. 3. Drift of the targets' polar coordinates colored according to their face (light blue and light red) and the mean drift of the coordinates for each face (dark blue and dark red)

The investigations regarding the stability of the range r indicate that individual coordinates drift up to half a millimeter in distance during the 2.5 days. However, the majority of the targets change their distance by less than 0.2 mm. Conspicuously, all drifts have the same sign; so that a rigid body movement of the scanner cannot explain the drift (see Sect. 4). Possible reasons can be an additive constant or a scale error, which change with time. The fact that the range of the second scanner face drifts more than that of the first face cannot be explained.

The drift of the horizontal direction h is small. On average, the directions drift by a maximum of 3" ($\widehat{=}$0.1 mm at 5 m), which is far below the specified angular accuracy of the scanner, but equal to the empirical precisions.

The biggest systematics and drifts can be noticed in the vertical angle v. Clear differences can be seen between the polar elements in the first and second face, indicating that they are based on time-varying misalignments of the scanner. The mean value of the first face has drifted by over 20" ($\widehat{=}$0.5 mm at 5 m) after 2.5 days, the mean value of the second face by about 8" ($\widehat{=}$0.2 mm at 5 m).

The reason for the large drifts cannot be determined from the data, but it is clear that it cannot be explained by a time-dependent vertical index error alone. For this, the magnitude of the drifts in the first and second face should be the same only with reversed signs.

Besides the polar observations, the measured air temperature and the internal scanner temperature are also included in the analysis. In Fig. 4, on the left side, both temperatures are shown dependent on time.

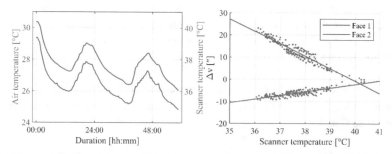

Fig. 4. Air and scanner temperatures over time (left) and relation between vertical angle drift and scanner temperature

It can be stated that the air and scanner temperature time series are very similar. However, the scanner temperature is always 10 °C higher than the air temperature. Furthermore, it can be seen that the two temperature curves show daily variations of about 2 °C, which is very probably caused by general ambient temperature change in the room. Because of the constant offset also the internal scanner temperature varies about 2 °C per day. These diurnal variations can also be guessed from the polar observations and are particularly evident for the drift of the vertical angle Δv. The correlations between the temperatures and polar observations are shown in Table 1.

Table 1. Absolute correlations between the mean drift of the polar observations and air as well as scanner temperatures

Correlations	Range r			Horizontal direction h			Vertical angle v		
	Face 1 (%)	Face 2 (%)	Mean (%)	Face 1 (%)	Face 2 (%)	Mean (%)	Face 1 (%)	Face 2 (%)	Mean (%)
Air temp.	21	69	45 ·	32	90	61	95	86	91
Scan. temp.	22	71	46	35	91	64	95	86	91

From the table of correlations, essentially three things can be seen:

1. The correlations between the air and scanner temperatures do not differ much. This is due to the similarity of the two temperature curves. The correlation between air

and scanner temperature is 99.7%. The time shift between the change of the ambient temperature and the scanner temperature is not detectable, because of the sampling interval of 20 min (see Sect. 2).

2. The correlations between temperature and distance drift are the smallest (about 46%), the correlations between temperature and vertical drift are the largest (about 91%). The latter suggests a strong correlation between temperature and vertical drift, so we investigate this more closely. Figure 4, on the right side, shows the vertical drift as a function of the scanner temperature. From the data points, a linear trend between drift and scanner temperature can be inferred. For the first face a drift of -5"/ 1 °C is estimated, for the second face the drift is only 2"/ 1 °C. The reason for the different gradients of the lines is based on the different magnitudes of the drifts between the first and second face.

3. A reason why the correlation values of temperature and the range drift and temperature and horizontal drift is clearly greater in the second face than in the first face cannot be found.

From the analysis of data set A, it can be stated that, in the case of long-time measurements over a period of 59 h (repeated scans), the coordinates of the targets move by about 1 mm on average. This is only true for scanners that are already warmed up. The movement by 1 mm can be essentially attributed to distance measurement with a continuous drift and a temperature-dependent drift in the vertical angle measurement. The horizontal direction proves to be very stable.

3.2 Analysis of Data Set B

Data set B is used to analyze measurements over a period of 5 h, which also includes the warm-up phase of the scanner. As for data set A, the temporal drift of the polar observations (Δr, Δh, Δv) related to the first scan is examined first (Fig. 5). In contrast to dataset A, only the mean of one face is plotted in the figure, not the individual targets. In addition, each day on which the experiment was repeated is represented by an individual line in the figure.

From the graphs, it can be derived that all three polar elements are influenced by the warm-up phase of the scanner, but to different extents. The range drifts up to 0.2 mm at the beginning of the measurements, the horizontal direction up to 10" and the vertical angle more than 40" in the worst case.

From the analysis of data set A (Sect. 3.1), we know which magnitudes of drift can occur in a warmed-up scanner. Thus, if the values for the drift are larger, we assume that these parts belong to the warm-up phase. With this approach, the lengths of the warm-up phase can also be estimated. The calculated warm-up times (WUT) are shown for all polar observations and each day in Fig. 5 with black circles. The mean WUT for both faces is given in Table 2. If the standard deviations of the warm-up times are considered, it becomes apparent that the exact values for the WUT are quite uncertain. The WUT of the distance measurement and horizontal direction are estimated to be approximately half an hour; the WUT of the vertical angle is estimated to be about one hour.

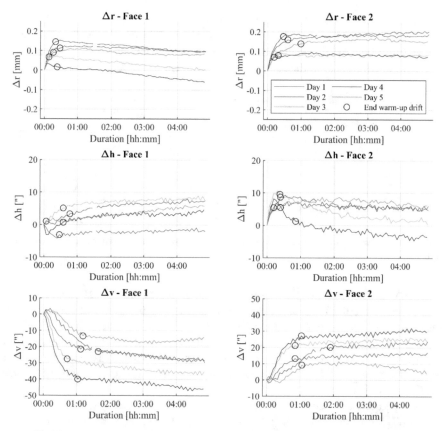

Fig. 5. Mean Drift of the polar observations during the scanner's warm-up phase

Table 2. Warm-up times (WUT) for polar observations and scanner temperature and the temperature magnitude $\Delta\Delta T$ that the scanner warmed up

	WUT Δr	WUT Δh	WUT Δv	WUT ΔT	$\Delta\Delta T$
Day 1	00:32	00:20	00:59	01:15	8,0
Day 2	00:24	00:36	01:46	01:31	11,6
Day 3	00:14	00:29	00:47	01:05	7,6
Day 4	00:18	00:43	01:02	01:22	11,3
Day 5	00:38	00:14	01:07	01:11	7,5
Mean	00:25	00:28	01:08	01:17	–
Std.	00:10	00:12	00:22	00:10	–

Since the WUT of the individual days varies strongly, it will be investigated whether there is a link to temperature. For this purpose, the scanner temperature in Fig. 6 (left) is examined first. It turns out that the temperatures differ between the individual days. This is explained by the fact that the scanner heats up to a certain value above the ambient air temperature (about 11 °C, see Sect. 3.1). As also the ambient temperature varies between the days, the differences in the scanner temperature are plausible.

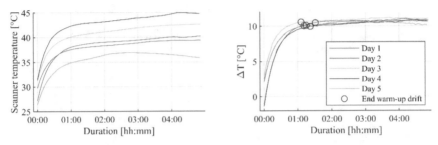

Fig. 6. Scanner temperature over time (left) and difference between scanner and air temperature (right)

In order to calculate the WUT from the temperature, the differences between air and scanner temperature ΔT are calculated and shown in Fig. 6 (right). All curves seem to converge over time to approximately 10–11 °C. However, even here the WUT differs between the days. It is assumed that the WUT of the temperature is related to temperature differences $\Delta \Delta T = \Delta T_{End} - \Delta T_{Start}$, which the scanner has to warm up in order to reach the target temperature (about 11 °C above air temperature). The temperature differences $\Delta \Delta T$ are also listed in Table 2.

If temperature differences $\Delta \Delta T$ are compared with the warm-up time of the temperatures (WUT ΔT) in Table 2, it can be seen that the higher the temperature magnitude the scanner warms up, the longer the WUT for the temperature.

The quintessence of the analysis of data set B is that the warm-up phase of the scanner affects all polar observations. The duration (approx. 30 min) and effect on range and horizontal angle are clearly smaller than for the vertical angle. The warm-up time of the vertical angle is about 1 h. Besides these approximate values, the warm-up times seems to be slightly different depending on the scanner and ambient temperatures.

3.3 Verifying the Target Stability Using Keypoints

As revealed by the analysis of the target stability (see Sects. 3.1 and 3.2), polar observations drift differently between the first and second face. We want to verify that these differences are not caused by the different ways of mounting the targets between the two faces (see Sect. 2). For this, the drifts of 2D keypoints in the intensity image is calculated. The procedure is similar to the 2D keypoint calibration of Medic et al. [9]:

1. By means of the Förstner operator [10], keypoints are calculated in 10° x 10° segments of the panoramic intensity image.
2. From corresponding image segments of two scans, the keypoints are matched and the drifts are calculated by comparing the polar coordinates of the matched 2D keypoints.

Figure 7 shows the keypoints and drifts for three exemplary image sections. As an example, the drifts between the first and 75th scan of the fourth day from data set B are depicted. It can be seen that the majority of the arrows point in a similar direction.

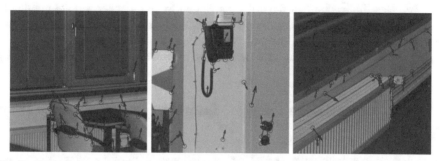

Fig. 7. Keypoints and calculated drifts of three exemplary image segments, two segments from face 1 (left and middle) and one segment from face 2 (right), all drifts are magnified 50 times

Since the precision of the individual keypoints or drifts is lower than the precision of the target centers [9], the median drift is calculated for each image segment. The median drifts (red arrows) are visualized together with the drifts of the target centers (blue arrows) in Fig. 8.

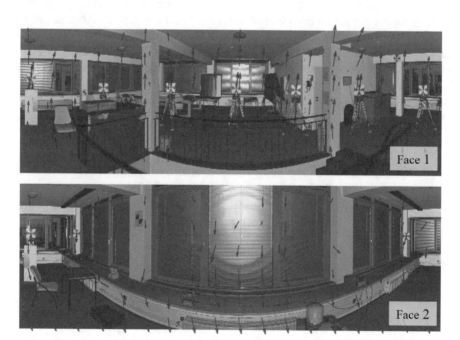

Fig. 8. Median drifts of the keypoints for all image segments (red arrows) and drifts of the target centers (blue arrows), all drifts are magnified 500 times

From a visual analysis of Fig. 8 two conclusions can be drawn: Firstly, the drifts of the keypoints (red arrows) confirm the drifts of the target center coordinates (blue arrows) very well. Secondly, the drifts differ mainly between the two scanner faces. All drifts of the first face point to the top and all drifts of the second face point to the bottom. This is a clear indication that the drifts are caused by the scanner and are not due to instability of the measurement environment, individual components of the room or ways of mounting the targets. Thus, the targets are further considered stable.

A processing of all days and scans with keypoints was avoided due to the complexity of the calculation.

4   Consequences of the Temporal Instability on the Measured Point Cloud

From the previous analysis, we know the temporal instability of the polar observations during long-term measurements and during the warm-up phase of the scanner. In this section we analyze if the drifts are causing a rigid body movement of the point cloud (Sect. 4.1) or if the drifts deform the inner shape of the point cloud (Sect. 4.2).

4.1   Rigid Body Movement of the Point Cloud

In order to analyze if the drifts cause a rigid body movement in the point cloud, we transform the consecutive scans to the last scan of the measurement series. As transformation equation we choose a rigid body transformation with six parameters. Three parameters describe the rotations of the point cloud around the coordinate axes and three parameters refer to the translations along the axes. The parameters are visualized in Fig. 9. If there was no drift, warm-up phase, or random measurement errors, we expect all six parameters to be exactly zero.

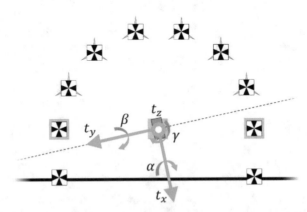

Fig. 9. Visualition of the transformation parameters α, β, γ, t_x, t_y and t_z

In the previous analysis we already found out that many drifts differ between the first and second face. Therefore, we applied the rigid body transformation once for each scan with the targets of each face separately and once for each scan with all targets of both faces. The obtained transformation parameters t_z and β for the data set B are exemplary shown in Fig. 10. Contrary to the figures in Sect. 3, the values are zero at the end and not at the beginning since all scans are transformed to the last scan.

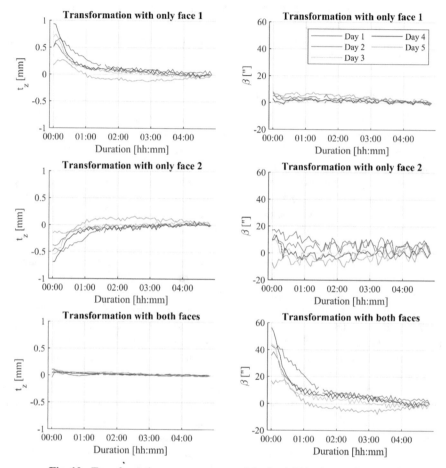

Fig. 10. Transformation parameters t_z and β of a rigid body transformation

From the left column of Fig. 10 it can be seen that the transformation parameter t_z differs clearly in the first hours. The reason for this drift is the drift of the vertical angle during the warm-up phase. The drifts of the vertical angle correspond to a height change of the targets. Since the vertical drifts differ in their sign between the first and the second face it seems that the targets in face 1 lift up and the targets in face 2 go down. This apparent movement in the target coordinates is compensated in the translation t_z along the z-axis corresponding to a rigid body movement of the point cloud in height direction.

If both faces are used for the rigid body transformation, the vertical drifts between the first and second face are contradictory. The translations of the two faces cancel each other out. Instead, the drifts of the targets in both faces can be better explained by the rotation β, a tilting around the y-axis (see Fig. 10, right column). This means that the drift of the vertical angle during the warm-up phase can be interpreted as a rigid body rotation of the scanner around the y-axis.

The higher noise of the transformation parameters in the second face compared to the first face can be explained by the fact that only three targets were scanned in the second face instead of seven in the first face.

As the analyses of the other transformation parameters and data set A has not yielded any further findings, the results are not discussed here.

4.2 Shape Deformation of the Point Cloud

It has to be checked whether the drifts cause only rigid body movement of the point cloud or also an inner deformation of the point cloud geometry, i.e. a shape deformation. For this purpose, the drifts of the polar elements are considered again, as already in Fig. 5, but now both before and after applying the rigid body transformation.

We assume the following: If the drifts in the polar elements were to cause only a rigid body movement, the drifts would have to be zero after applying the rigid body transformation. Remaining systematics in the drifts are due to drifts that cannot be described by the 6-parameter transformation. Thus, they cause a deformation of the inner shape of the point cloud.

Figure 11 shows the drifts before and after the rigid body movement with all targets of both faces. It should be noted that, in contrast to Fig. 5, all values refer to the last scan, since we assume that this scan is not influenced by the warm-up phase.

From the figure it is evident that the drifts become clearly smaller by applying the rigid body transformation. This is especially obvious for the vertical angle, but also for the range and horizontal drift in the first face. The reason for the clear improvement of the vertical angle drift lies in the rotation around the y-axis described in the section above, which compensates for big parts of the drift.

It is also apparent from Fig. 11 that not all drifts are balanced completely. Especially the drifts during the first hour (warm-up phase) in the range, the horizontal direction and in the second face are not compensated by the rigid body transformation. This is a clear indication that these drifts lead to an inner shape deformation of the point cloud.

In addition, the different number of targets in the first and second face leads to a stronger weighting of the first face, whereby the drifts of the first face are better compensated by the transformation than those of the second face.

In order to determine the ratio between rigid body movement and inner shape deformation, the mean absolute errors (MAE) between the coordinates of each scan and the last scan are calculated, both with and without rigid body transformation. From the values shown in Fig. 12 it is apparent that the MAE in the vertical angle can be clearly minimized by a rigid body transformation. As in Fig. 11 the improvement is smaller for the range and is not notable for the horizontal direction.

If the MAE before and after rigid body transformation are put into relation, the ratio of rigid body movement and internal shape deformation can also be calculated. In mean 23% of the range MEA can be reduced by a rigid body transformation. This also means that 23% of the range drift can be understood as a rigid body movement, whereas the

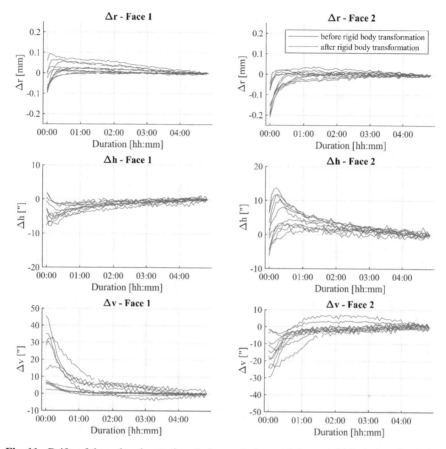

Fig. 11. Drifts of the polar observations before and after applying a rigid body transformation

remaining 79% of the range drift causes an inner shape deformation of the point cloud. For the horizontal direction only 9% of the drift is explainable by a rigid body movement. On average 52% of the vertical angle drift is, as from the previous analysis expected, a rigid body movement, but still 48% of the drift causes inner shape deformations.

5 Relevance of Re-Stationing of the Scanner During Long-Term Measurements

From the analyses in Sect. 3 we know that the polar elements of laser scanners drift during the warm-up phase and during long-term measurements. But what can we practically do against these drifts?

One way to minimize drifts caused by the warm-up phase would be to warm up the scanner for at least one hour before taking measurements. This, however, is not very efficient.

Another way to reduce both the warm-up and the long-term drifts is to re-station the scanner at regular time intervals. Re-stationing, also known as registration, in laser

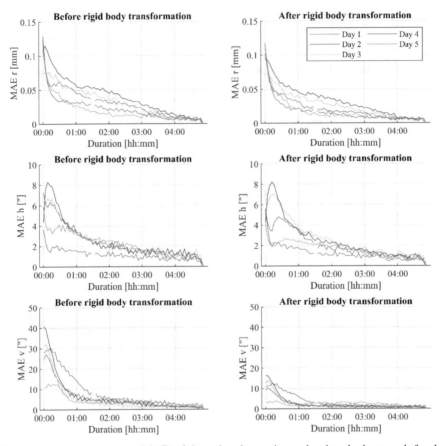

Fig. 12. Mean absolute error (MAE) of the polar observations related to the last epoch for the raw polar observations (left) and for re-stationed scanner (right)

scanning usually equals a rigid body transformation as it was performed and analyzed in Sect. 4. Therefore, the advantages of re-stationing can be taken directly from Sect. 4: About 20% of the range drift, 10% of the horizontal drift and 50% of the vertical drift can be compensated by re-stationing the scanner during long-term measurements. Analyzing the drifts after the hour-long warm-up phase depicted in Fig. 9, it can be seen that in the subsequent long-term measurement a stability of less than 0.05 mm for the distance and less than 8" ($\hat{=}$0.2 mm at 5 m) for the angles is achieved by re-stationing.

Despite this improvement, especially in the warm-up phase, drifts are still clearly visible leading to an inner shape deformation of the scans. These inner deformations cannot be absorbed by a simple re-stationing of the scanner. However, they can be minimized, for example, by a time variable calibration of the laser scanner [11]. Thus, for instance, also shear and scaling of the point cloud, which are caused by temporally unstable calibration parameters, can be compensated.

6   Conclusion

The findings from the analyses of Sect. 3–5 have provided the following answers to the questions posed in Sect. 1:

1. During long-term measurements drifts occur especially in the distance measurements and the vertical angle measurements. During the warm-up phase clearly increased drifts can occur. In the data shown in this paper, the range drifts up to 0.2 mm and the vertical angle up to 45 ".
2. The analysis in Sect. 4 shows that the drifts are a combination of rigid body movement and internal shape deformation. While the drifts in the distance and in the horizontal angle can only be explained between 10 and 20% by a rigid body movement of the scanner, for the vertical angle this is on average around 50%. The remaining drifts are due to a shape deformation caused by calibration parameters that are temporally unstable.
3. A continuous re-stationing of the scanner can clearly minimize drifts. It can compensate up to 80% of the vertical drifts during the warm-up phase, and even after the warm-up phase it is possible to achieve stabilities of less than 0.05 mm for the distance and 8" for the angle measurements.

With the results presented here, orders of magnitude for the stability of polar observations during long-term measurements and the warm-up phase of the scanner are known for the first time. This is an important step on the way to a better analysis and understanding of deformation measurements and long-term measurements with laser scanners.

References

1. Woźniak, M., Odziemczyk, W.: Monitoring of the wut grand hall roof in conditions of high temperature changes. Reports on Geodesy, 105–110 (2009).
2. Odziemczyk, W.: Stability test of TCRP1201 + total station parameters and its setup. In: XXIII. Autumn School of Geodesy, E3S Web of Conferences, Vol. 55. EDP Sciences, Les Ulis Cedex, France (2018).
3. Gučevic, J., Delčev, S., Ogrizović, V.: Determining temperature dependence of collimation error of digital level Leica DNA 03. In: FIG Working Week 2011-Bridging the Gap between Cultures in Marrakech, Morocco (2011).
4. Elias, M., Eltner, A., Liebold, F., Maas, H.-G.: Assessing the influence of temperature changes on the geometric stability of smartphone- and raspberry pi cameras. Sensors 20(3), 643 (2020).
5. Reshetyuk, Y.: Calibration of terrestrial laser scanners for the purposes of geodetic engineering. In: 3rd IAG Symposium on Geodesy for Geotechnical and Structural Engineering and 12th FIG Symposium on Deformation Measurements, Baden, Austria (2006).
6. Janßen, J., Medić, T., Kuhlmann, H., Holst, C.: Decreasing the uncertainty of the target center estimation at terrestrial laser scanning by choosing the best algorithm and by improving the target design. Remote Sensing 11(7), 845 (2019).
7. Zoller + Fröhlich: Reaching new levels Z + F Imager 5016 User Manual V 1.8. User manual, Wangen im Allgäu, Germany (2018).

8. Medić, T., Holst, C., Janßen, J., Kuhlmann, H.: Empirical stochastic model of detected target centroids: Influence on registration and calibration of terrestrial laser scanners. Journal of Applied Geodesy 13(3), 179–197 (2019).

9. Medić, T., Kuhlmann, H., Holst, C.: Automatic in-situ self-calibration of a panoramic TLS from a single station using 2D keypoints. In: ISPRS Annuls of the Photogrammetry, Remote Sensing and Spatial Information Sciences, Vol. IV-2W5, ISPRS Geospatial Week 2019, 10–14 Junge 2019, Enschede, Netherlands (2019).

10. Förstner, W., Gülch, E.: A fast operator for detection and precise location of distict point, corners and centres of circular features. In: Proceedings of the ISPRS Conference on Fast Processing of Photogrammetric Data, 281–305, Interlaken, Switzerland (1987).

11. Medić, T., Kuhlmann, H., Holst, C.: A priori versus in-situ terrestrial laser scanner calibration in the context of the instability of calibration parameters. In: Proceedings of the 8th International Conference on Engineering Surveying & 4th SIG Symposium on Engineering Geodesy (INGEO & SIG 2020), Dubrovnik, Croatia, Also a chapter of this proceeding book (2020).

Using the Resolution Capability and the Effective Number of Measurements to Select the "Right" Terrestrial Laser Scanner

Berit Schmitz[✉], Daniel Coopmann, Heiner Kuhlmann, and Christoph Holst

Institute of Geodesy and Geoinformation, University of Bonn, Nussallee 17, 53115 Bonn, Germany
{schmitz,c.holst}@igg.uni-bonn.de,
{s7dacoop,heiner.kuhlmann}@uni-bonn.de

Abstract. The point-to-point distance, the spot size and its shape limit the minimum size of objects that can be spatially resolved in a TLS point cloud. As the laser beam has a footprint of at least a few millimeters, adjacent laser spots overlap if the sampling interval is chosen small. Thus, they do not provide individual information about the object surface and they are correlated. To evaluate the performance of different terrestrial laser scanners to resolve small objects spatially, we investigate their resolution capabilities. Our results show that the expansion and the magnitude of the resolution capability vary between the scanners due to the different focusing and shape of the laser beam, and the rotational speed of the deflecting mirror. Furthermore, we use the resolution capability to assess which scanner provides the most uncorrelated information. Thus, this study provides a measure to judge the scanners' usability for specific applications, such as finding a crack in a wall.

Keywords: Laser spot · Resolution · Correlation · Stochastic model · Level-of-detail

1 Motivation

The fast and dense data acquisition of terrestrial laser scanners (TLSs) gives many opportunities to sample objects with many details. Nowadays, TLSs are capable to sample up to two million points per second with a point spacing of less than a millimeter on ten meters, and a point accuracy of a few millimeters or even less [1, 2]. However, the size of objects that can be spatially resolved in a point cloud varies between different scanners as it depends on the point distance as well as on the laser spot size [3]: If the object is scanned with a high resolution, neighboring laser spots overlap as the laser beam is at least a few millimeters. Hence, they do not provide individual information about the object and the resolution does not equal the resolution capability [4]. Furthermore, the range measurement is averaged over the whole illuminated area, which causes a smoothing of the surface [5].

A. Kopáčik et al. (Eds.): *Contributions to International Conferences on Engineering Surveying*,
SPEES, pp. 85–97, 2021. https://doi.org/10.1007/978-3-030-51953-7_7

The following example gives an impression of this issue. Figure 1 shows a crack in a rock, once in a photo (left) and once in a point cloud (right). It is obvious that the point cloud does not resolve the crack as good as the photo. This is induced by the finite size of the laser spot, which leads to a smoothing of the rock roughness [5].

Fig. 1. Crack in a rock, left: photo, right: point cloud

Since neighbored measurements cannot be treated as independent, the number of uncorrelated measurements is reduced with respect to the number of all measured points. A measure for the number of uncorrelated points is given by the effective number of measurements [4].

The resolution capability varies between different TLSs as the resolution and the spot size differ between scanner models. In order to find the "right" scanner for a special application, we investigate the resolution capability and the effective number of measurements of nine different TLSs in order to answer the following questions:

- Which size of objects can I resolve with my scanner?
- Which is the right scanner to resolve the object?
- How much individual information do I obtain with my scanner?

Section 2 introduces the terms resolution capability and effective number of measurements and describes the methodology how to determine both. Section 3 describes the data collection for all scanners and Sect. 4 analyzes the resolution capability and the effective number of measurements. Section 5 discusses the results and Sect. 6 concludes our findings.

2   Resolution Capability of TLS

The ability of TLSs to resolve two objects on adjacent lines-of-sight is called resolution capability in angular direction [3]. In contrast, the ability to resolve two objects on the same line-of-sight is defined by the resolution capability in distance direction [6]. This section explains both kinds (Sect. 2.1 and 2.2) and shows how to determine the effective number of measurements (Sect. 2.3).

2.1 Resolution Capability in Distance Direction

The resolution capability in distance direction mainly depends on the range precision as more precise measurements can be distinguished faster. Considering a scanner that is shifted by very small distance intervals, the measurements at each interval have a probability distribution function depending on the variance. With these functions, it is possible to test at which size of interval the measurements significantly differ from each other, so that we can distinguish them. The result is the resolution capability in distance direction.

Hence, the determination depends solely on the precision of the rangefinder. Wujanz et al. [7] invented a method to determine the range precision depending on the intensity. This method is also applied to other scanners (e.g. [8]). Since the range precision is completely described by the intensity, the distance resolution capacity likewise depends on the intensity. As this method already exists and users can determine the resolution capability in distance direction of their scanners using the range precision, we do not further focus on it.

2.2 Resolution Capability in Angular Direction

Herein, we focus on the resolution capability in angular direction. According to Lichti and Jamtsho [3], it depends on the point distance and the spot size. Schmitz et al. [4] discovered that the resolution capability is improved using higher quality levels. A higher quality means that subsequently measured points are averaged to reduce the range noise. This leads to a slower rotation of the deflecting mirror. Hence, the best resolution capability is achieved using the highest resolution and the highest quality. Both lead to slowing down the rotation speed of the mirror, which could also be a potential influence on the resolution capability.

The resolution capability is calculated according to the method proposed in Schmitz et al. [4]. A test specimen called Böhler-Star (Fig. 2) is scanned. This test specimen has a fore- and a background plane. The distance between both planes is 25 cm. The foreground plane has recesses beginning in the middle of the star with an opening angle of 15°. The overall size of the Böhler-Star is 1.25 m × 1.25 m. If a laser spot hits both fore- and background plane, the signal is averaged and mixed pixels occur. The proposed methodology quantifies the length of the transition where the spot hits both surfaces. The length of the transition equals the resolution capability. Since the Böhler-Star contains many edges, the resolution capability can be determined in twelve angular directions. For this method, it is important that the mixed pixels filter of the import software is turned off. A detailed explanation of the methodology can be found in Schmitz et al. [4].

2.3 The Effective Number of Measurements

Not only does the resolution capability affect the level-of-detail of the point cloud, it also influences its stochastic model. Correlations in terrestrial laser scans occur due to systematic errors caused by internal scanner misalignments, scan geometry, atmospheric conditions, and object surface properties [9] that cannot be reduced in the functional

Fig. 2. Test specimen Böhler-Star

model. Considering the resolution capability, the object surface is the most important factor for the correlations.

So far, physical correlations are neglected in the stochastic model of TLSs as they are not known [10, 11]. Schmitz et al. [4] proposed an approach to derive correlations of range measurements from the resolution capability and to determine the amount of individual information given by each point by estimating the effective number of measurements:

$$n_{eff} = \frac{n}{1 + 2\sum_{k=1}^{m_x}\sum_{j=0}^{m_y} \frac{n_x-k}{n_x} \cdot \frac{n_y-j}{n_y}\mathbf{K}(k,j)}, \tag{1}$$

where n denotes the total number of points, n_x the amount of points in horizontal direction and n_y the ones in vertical direction with $n = n_x \cdot n_y$. The correlation is proportional to the spatial distance between points on the surface. The spatial distance between two neighbored points is denoted as one interval. The number of intervals in horizontal and vertical direction are defined by k and j. Since the correlation function is only stable for point distances up to 10% of the maximum distance [12], m_x and m_y define the number of considered intervals included in the computation, which equal $m_x = n_x/10$ and $m_y = n_y/10$. Furthermore, $\mathbf{K}(k, j)$ describes the value of correlation for certain intervals. Herein, only range measurements are considered to be correlated. Correlations of angular measurements are neglected.

3   Data Collection

We compare the resolution capability of nine different TLSs: Leica ScanStation P40, Leica RTC360, Leica Nova MS60, Faro Focus 3D X130, Leica BLK360, Leica HDS6100, Leica ScanStation P20, Leica ScanStation P50 and Zoller + Fröhlich Imager 5016. Table 1 gives the corresponding specifications defined by the manufacturers [1, 2, 13–19]. The technology of the distance measurement is important for the rangefinder:

phase-based or time-of-flight (here time-of-flight enhanced by waveform digitizing technology – WFD), the wavelength, the spot size, and the highest achievable resolution plays an important role for the resolution capability.

Table 1. Scanner specifications

Scanner	Distance measurement	Wave-length [mm]	Spot size at front window [mm]	Highest resolution [mm@10 m]
P20	WFD	808	2.8	0.8
P40	WFD	1550	3.5 (FWHM)	0.8
P50	WFD	1550	3.5 (FWHM)	0.8
RTC360	WFD	1550	6 ($1/e^2$)	3.0
BLK360	WFD	830	2.25 (FWHM)	5.0
HDS6100	Phase-based	690	3	1.6
MS60	WFD	658	8×20 @ 50 m	1.0
Focus X130	Phase-based	1550	2.25 ($1/e^2$)	1.6
Imager 5016	Phase-based	1500	3.5 ($1/e^2$)	0.6

Attention must be paid on the definition of the spot size given in the specifications since manufacturers use different ones and sometimes they do not declare which one they use: FWHM (full width half maximum) is the beam width where the signal contains at least half of the emitted intensity. The Gaussian beam ($1/e^2$) is where the beam contains the entire signal that has an intensity of at least 13.5% of the maximum intensity. For more details, we refer to Jacobs [20]. These values show that it is not straightforward to judge the scanner performances from the manufacturers' specifications since different definitions are used and we do not know which part of the spot is actually used for the range measurement. Since this also influences the specifications of the angle of divergence, we refrain from stating these values as they are not a good measure for comparability.

The Böhler-Star (Fig. 2) was scanned with all the above mentioned scanners at four different distances (10, 20, 30 and 40 m). We selected these distances as not all scanners can measure longer ranges and for tasks with high demands on the level-of-detail, usually shorter distances are chosen. Furthermore, we selected two scanner settings for each setup. First, we scanned with 3.1 mm @ 10 m resolution and with the lowest quality level. This resolution is possible to achieve with each of the tested scanners besides BLK360, which can only achieve 5 mm @ 10 m. Additionally, for those scanners that can perform better, the best possible settings were chosen. As Schmitz et al. [4] found out that the resolution capability is better while using a higher quality level, we used the highest resolution (see Table 1) and the highest quality as the best settings. Hence, subsequently measured points are averaged, and the rotational speed of the deflecting mirror is reduced.

For most of these scanners, it is possible to deactivate the mixed pixels filter during the import process. Only for BLK360, Faro Focus 3D X130 and MS 60, there was no possibility to choose. From the point clouds, we conclude that mixed pixels were only eliminated during the import of the BLK data. Hence, they need to be judged carefully.

4 Analysis of the Resolution Capability

We determined the resolution capability according to the description in Sect. 2.1. The results are analyzed in the following. Section 4.1 compares the shape of the resolution capability of all scanners. In Sects. 4.2 and 4.3, the magnitude of the resolution capability is described for the same and the best scanner settings. Afterwards, the effective number of measurements is computed and compared in Sect. 4.4.

4.1 Comparing the Spatial Expansion in Different Angular Directions

We estimated the resolution capability in different angular directions in order to determine its spatial expansion. Figure 3 shows the spatial expansion of the resolution capability of all scanners at different distances for the same scanner settings (lowest quality, 3.1 mm @ 10 m, BLK360: 5 mm @ 10 m).

First of all, it is obvious that the resolution capability strongly depends on the measuring distance as the area of the polygons increases with the distance. Considering the shape of the polygons, the spatial expansion varies between most scanners. P20, HDS6100 and Imager 5016 show an elliptical expansion of the resolution capability, whereas P40, P50 and RTC360 have a circular expansion. Faro Focus 3D and MS60 show a slight elliptical expansion. The resolution capability of BLK360 also expands circularly, but its determination is more noised compared to the other scanners since the maximum range of the BLK360 is much shorter and the accuracy is worse. Reasons for the different spatial expansions could be the fast rotation of the deflecting mirror and the vertical sampling of points. Another aspect is the focusing of the laser beam, which could lead to different shapes of the polygons.

Comparing the newer scanners such as P40, P50, RTC360 and Imager 5016, the scanners using WFD have a circular expansion and the phase-based scanner has an elliptical expansion. Furthermore, the old WFD scanner P20 is also elliptical, which demonstrates an improvement of the rangefinder technology for the new scanners.

P40 and P50 use the same technology and they provide the same results. Hence, the methodology is transferable to the same scanner types. Since the RTC360 also shows similar results, it is obvious that Leica uses the same technology for all three scanners.

4.2 Comparing the Magnitude with the Same Scanner Settings

This section compares the magnitude of the resolution capability of all nine scanners. Figure 4 shows the values in horizontal direction (top) and in vertical direction (bottom). All values are obtained for the same point spacing (3.1 mm @ 10 m; BLK360: 5 mm @ 10 m) and the lowest quality level. The resolution capability increases with longer distances. For the main part of the scanners, this increase is almost linear. Considering

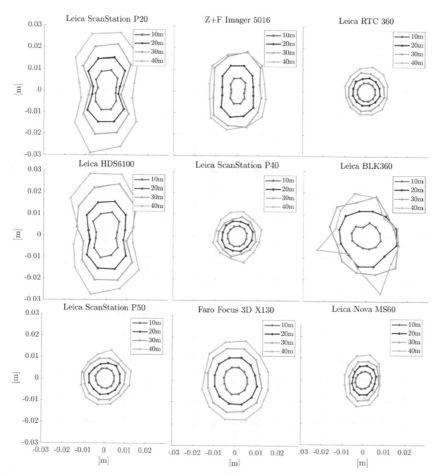

Fig. 3. Spatial expansion of the resolution capability at different distances, lowest quality and a resolution of 3.1 mm @ 10 m (5 mm @ 10 m for BLK 360)

the resolution capability in horizontal direction, the Leica MS60 provides the smallest values. The Imager 5016 has the highest resolution capability on 10 m, but it deteriorates faster than it does for the other scanners. P20, P40, P50 and RTC360 provide almost similar values, which induces the same technology introduced to all these scanners from Leica. The lowest resolution capability is obtained for the HDS6100, Faro Focus 3D and the BLK360, which represent the rather older or cheaper scanners.

For the vertical direction, the results differ. Again, the MS60 has the lowest values, but P40, P50 and RTC360 obtain the same results. Imager 5016, Faro Focus 3D and BLK360 can be grouped as the middle part. HDS6100, Imager 5016 and P20 provide the lowest resolution capability as they have high values compared to the horizontal direction caused by their elliptical expansion of the resolution capability. Either their spots are elliptical or they are strongly affected by the rotational speed of the deflecting mirror in vertical direction.

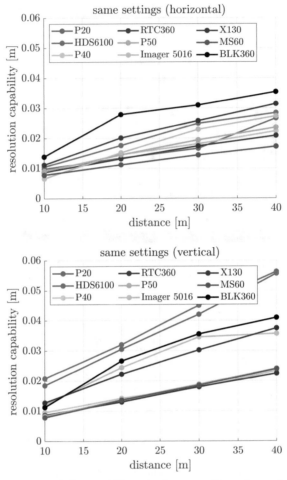

Fig. 4. Magnitude of the resolution capability using lowest quality and a resolution of 3.1 mm @ 10 m (5 mm @ 10 m for BLK 360)

4.3   Comparing the Magnitude with the Best Possible Settings

Schmitz et al. [4] found out that the highest resolution combined with the highest quality level leads to the highest resolution capability. For this reason, we estimate the resolution capability with the best possible settings (Fig. 5).

Due to the different possibilities to choose the sampling interval and quality level, the results differ from the previous analysis. Obviously, the BLK has the lowest resolution capability in both horizontal and vertical direction. Due to the lowest resolution, lowest maximum range and lowest price of all scanners, this is not surprising.

We can achieve the highest possible resolution with the P20, HDS6100 and MS60. Those scanners provide the smallest values for horizontal and vertical direction. This induces that the resolution capability of P20 and HDS6100 is affected by the rotational speed since the shape of the spot does not change using different sampling intervals.

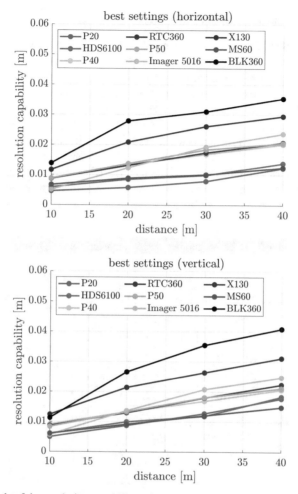

Fig. 5. Magnitude of the resolution capability using the highest sampling interval and the highest quality level

Nevertheless, the eccentricity of vertical and horizontal resolution capability is much more pronounced for a larger point distance and hence, for a faster rotation of the mirror. Thus, the rotational speed of the deflecting mirror has a big impact on the results.

This does not hold for newer Leica scanners. Surprisingly, RTC360, P40 and P50 again obtain the same results even though the highest resolution of RTC360 is 3 mm @ 10 m and P40 and P50 can obtain a point spacing of 0.8 mm @ 10 m. This shows that especially on the shorter ranges, the spot size is more important than the point spacing for these scanners. Furthermore, the rangefinder technology can cope better with the rotation of the mirror.

4.4 Comparing the Effective Number of Measurements

As described in Sect. 2.2, the effective number of measurements gives a value for the number of uncorrelated measurements in the point cloud. Since we only consider the range observations to be correlated, the number of range measurements equals the number of points. Using Eq. (1), we estimated the effective number of measurements for a 2 m x 2 m surface at a distance of 30 m. This is a common range and the effect of the incidence angle is not that high on this surface. The object is sampled with a point distance of 9.3 mm, which corresponds to the resolution of 3.1 mm @ 10 m. The results of the resolution capability from Sect. 4.2 are taken to compare the results. Table 2 presents the number of measured points n as well as the effective number of measurements n_{eff}, and the ratio of these values. This ratio represents a measure of how many points provide individual information about the object.

Table 2. Effective number of measurements for each scanner

Scanner	n	n_{eff}	n_{eff}/n [%]
ScanStation P20	46,225	17,354	37.6
ScanStation P40	46,225	22,405	48.5
ScanStation P50	46,225	20,667	44.7
RTC360	46,225	23,106	50.0
BLK360	46,225	8,744	18.9
HDS6100	46,225	9,788	21.2
MS60	46,225	27,932	60.4
Focus X130	46,225	11,943	25.8
Imager 5016	46,225	12,835	27.8

The effective number of measurements varies for all scanners since the resolution capability differs as well (see Table 2). Obviously, point clouds of some scanners provide more individual information of the measured surface than others. The most information is obtained with the MS60. Here, the resolution capability is the highest. Thus, laser spots overlap less and the smoothing effect is small. Nevertheless, only 60% of the points provide individual information even though not the highest resolution is chosen in the scanner settings. This percentage decreases using other scanners. According to the amount of individual information, the scanners can be sorted as follows: MS60, RTC360, P40, P50, P20, Imager 5016, Faro X130, HDS6100, BLK360. Hence, to get the most detailed point cloud, the MS60 should be chosen as scanner.

5　Discussion

The previous investigations determine the magnitude and the spatial expansion of the resolution capability of different TLSs as well as the effective number of measurements obtained for a 2 m x 2 m surface at a distance of 30 m and a resolution of 3.1 mm @ 10 m. Our investigations show that both magnitude and spatial expansion vary between most scanners. Especially older scanners show an elliptical expansion of the resolution capability since the rotational speed of the mirror around the horizontal axis and the focusing of the laser beam influence them.

Newer scanners that use WFD rangefinder technology, such as P40, P50 and RTC360, show a rather circular expansion and they are not highly influenced by the change of point distance and rotational speed of the mirror. Furthermore, they yield similar results, which shows that they are produced with the same rangefinder technology, and that the results can be transferred to the same scanner types. Hence, they provide the best results using the same scanner settings with a resolution of 3.1 mm @ 10 m, which leads to a much more economic data collection since it is faster and the point cloud files are smaller, but the level-of-detail of the point cloud remains almost equal.

The highest resolution capability can be achieved with MS60, P20 and HDS6100, but only if the rotational speed is low and the point distance small for the two latter mentioned scanners. Hence, the old scanners can acquire very detailed scans but the data collection is less efficient than with newer scanners such as P40, P50 or RTC360.

More accurate and more expensive scanners do not necessarily lead to better results regarding the resolution capability. Using the same scanner settings, the BLK360 as the cheapest scanner can provide a higher resolution capability in the vertical direction than the older scanners HDS6100 or P20. Furthermore, the two newest and best scanners on the market regarding the accuracy (Imager 5016 and P50), do not provide the best resolution capability. Especially the Imager 5016 resolves very well on short distances, but on higher distances, the resolution capability deteriorates faster than for other scanners.

The scanner performances do not correspond to the magnitude of spot sizes given in Table 1 where Faro Focus 3D X130 had the smallest one, which would potentially lead to the highest resolution capability. This shows that it is not sufficient to just take a look at the manufacturers' specifications since the rangefinder's technology and the handling of the incoming signal are different and hence, they influence the resolution capability.

The amount of individual information given by each scan point is quantified using the effective number of measurements. The results are directly related to the size of the resolution capability and hence, the MS60, which has the highest resolution capability, provides the highest amount of individual information in a point cloud.

Referring to the questions mentioned in the motivation (Sect. 1), users can judge the performance of their scanners with the given results. Figures 4 and 5 show the minimum size of objects that can be resolved in the point cloud for different scanners and scanner settings. Thus, it is possible to check the own scanner's capabilities and to judge its usability for the application. Furthermore, we can compare different scanner performances also regarding the efficiency and the amount of individual information given in a point cloud. Hence, we can choose the right scanner for our application.

6   Conclusion

This study determined the magnitude and spatial expansion of the resolution capability of nine different TLSs. Furthermore, it compared the effective number of measurements, i.e. the number of uncorrelated points in a point cloud to judge the scanners' performances regarding the level-of-detail that can be achieved in the point cloud. The major findings are summarized as follows:

- Older scanners such as HDS6100 and P20 achieve the highest resolution capability if the best possible settings are used, but the data acquisition is less efficient than with newer scanners such as P40, P50 or RTC360.
- Scanners that have the same rangefinder technology such as P40 and P50 provide the same results.
- For some scanners, the resolution capability expands elliptically. Hence, it is not sufficient to determine one value for the resolution capability. It is always important to consider the shape.
- To get a detailed scan, it is not always necessary to scan with the highest resolution. Due to the overlap of spots and the resulting smoothing effect, the resolution capability is not necessarily higher.
- Not only do point-to-point distance, spot size and shape influence the resolution capability, the rotational speed as well as the rangefinder technology play an important role.

According to the estimated values, scanners can be chosen for certain applications. If, for example, objects of at least 4 cm should be resolved in the point cloud, it is sufficient to scan with the BLK360 if ranges up to 40 m are used. Furthermore, it demonstrates the limits of each scanner. With these values, TLS users can judge the utility of their scanners before scanning. In future studies, these investigations can be extended by using longer distances, more scanners and scanner settings and different incidence angles to get a broader analysis about the resolution capability of terrestrial laser scanners. Furthermore, these results should be combined with the resolution capability in distance direction to get a complete scanner evaluation considering the resolution capability.

Acknowledgements. We thank PILHATSCH INGENIEURE for providing us P40 and RTC360 for this experiment and Marek Dymel for carrying out the measurements with these scanners.

References

1. Leica Geosystems: Leica ScanStation P50 Because every detail matters. Data sheet, Heerbrugg, Switzerland (2017), available online: www.leica-geosystems.com, last accessed 2020/01/10.
2. Leica Geosystems: Leica RTC360 User Manual Version 1.0. User manual, Heerbrugg, Switzerland (2018).
3. Lichti, D., Jamtsho, S.: Angular resolution of terrestrial laser scanners. Photogramm. Rec., 21 (114), 141–160 (2006).

4. Schmitz, B., Kuhlmann, H., Holst, C.: Investigating the resolution capability of terrestrial laser scanners and its impact on the effective number of measurements, ISPRS J. Photogramm. Remote Sens. (159), 41–52 (2020).

5. Bitenc, M., Kieffer, D. S., Khoshelham, K.: Range versus surface denoising of terrestrial laser scanning data for rock discontinuity roughness estimation. Rock Mech. Rock Eng. 52(9), 3103–3117 (2019).

6. Kamerman, G.W.: Laser radar. In: The Infrared & Electro-Optical Systems Handbook, Volume 6, Chapter 1, 1–76, Infrared Information Analysis Center, Ann Arbor (USA) and Spie Optical Engineering Press, Bellingham (USA) (1993).

7. Wujanz, D., Burger, M., Mettenleiter, M., & Neitzel, F. (2017). An intensity-based stochastic model for terrestrial laser scanners. ISPRS Journal of Photogrammetry and Remote Sensing, 125, 146–155.

8. Schmitz, B., Holst, C., Medic, T., Lichti D. D., Kuhlmann, H. (2019) How to Efficiently Determine the Range Precision of 3D Terrestrial Laser Scanners, Sensors, 19 (6), 1466.

9. Soudarissanane, S., Lindenbergh, R., Menenti, M., Teunissen, P.: Scanning geometry: influencing factor on the quality of terrestrial laser scanning points. ISPRS J. Photogramm. Remote Sens. 66 (4), 389–399 (2011).

10. Kauker, S., Holst, C., Schwieger, V., Kuhlmann, H., Schön, S.: Spatio-temporal Correlations of Terrestrial Laser Scanning, Allgemeine Vermessungs-Nachrichten (AVN), 6/2016, 170–182 (2016).

11. Jurek, T., Kuhlmann, H., Holst, C.: Impact of spatial correlations on the surface estimation based on terrestrial laser scanning. J. Appl. Geodesy 11 (3), 143–155 (2017).

12. Heunecke, O., Kuhlmann, H., Welsch,W., Eichhorn, A., Neuner, H.: Handbuch Ingenieurgeodäsie: Auswertung geodätischer Überwachungsmessungen (2., neu bearbeitete und erweiterte Auflage); Wichmann Verlag, Berlin, Offenbach (Germany) (2013), (in German).

13. Faro: Faro Laser Scanner Focus 3D X130. User manual (2015), available online: https://knowledge.faro.com, last accessed 2020/01/10.

14. Leica Geosystems: Leica HDS6100 Latest generation of ultra-high speed laser scanner. Data sheet, Heerbrugg, Switzerland (2009), available online: www.leica-geosystems.com/hds, last accessed 2020/01/10.

15. Leica Geosystems: Leica ScanStation P20 Industry's Best Performing Ultra-High Speed Scanner. Data sheet, Heerbrugg, Switzerland (2013), available online: www.leica-geosystems.com/hds, last accessed 2020/01/10.

16. Leica Geosystems: Leica Nova MS 60. Data sheet, Heerbrugg, Switzerland (2015), available online: www.leica-geosystems.com, last accessed 2020/01/10.

17. Leica Geosystems: Leica ScanStation P30/P40 Because every detail matters. Data sheet, Heerbrugg, Switzerland (2016), available online: www.leica-geosystems.com, last accessed 2020/01/10.

18. Leica Geosystems: Leica BLK360 Imaging Scanner 3D. Data sheet, Heerbrugg, Switzerland (2017), available online: www.leica-geosystems.com, last accessed 2020/01/10.

19. Zoller + Fröhlich: Reaching new levels Z + F Imager 5016 User Manual V 1.8. User manual, Wangen im Allgäu, Germany (2018).

20. Jacobs, G.: Understanding spot size for laser scanning. Professional Surveyor Magazine, October 2006, 1–3 (2006).

Quantification of Systematic Distance Deviations for Scanning Total Stations Using Robotic Applications

Finn Linzer[1]([✉]), Miriam Papčová[2], and Hans Neuner[1]

[1] TU Wien, Wiedner Hauptstraße 8-10, 1040 Vienna, Austria
{finn.linzer,hans.neuner}@tuwien.ac.at
[2] Research Institute of Geodesy and Cartography in Bratislava, Chlumeckého 4, 826 62 Bratislava, Slovakia
miriam.papcova@skgeodesy.sk

Abstract. The incidence angle (IA) of the measuring beam and influences due to different materials cause unacceptable systematic deviations when it comes to high-precision surveying tasks in the field of engineering geodesy. For the determination of these distance deviations, an approach is introduced, which is based on the absolute comparison between reference measurements and laser scans. However, the implementation of the developed method, which reveals this influence for scanning total stations, was conducted manually by an operator. Thus, the invested expenditure is considered to be very high. In this paper, a further development of the approach is presented which mainly introduces robotic-based automations in order to significantly reduce the required effort for data acquisition. We show that the automation has nearly no impact on the accuracy of the obtained error curves, as these are consistent to results obtained about three years ago. Furthermore, a higher discretization level of the error curves could be achieved.

Keywords: Incidence angle (IA) · Automations · ROS · Systematic deviations

1 Introduction

Previous investigations with total stations show an influence on the distance-dependent component at varying angles of incidence (IA) on differently tilted surfaces [1–3]. The measuring instrument used for this investigation is particularly suitable for the detection of effects that can occur in laser scanning due to changing surface geometry and structure. Scanning total stations are a modern development for many areas in the industry. With several thousand points per second, the scanning of a point cloud of the environment is relatively slow in comparison. However, unlike for established laser scanners, via the ocular it is possible to determine the instruments zero-point P_0 by spatial backwards intersection (SBI) with respect to a highly accurate geodetic network. This enables an accuracy in stationing that allows a direct comparison with higher-quality point clouds.

A. Kopáčik et al. (Eds.): *Contributions to International Conferences on Engineering Surveying*,
SPEES, pp. 98–108, 2021. https://doi.org/10.1007/978-3-030-51953-7_8

For the determination, reference distances are derived by a specially developed measuring method with superior accuracy compared to the measurement result and analyzed as distance deviations.

In the above-mentioned study, the deviations found were up to 0.8 mm between flat and steep sights, for many applications in the economy this is sufficient.

However, in order to be able to carry out measurement tasks in the field of engineering geodesy and laser scanning with highest accuracy, systematic deviations of this magnitude needs to be applied. Point clouds measured by terrestrial laser scanners (TLS) are influenced by instrumental mechanism, atmospheric conditions, the properties of the object's surface and the acquisition configuration [2]. In this context, it already has been discussed that under varying IA, the reflected signal strength varies and needs to be taken into account [4, 5].

With respect to the forthcoming challenge, the advantage of the scanning total station is used under laboratory conditions. The measurement setup proposed by Zámečníková et al. [3] is reproduced in this work, which is described in the second chapter. A methodically enhancement is explained in the third chapter. Previously observed deviations in the distance component can essentially be confirmed by a new epoch. Furthermore, the resolution with regard to the investigated IA was extended. Acquired results are discussed in the fourth chapter, as we come to a conclusion in the fifth chapter.

For the measurement setup a scanning total station, a laser tracker with T-Scan attachment and a stable setup for the object to be scanned are required. All measurement result computations and control commands concerning the acquisition of point clouds were implemented into the ROS (Robot Operating System) environment. The use of a robot arm further supports the measuring process. Thus, the procedure will be expanded with the help of a robot arm in order to be able to detect occurring effects in a faster sequence and more reliably in the future.

2 Methodology

Using the method proposed by Zámečníková et al. [3], the absolute distance deviations ΔD can be obtained by determining the difference between a highly accurate distance D_{ref} against the measured distance to single points of the laser scan D_{TLS}. For this purpose, the point clouds of a reference measurement are transformed into a joint coordinate system with the measured point clouds of the scanning total station via SBI. The coordinate description of all points related to the MS50 origin P_0 enables the direct computation of polar coordinates. Therefore, a Leica MS50 and a Leica laser tracker LTD800 were positioned in relation to the scanned measuring object (Fig. 1); expected standard deviations are shown in Table 1.

The demands on the measuring equipment and the observer are immense due to the high accuracy to be achieved and the duration. A single measuring epoch that examines up to 31 different IA requires up to 10 h. In addition, the measuring devices should be turned on 3 h in advance due to the recommended warm-up time. In order to relieve the observer and at the same time achieve results in a faster sequence, the individual tasks should be automated as far as possible and run autonomously.

For further automation, the measurement setup will be divided into four sections. According to the current status, a varying degree of automation will be achieved for

Fig. 1. The measuring instruments used, as well as the measuring object made of granite

Table 1. Technical parameters of MS50, LTD800 (+TS50A)

Technical parameter	Leica MS50 (σ)	Leica LTD800 + T-Scan TS50A (2σ)
Angle measurement	0.3 mgon	0.6 mgon (LTD)
Distance measurement with reflector	1 mm + 1.5 ppm	25 microns + 10 microns/m (LTD) 3 microns (Optics center CCR)
Distance measurement without reflector	2 mm + 2 ppm	/
Scanning-range noise	0.5 mm @ 25 m/62 Hz	80 microns + 3 microns/m (TS)
Laser dot size	7 × 10 mm @30 m	4.5 mm (LTD)

each task. The aim is to let as many tasks as possible run independently to generate more measurement results in less time, in order to gain a better understanding of the underlying processes.

(1) Network measurement | ca. 1.5 h | manually

(2) Scan process TLS + TS | 10 min/IA | semi-automated

(3) Control measurement | ca. 45 min | manually

(4) Evaluation | ca. 10 min | automated

Performing all four steps define a single measurement epoch, consisting of a network measurement and control, multiple scans of the measuring object for each IA and an evaluation of the result.

The network measurement establishes the spatial relationship between the laser tracker and the total station. Derived from the technical parameters shown in Table 1, a positioning accuracy of 0.1 mm for P_0 is expected to be achievable due to the high angle measurement accuracy of the total station and the coordinate determination with the laser tracker. The coordinates of the individual network points (31–73) are determined by the laser tracker by interferometry and in precise mode in each new measurement epoch. A Leica Corner Cube Reflector (CCR) is used for this procedure, which is magnetically held in the consoles provided for this purpose.

After determining the coordinates by laser tracking the reflector is rotated in the mount without removing it, so that it can also be aimed by the total station. With high overdetermination the pose of the measuring instruments to each other can be determined by means of SBI. To compensate for negative effects due to the shape of the laboratory, all available target consoles are taken into account. Any deviation along the distance-determining axis is fully taken into account in the subsequent examination of the distances (Fig. 2).

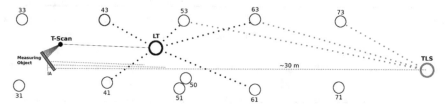

Fig. 2. Configuration of the reference network

The network measurement cannot be automated at present. For interferometric network measurements, the CCR must be moved from the laser tracker's birdbath to the respective network point without interrupting the sight. The subsequent aiming of the CCR by means of the total station must be carried out carefully due to the demand for high accuracy. As shown in Fig. 3, the origin of the Corner Cube is difficult to see even with additional lighting. With increasing distance, the demands on the operator also increase. A CCR more than 40 m away is practically unfeasible to target. Depending on experience, the entire process takes about 1.5 h to aim for 11 available network points in two sights. Due to a fixed data acquisition and the processing behind it, the evaluation can be done quickly.

Fig. 3. Target signaling, CCR with illumination

The third aspect of the methodology, the control of the network based on a statistical test [6], is carried out at the end of an epoch, for which the same care is demanded. The total station was mounted on a founded measuring pillar in order to keep movements as low as possible even over a longer period of time and despite a lot of usage. In order to be able to control the stability of the net configuration, even during the current measurement period, several prisms are placed on the net points. This is of importance as soon as the second point of the methodology, the scanning processes at different IAs, changes from semi-automatic to fully automatic operation in time.

The main amount of time spent by the observer currently arises at the second sub-item. For each investigated IA the measuring object must be rotated accordingly and the situation must be recorded with two different point clouds by two different measuring devices. Automation of this procedure could lead to the greatest time saving of the method. Using the T-Scan, a line laser scanner belonging to the laser tracker, the test object is measured on an area of about 10 cm × 20 cm within about 2 min measuring time acquiring about 1 million points. The material composition must be selected so that it is suitable for laser scanning. A laser scan is then triggered by the total station for the same area section. With an average point density of 2 mm on the object and a frequency of 62 Hz, the scan takes 3–5 min. Measured point clouds can now be related to one another to find individual points that are of corresponding relationship. Zámečníková et al. [3] therefore introduced the N-method, where an area-wise reference determination without direct signalizing is realized.

In Fig. 4 the scanned distance D_{TLS} as well as the reference D_{ref} refer to the same origin P_0. As the distance crossed is determined by a constellation where both distances refer to the same spot on the measuring object, ΔHz and ΔV in both cases are therefore required to be very similar. Distance differences ΔD_i are determined (1) for every point pair (200–600 per IA). The mean (2) is computed and can be seen as reliable if it does not differ from the median and if its standard distribution (3) is about one potency lower than the manufacturer's specifications.

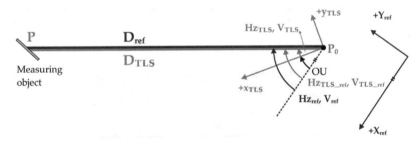

Fig. 4. Absolute differences of a distance measurement [3]

$$\Delta D_i = D_{ref\_i} - D_{TLS\_i}, \ \Delta D_m = \sum_{i=1}^{n} \frac{\Delta D_i}{n}, \ \sigma_{\Delta D_m} = \sqrt{\sum_{i=1}^{n} \frac{(\Delta D_m - \Delta D_i)^2}{n(n-1)}} \quad (1, 2, 3)$$

The scanning process must be relaunched from a computer for each IA, as it is not part of the automated methodology as of now (Fig. 5). The triggering of the process is currently still done by the observer, as he has to twist the sample himself.

Apart from this, no further intervention is currently necessary for the second sub-item in the event of a flawless process.

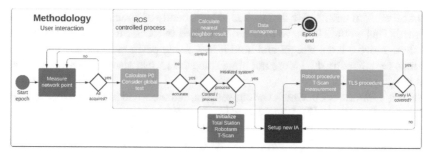

Fig. 5. Flow diagram of extended methodology

3   Autonomous Implementation

The newer approach to implementing the method involves the use of a robot arm. The robot arm of Universal Robots UR5 is capable of lifting loads up to 5 kg. A special design allows the T-Scan TS50A to be mounted on the end effector (Fig. 6). The path of movement is planned individually for all angles of incidence in such a way that the scanner can pick up the sample in the area mentioned. As soon as 1 mio. points have been measured so that, from experience, about 400 pairs of points can be identified, the robot arm automatically moves out of the field of view of the total station so that the latter can then perform its measurement.

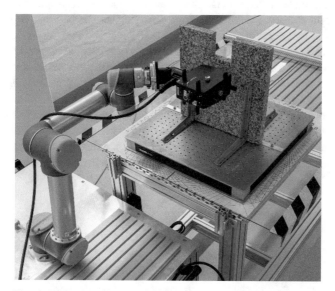

Fig. 6. Robot arm, Leica T-Scan and measuring object (IA: 0 gon)

The robot arm can be controlled via a supplied input panel, via the RoboDK control software or via the Robot Operating System (ROS) driver supported by the manufacturer. The object itself cannot be rotated by computer control yet. In the near future, it would be conceivable to have a disc that moves independently as soon as a single IA has been recorded. This would also have to be implemented in ROS. Such a disk would have to be able to move a large weight and hold it in a stable position (granite). The accuracy demands for adjusting the IA are not high, since the IA can also be determined directly from the point cloud.

ROS is a framework in which various sensors, actuators and their control software are abstracted for operation [7]. Widely used in computer science, e.g. for controlling entire robot systems, its advantages can also be used in geodetic measurement technology. With Python and C++, ROS covers the programming languages commonly used in engineering geodesy. In addition, MATLAB also provides a suitable and easy-to-use interface. Through quasi-standardized interfaces, between the instruments themselves and with the user, individual processes can be optimized and coordinated. Rejchrt et al. [8] and Linzer et al. [9] have shown that ROS is particularly suitable for setting up geodetic measurement projects if modular components are used for the setup. In the preliminary stages of this investigation, the measuring instruments were prepared for automated operation. In addition to the use of a total station, the control of the Leica LTD800 laser tracker was also implemented. In this project, the sampling of scan lines using the T-Scan were also programmed in C++, which to our knowledge is now possible in ROS for the first time.

From the instruments origin P_0, the distance D_{TLS} is determined for each of the 4000–8000 points and is valid in relation to the unambiguous direction. The determination of the reference distance D_{ref} is calculated for about 1 million points each. The dense cloud approach described earlier should facilitate the assignment of corresponding points later. If a higher point density is desired, the measuring object can be scanned several times in the same way by the robot arm.

All measuring instruments are integrated into a specially set up network in the laboratory and controlled via WLAN/LAN. The triggering of the total station in ROS was solved via the geocom interface [10]. For the laser tracker, the emScon interface is available [11]. Scanlines are accessed via a module written in C++ which enables the transmission of incoming data acquired from a windows platform to the ROS environment via network protocol (UDP). Due to the defined interfaces an analysis can also be directly realized in ROS. For the calculation of corresponding pairs of points ΔD_i, those points are iselected that are in a similar direction from the perspective of the total station (Fig. 7).

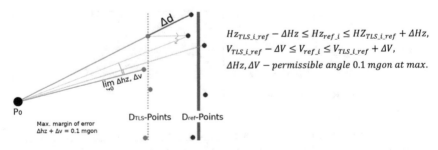

$$Hz_{TLS\_i\_ref} - \Delta Hz \leq Hz_{ref\_i} \leq HZ_{TLS\_i\_ref} + \Delta Hz,$$
$$V_{TLS\_i\_ref} - \Delta V \leq V_{ref\_i} \leq V_{TLS\_i\_ref} + \Delta V,$$
$$\Delta Hz, \Delta V - permissible\ angle\ 0.1\ mgon\ at\ max.$$

Fig. 7. The corresponding point is assessed according to the deviation Hz, V

The search for a corresponding point is accelerated by a K-D tree. A predefined maximum value for ΔHz and ΔV respectively must not be exceeded. Due to the massive amount of data, the calculation for all IAs of a measurement epoch takes about 10 min on a standard computer. The calculation follows the strict scheme of data acquisition and requires little or no user control. It may be necessary, for example, to uncover errors which, however, occur less frequently in an automated process or which can be taken into account in further development. As soon as an autonomously rotating panel has been implemented, the observer can devote himself to other tasks during the scanning processes of all IA.

This allows more time to be spent on the analysis of emerging issues. In addition, the influence of the observer on the result is minimized, overall resources and time are saved. In contrast, the determination of the network points still requires an experienced geodesist. Automatic measuring functions (such as ATR) do not meet the required standards yet.

The developed automation of the most time-consuming part of the methodology allows to generate results in a faster sequence and with increased IA resolution as well. Except for the rotation of the measuring object, the setup and execution could be simplified, extended knowledge for the execution is only necessary at the beginning and end for measurements of the geodetic network. Due to this, a higher degree of discretization in the range 0–20 gon could be generated without any added difficulty. The interfaces for input and output data defined in ROS allow a quick evaluation, which can be verified right away.

4 Results

The epoch with extended methodology from January 2020 is carried out more than three years after previous investigations. During this time, no maintenance was performed on the scanning total station, so that the results of the error curves obtained remain directly comparable. Encountered deviations are possibly influenced by the eccentricity between the collimation axis and the distance measurement axis, the estimation of the orientation unknown and the calculation of the reference distance D_{ref} [3].

The results shown in Fig. 8a for the range −60 to +60 gon show the course of the determined mean ΔD_m against the IA. The previously made assumption that the strongest deviations occur around the 0 gon range can now be made more accurate. The range from 0 to 20 gon was discretized to both sides of the IA by 10 further observations. In this range the curve falls more steeply than previously assumed. A flattening starting at 12 gon instead of 20 gon could be revealed.

An undesired effect is that distances are systematically measured too short or too long as the twist of the measuring object increases. As mentioned, this effect occurs when an eccentricity of the collimation axis with respect to the distance measuring axis occurs. It is eliminated by averaging both directions of rotation. The magnitude of the determined deviation is given in Fig. 8b for each IA. For D_{TLS} deviations of up to 0.8 mm were found. Additionally, the values for the median in relation to the means are given. As can be seen, the median and mean differ less than 0.1 mm, which indicates that possible outliers of single distance differences do not significantly influence the reported

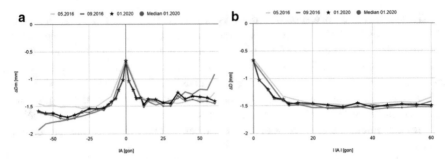

Fig. 8. Curves dated 2016 belong to Zámečníková et al. (2018). **a** Quantified distance deviations under IA; **b** corrected by averaging

mean values. For every IA in Fig. 8a the standard deviation was computed to less than 0.06 mm. Figure 9 shows the resulting standard deviation in respect to the corrected average (Fig. 8b). It is highest for 0 gon but still ten times below the standard deviation given by the instruments description. Differences likely result from the different number of pairs of points found.

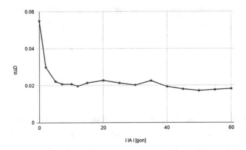

Fig. 9. Standard deviation of corrected deviations

Figure 10 shows the histograms of occurring distance deviations for 0 and 15 gon. While the deviation distribution for 0 gon is skewed more to the left, the distribution for 15 gon is skewed more to the right and is shifted by the corresponding mean value. The standard deviation is just under 0.5 mm and was computed to meet the manufacturer's specifications for all IA. Supported by the impression of the shown graph (Fig. 10c) the computed skewness switches from left and stays on stays right for every IA off above approximately 10 gon. With a skewness of around 0.5 the deviation distribution can be considered fairly symmetrical, supporting the statement made about the median above.

A direct comparison of all epochs shows a very good consistency and a high degree of repeatability. Instead of the T-Scan 5 used in previous epochs, the older version TS50A was used. The T-Scan 5 measures up to 15 times more points than the TS50A in the same period. This, however, with the same accuracy of about 80 micrometres. Thus, there is obviously no influence of the chosen measuring instrument to generate the reference. Since the same measuring object made of granite was used as in the past, the good agreement of the obtained results with those from previous investigations

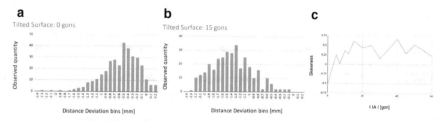

Fig. 10. **a** Deviation histogram 0 gon (350 Pairs); **b** 15 gon (260); **c** skewness over all IA

performed about 3 years ago. It underlines the high degree of repeat accuracy in the implementation of the developed method.

5   Conclusion

In this paper an extended method for quantifying distance deviations under IA was presented, which eases the problem of reproducibility of TLS scanned point clouds by partial automation. By identifying corresponding point pairs in regard to a reference scan, absolute values could be computed. By dividing the chosen approach into four parts, a structure was created which allows task-specific improvements to be made. By means of ROS, both the methodology and the evaluation of data sets could be improved. This includes the acquisition and processing of point clouds, e.g. with the Leica T-Scan as it works with a robot arm. Since ROS abstracts individual processes, the proposed setup gets a modular structure, which leads to the fact that individual problems can be specifically addressed. As soon as the targeted surface is able to rotate automatically, the individual processes can be thought of as an integrated solution. The total time duration for one epoch as well as the necessary requirements for the observer has been reduced. The measurement results shown are comprehensible in view of previous examinations. As in the past, the consistent results can be considered a success. However, specific effects related to the measuring object such as roughness and material penetration still may superimpose the obtained results. In future work the measurement setup implemented in ROS provides faster access to such results thus allowing the investigation of other materials.

By using the defined interfaces it is conceivable to implement other TLS as well. The investigation of the IAs impact for full scanners in future with the developed method is challenging, hence requiring a solution for the highly accurate determination of P_0. Future studies will be carried out in that sense. Attention must also be paid to the signal strength related to the examined distance. The extended method provides an advanced framework for future studies to get a sophisticated understanding for variations of diverse influences in respect to tilted surfaces.

References

1. Gordon, B.: Zur Bestimmung von Messunsicherheiten terrestrischer Laserscanner. PhD Thesis. TU Darmstadt (2008).
2. Soudarissanane, S., Lindenbergh, R., Menenti, M., Teunissen, P.: Scanning geometry: Influencing factor on the quality of terrestrial laser scanning points. In: ISPRS Journal of Photogrammetry and Remote Sensing, 66, 389–399 (2011).
3. Zámečníková, M., Neuner, H.: Methods for quantification of systematic distance deviations under incidence angle with scanning total stations. In: ISPRS Journal of Photogrammetry and Remote Sensing, 144, 268–284 (2018).
4. Wujanz, D., Burger, M., Mettenleiter, M., Neitzel, F.: An intensity-based stochastic model for terrestrial laser scanners. In: ISPRS Journal of Photogrammetry and Remote Sensing, 125, 146–155 (2017).
5. Zámečníková, M., Wieser, A., Woschitz, H., Ressl, C.: Influence of surface reflectivity on reflectorless electronic distance measurement and terrestrial laser scanning. In: Journal of Applied Geodesy, 8(4), pp. 311–326. Retrieved 28 Jan. 2020, from https://doi.org/10.1515/jag-2014-0016 (2014).
6. Niemeier, W.: Ausgleichungsrechnung, second ed. Walter de Gruyter, Berlin (2008).
7. ROS Wiki, Open Source Robotics Foundation. http://wiki.ros.org. (retrieved 21.01.2020).
8. Rejchrt, D., Thalmann, T., Ettlinger, A., Neuner, H.: Robot Operating System – A Modular and Flexible Framework for Geodetic Multi-Sensor Systems. In: AVN Ausgabe 6-7, Wichmann Verlag im VDE Verlag Gmbh, Berlin (2019).
9. Linzer, F., Barnefske, E., Sternberg, H.: Robot Operating System zur Steuerung eines Modularen Mobile-Mapping-Systems – Aufbau, Validierung und Anwendung. In: (peer-reviewed) AVN Ausgabe 126. Wichmann Verlag im VDE Verlag Gmbh, Berlin (2019).
10. LEICA: GeoCOM Reference Manual. Leica Geosystems Heerbrugg, Schweiz (2013).
11. HEXAGON: emScon Reference Manual. Hexagon AB (schweden) https://support.hexagonmi.com/s/article/What-is-emScon (retrieved 21.01.2020).

Testing Capabilities of Locata Positioning System for Displacement Measurements

Igor Grgac$^{(\boxtimes)}$ ⓘ, Rinaldo Paar ⓘ, Ante Marendić ⓘ, and Ivan Jakopec ⓘ

Faculty of Geodesy, University of Zagreb, Kačićeva 26, 10000 Zagreb, Croatia
igor.grgac@geof.unizg.hr

Abstract. GNSS has limitations or cannot be applied in specific environments with poor geometry like city streets, tunnels, bridges, quarries, mines, ports or in indoor environment in general. In 2003 Locata Corporation from Australia began with the development of a new, completely independent technology called Locata, which was designed to overcome the limitations of GNSS. The paper will display capabilities of Locata positioning system for determining simulated displacements. For that purpose, LocataNet consisting of six LocataLite signal transmitters was established in an open field. As all LocataLite transmitter were set up on the ground, the network geometry was not adequate for vertical positioning and only horizontal displacements were determined. After initialization of the Locata rover, different horizontal displacements were introduced by moving rover antenna. Displacements of magnitudes from 0.5 to 2.0 cm and in different directions were introduced. Positions of the rover after each introduced displacement were determined from single difference carrier phase measurements. From determined positions, displacements were calculated and compared to the true value of displacements. The results shown that Locata positioning system is capable to determine sub-centimeter level displacements.

Keywords: Displacement measurements · Locata positioning system · LocataLite · LocataNet

1 Introduction

Although the Global Navigation Satellite System (GNSS) is widely used in many areas of positioning and navigation, it is well known that this technology, in unfavorable observing environments, has its limitations or cannot be applied (indoors or underground). To overcome these limitations many researchers complemented GNSS with other, mainly terrestrial technologies. One of the solutions was the application of pseudolites (pseudo-satellites), ground-based generators and transmitters of GPS-like signals, for use in the local area (e.g. [1–3]).

However, the technology based on pseudolites has also its limitations. Extensive research and testing has concluded that pseudolites have fundamental technical problems that are very difficult to overcome in the real world: e.g. controlling transmission power levels, near/far problems, configuring special antennas, designing the "field of

A. Kopáčik et al. (Eds.): *Contributions to International Conferences on Engineering Surveying*,
SPEES, pp. 109–117, 2021. https://doi.org/10.1007/978-3-030-51953-7_9

operations" such that GNSS and pseudolites can work together [4, 5] and the basic problem—time synchronization. Also, one of the biggest problem of pseudolites is that they are restricted by law in many countries, due to transmitting on restricted GPS frequency.

In 2003 Locata Corporation (Canberra, Australia) began with the development of a new terrestrial radio frequency (RF)-based positioning technology, known as Locata, which was designed to overcome the limitations of GNSS and other pseudolite-based positioning systems. Locata positioning system is designed to enhance GNSS with additional positioning signals, but also can work completely independent where tracking of GNSS signals is not possible.

Locata primarily found its application in navigation applications [6–8], especially for navigation of the machinery in open cut mines [9, 10], where it is integrated in Leica Jigsaw Positioning System [11, 12]. Despite that, it was also researched for measurement of simulated displacements [13–15], and tested for measuring displacement of a bridge [16] and dam [17–19].

Four years ago, Locata positioning system was acquired and implemented for the first time in Croatia within the project "Wearable Outdoor Augmented Reality System for the Enrichment of Touristic Content" [20]. After the project ended, acquired Locata positioning system was tested to get insight into possibilities of the Locata system for displacement measurement.

2 Locata Positioning Technology

Locata positioning system consists of time-synchronized transceivers called LocataLites. Four or more LocataLite transceivers forms LocataNet which transmits signals that allow carrier-phase point positioning of mobile rovers.

Main invention of Locata positioning technology is patented, wireless technology of time synchronization between LocataLites called TimeLoc. Therefore, there is no need for base station, the connection for data transfer from the base to the mobile receiver, and no requirement for measurement double differencing.

Unlike pseudolites, which mostly transmit signals at the GPS frequency bands which causes many difficulties [4], Locata incorporates proprietary signal transmission structure that operates on the 2.4 GHz Industrial Scientific Medical (ISM) license free frequency band [5]. This allows high power of transmitted signals so that Locata can work both indoor and outdoor environments.

2.1 Core Components of Locata Positioning System

Locata system consists of two core components [21] (Fig. 1):

- LocataLite—transmitter,
- Locata—receiver (rover).

The LocataLite transmitter generates a carrier-phase signal modulated with a proprietary ranging code in the 2.4 GHz (ISM) band. A LocataLite currently transmits four

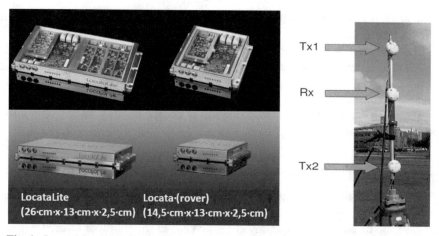

Fig. 1. LocataLite transceiver and Locata rover (left) [21], LocataLite antennas (right) [22]

PRN-style signals on two frequencies and from two spatially separated antennas, producing four usable ranging signals. The receiver and the transmitter share the same clock which is a cheap temperature-compensated crystal oscillator (TCXO).

Locata receiver (rover) utilizes signals broadcast by LocataLites to position itself, using either code or carrier-based techniques.

2.2   Time Synchronization of Locata Positioning System

The synchronization of transmitters that are broadcasting a positioning signal is the fundamental requirement for radio-positioning systems. The required level of synchronisation is extremely high, considering a one nanosecond error in time equates to a range error of approximately thirty centimetres [23].

A patented wireless time synchronisation procedure of one or more LocataLite devices is a key innovation of the Locata technology and is known as TimeLoc. The TimeLoc procedure to synchronise one LocataLite (B) to another LocataLite (A) can be described in the following steps [24]:

1. LocataLite A transmits a C/A code and carrier signal on a particular PRN code.
2. The receiver section of LocataLite B acquires, tracks and measures the signal (C/A code and carrier-phase measurements) generated by LocataLite A.
3. LocataLite B generates its own C/A code and carrier signal on a different PRN code to A.
4. LocataLite B calculates the difference between the code and carrier of the signal received from LocataLite A and its own locally generated signal. Ignoring propagation errors, the differences between the two signals are due to the difference in the clocks between the two devices, and the geometric separation between them.
5. LocataLite B adjusts its local oscillator (using Direct Digital Synthesis (DDS) technology) to bring the code and carrier differences between its own signal and LocataLite A to zero. The signal differences between LocataLite A and B are continually

monitored and adjusted so that they remain zero. In other words, the local oscillator of B follows precisely that of A.

6. The final stage is to correct for the geometrical offset (range) between LocataLite A and B, using the known coordinates of the LocataLites, and after this TimeLoc is achieved.

Theoretically, there is no limit to the number of LocataLites that can be synchronized together using TimeLoc.

2.3 Locata Positioning Network

When four or more LocataLites are deployed, they form a positioning network called a LocataNet. This positioning network is time-synchronous, so that Locata receiver can compute its position without any correctional data. At least four LocataLites in a LocataNet are required since Locata rover needs to determine four unknown parameters: three coordinates and receiver's clock error.

When forming a LocataNet there are two basic considerations for the position of the LocataLites. First, the LocataLites must be able to receive the signal from at least one other LocataLite. The other basic consideration is that the geometry of the network is suitable for the positioning precision requirements of specific task, which can be analyzed by calculating the dilution of precision (DOP) values for any LocataNet configuration.

A LocataNet is a typical Master-Slave structure. A Master is first selected among the LocataLites and then all the others are synchronized to its clock during a TimeLoc process [25].

3   Measurement of Displacements Using Locata

For the purpose of testing Locata positioning system for measurement of simulated displacements LocataNet configuration is established in the open field without any obstructions near the Faculty of Geodesy building in Zagreb. LocataNet consists of six LocataLite transceivers arranged in rectangular shape (Fig. 2). LocataLite LL3 was chosen as a master LocataLite and all other LocataLites were time synchronized to it.

Coordinates of LocataLite antennas were determined by total station measurements using Leica TPS1201 instrument. Locata rover antenna was set up on a tripod in the center of the LocataNet. Initial coordinates of the Locata rover antenna were also determined by total station measurements.

Since all LocataLite transceivers were set up approximately in one plane, the network geometry was not adequate for vertical positioning, therefore only horizontal displacements were determined.

Total 13 measurement sessions were conducted. In the first measurement session Locata receiver was placed on the known point in the center of the field. In the following 11 session horizontal displacements of different magnitudes (5, 10 and 20 mm) and different direction (East and North) were introduced to the rover antenna. In the final session rover receiver was returned to the initial position. Displacements were introduced by centering the antenna on the benchmark with marked true values of displacements. Considering that centering using an optical plummet of the tribrach have accuracy of 0.5 mm, given by the manufacturer, it is estimated that accuracy of the introduced displacements is better than 1 mm. Each measurement session has duration of 60 s. All measurements were conducted with registration rate of 1 Hz, which gives 60 measurement epochs for each session.

4 Determination of Displacements and Quality Analysis

Positioning solutions were determined in a local coordinate system with the origin in the center point of the LocataNet and coordinate axis aligned in East and North direction.

In the first step single difference (SD) carrier phase measurements were calculated to eliminate receiver's clock error. Ambiguities of SD carrier phase measurements are estimated through Known Point Initialization (KPI) process from the first 20 measurement epochs when rover was without movement.

Positioning solutions were estimated by least square adjustment independently for each measurement epoch.

Since the origin of coordinate system coincidence with initial position of the rover (center point of the LocataNet) estimated positioning solutions represents displacements. Positioning solutions of all 13 measurement sessions are shown on the Fig. 3.

Fig. 2. Configuration of established LocataNet

Mean values of displacements in each measurement session determined from Locata positioning solutions were compared to the true values of introduced displacements

(Table 1). Errors of mean values of determined displacements are in a range between −0.7 and 0.7 mm in East direction and in a range between −1.6 and 1.1 mm in North direction.

Table 1. Comparison of true and measured displacements with error ellipses

Session	True		Mean		Error ellipses		
	E [mm]	N [mm]	E [mm]	N [mm]	a	b	θ [°]
1	0	0	−0.4	−0.3	1.7	1.5	41.6
2	0	5	0.7	3.6	1.9	1.3	145.6
3	0	10	−0.7	9.4	2.1	1.4	125.7
4	0	20	−0.1	18.4	1.4	1.4	3.0
5	5	20	5.4	19.5	1.9	1.4	129.4
6	10	20	10.2	20.2	1.7	1.6	1.3
7	10	10	10.4	9.8	1.7	1.3	54.0
8	20	10	19.9	11.1	1.8	1.5	47.5
9	20	5	20.5	4.7	1.5	1.3	24.4
10	20	0	20.1	0.2	1.6	1.3	6.0
11	10	0	10.2	−0.5	1.6	1.3	135.0
12	5	0	4.9	−0.2	1.4	1.4	96.3
13	0	0	−0.3	−1.1	1.4	1.3	164.9

To indicate precision of determined displacements, error ellipses were calculated for each measurement session. Major semi-axes (a) are in a range between 1.4 and 2.1 mm. From these results it can be concluded that similar precision of determined displacements is achieved for all measurement sessions.

Achieved results shows that Locata positioning system is capable to detect sub-centimeter level displacements from one-minute long measurements.

In the next step, displacement detection from positioning solution of one epoch measurement is analyzed. Based on least square estimation of positioning solution in the very first measurement epoch and positioning solutions in other measurement epochs congruency test is conducted with different levels of test significance. For each session percentage of measurement epochs with detected displacements is calculated (Table 2).

For the 5% significance level of congruency test, in almost all measurement epochs containing displacements (from 2nd to 12th session) were detected (except in two measurement epochs in second session). Nevertheless, in the first and last session, when the rover was on initial position in 27% and 23% respectively, measurement epochs displacement was detected, although rover did not have any displacement.

This is the reason why congruency test was performed for different level of significance. Only when significance level for congruency test was set to 0.1%, in first and last session displacements were not detected. In that case displacements of 5 mm (in

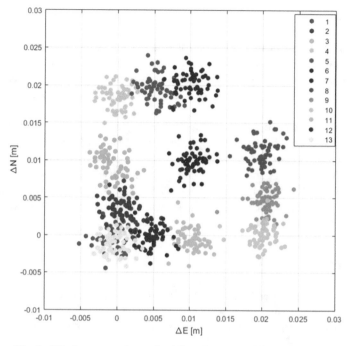

Fig. 3. Displacements determined from Locata positioning solutions

2nd and 12th session) were detected in only 18 and 68% measurements epoch, while displacements of 10 mm and higher where not detected only in 7% of measurement epochs of 3rd session.

Table 2. Percentage of detected displacements by congruency test for each measurement session with different levels of significance

Session	$\alpha = 5\%$ (%)	$\alpha = 1\%$ (%)	$\alpha = 0.5\%$ (%)	$\alpha = 0.1\%$ (%)
1	27	10	0	0
2	97	68	58	18
3	100	100	98	93
4–11	100	100	100	100
12	100	98	95	68
13	23	5	2	0

Described results lead to the conclusion that when displacements are calculated from positioning solutions from one measurement epoch centimeter level displacements can be detected, while any sub-centimeter level displacements cannot be detected reliably.

5 Conclusion

The biggest achievement of Locata positioning system is the wireless technology of time synchronization between signal transmitters. Time synchronization enables real time carrier phase positioning of Locata rover without the need for any corrections from base station, which is one of the biggest shortcomings of pseudolite based terrestrial positioning systems.

Ability of Locata positioning system for detecting displacement is tested within the LocataNet network established in an open field near the Faculty of Geodesy in Zagreb.

Research shows that calculation of displacements between two measurement epochs is sufficient for detecting centimeter level displacements, while any sub-centimeter level displacements could not be detected reliably. Nevertheless, calculation of displacements between two one-minute (60 epochs) long measurement sessions enables determination of sub-centimeter level displacements.

References

1. Wang, J.: Pseudolite Applications in Positioning and Navigation: Progress and Problems. Journal of Global Positioning Systems 1(1), 48–56 (2002).
2. Novaković, G., Đapo, A., Mahović, H.: Development and Pseudolite Applications in Positioning Navigation. Geodetski list 63(3), 215–241 (2009).
3. Novaković, G., Marendić, A., Grgac, I., Paar, R., Ilijaš, R.: Locata – A New Technology for High Precision Outdoor and Indoor Positioning, Geodetski list 69(4), 279–304 (2015).
4. Rizos, C., Li, Y., Politi, N., Barnes, J., Gambale, N.: Locata: A new constellation for high accuracy outdoor and indoor positioning. In: FIG Working Week, Marrakech, Morocco (2011).
5. Rizos, C.: Locata: A positioning system for indoor and outdoor applications where GNSS does not work. In: 18th Annual Conf., Association of Public Authority Surveyors, pp. 73–83. Canberra, Australia (2013)
6. Trunzo, A., Benshoof, P., Amt, J.: The UHARS Non-GPS Based Positioning System. In: 24th International Technical Meeting of the Satellite Division of The Institute of Navigation (ION GNSS 2011), Portland, Oregon, USA (2011).
7. Craig, D., Ruff, D., Hewitson, S., Barnes, J., Amt, J.: The UHARS Non-GPS Based Positioning System. In: 25th International Technical Meeting of the Satellite Division of The Institute of Navigation (ION GNSS 2012), Nashville, Tennessee, SAD, (2012).
8. Perrone, P., Hoekstra, G., Zuby, D., Rader, R.: Locata Positioning Used for World's First Fully-Autonomous Robotic Testing in Vehicle Collision Avoidance Systems. In: 27th International Technical Meeting of the Satellite Division of The Institute of Navigation (ION GNSS + 2014), Tampa, Florida, USA (2014).
9. Barnes, J., Rizos, C., Kanli, M., Pahwa, A.: A Solution to Tough GNSS Land Applications Using Terrestrial-Based Transceivers (LocataLites). In: 19th International Technical Meeting of the Satellite Division of The Institute of Navigation (ION GNSS 2006), Fort Worth, Texas, USA (2006).
10. Barnes, J., Lamance, J., Lilly, B., Rogers, I., Nix, M., Balls DeBeers, A.: An Integrated Locata & Leica Geosystems Positioning System for Open-Cut Mining Applications. In: 20th International Meeting of the Satellite Division of The Institute of Navigation (ION GNSS 2007), Fort Wort, Texas (2007).

11. Rizos, C., Lilly, B., Robertson, C., Gambale, N.: Open Cut Mine Machinery Automation: Going Beyond GNSS with Locata. In: 2nd International Future Mining Conference, Sydney, Australia (2011).
12. Rizos, C., Gambale, N., Lilly, B.: Mine Machinery Automation Using Locata-Augmented GNSS. In: ION 2013 Pacific PNT Meeting, Honolulu, Hawaii, USA (2013).
13. Barnes, J., Rizos, C., Pahwa, A., Politi, N., van Cranenbroeck, J.: The Potential of Locata Technology for Structural Deformation Applications. Journal of Global Positioning Systems 6(2), 166–172 (2007).
14. Barnes, J., van Cranenbroeck, J., Rizos, C., Pahwa, A., Politi, N.: Long Term Performance Analysis of a New Ground-Transceiver Positioning Network (LocataNet) for Structural Deformation Monitoring Applications. In: FIG Working Week 2007, Hong Kong, China (2007).
15. Barnes, J., Rizos, C., Pahwa, A., Politi, N., van Cranenbroeck, J.: The Potential of a Ground Based Transceiver (LocataLite) Network for Structural Monitoring of Bridges. In: Bridge design, construction and maintenance, Beijing, China (2007)
16. Barnes, J., Rizos, C., Kanli, M., Small, D., Voigt, G., Gambale, N., Lamance, J.: Structural Deformation Monitoring Using Locata. In: FIG International Symposium on Engineering Surveys for Construction Works and Structural Engineering, Nottingham, United Kingdom (2004).
17. Choudhury, M., Rizos, C.: Slow Structural Deformation Monitoring Using Locata - A Trial at Tumut Pond Dam. Journal of Applied Geodesy 4(4), 177–187 (2010).
18. Choudhury, M., Harvey, B., Rizos, C.: Mathematical Models and A Case Study of The Locata Deformation Monitoring System (LDMS). In: FIG Congress: Facing the Challenges - Building the Capacity, Sydney, Australia (2010).
19. Choudhury, M., Politi, N., Rizos, C.: Slow Structural Deformation Monitoring Using Locata - A Case Study at The Tumut Pond Dam. In: 5th World Conference on Structural Control and Monitoring, Tokyo, Japan (2010).
20. Grgac, I., Novaković, G., Ilijaš, R.: First Application of Locata Positioning Technology in Croatia. In: International Symposium on Engineering Geodesy, Varaždin, Croatia (2016).
21. Locata | Your Own GPS, http://www.locata.com, last accessed 2020/01/15.
22. Bonenberg, L. K., Roberts, G. W., Hancock, C. M.: Using Locata to augment GNSS in a kinematic urban environment. Archives of Photogrammetry, Cartography and Remote Sensing 22, 63 – 74 (2011).
23. Barnes, J., Rizos, C., Wang, J., Small, D., Voigt, G., Gambale, N.: High precision indoor and outdoor positioning using LocataNet. Journal of Global Positioning Systems 2(2), 73 – 82 (2003).
24. Barnes, J., Rizos, C., Wang, J., Small, D., Voigt, G., Gambale, N.: Locatanet: Intelligent time-synchronized pseudolite transceivers for cm-level stand-alone positioning. In: 11th Int. Assoc. of Institutes of Navigation (IAIN) World Congress, Berlin, Germany (2003).
25. Roberts, G. W., Montillet, J. P., de Ligt, H., Hancock, C., Ogundipe, O., Meng, X.: The Nottingham Locatalite Network. In: IGNSS Symposium 2007, The University of New South Wales, Sydney, Australia (2007).

First Step Towards the Technical Quality Concept for Integrative Computational Design and Construction

Laura Balangé$^{(\boxtimes)}$, Li Zhang , and Volker Schwieger

Institute of Engineering Geodesy, Geschwister-Scholl-Straße 24D, 70174 Stuttgart, Germany
{laura.balange,li.zhang,volker.schwieger}@iigs.uni-stuttgart.de

Abstract. In a world with a growing population, the development of new construction forms is becoming increasingly important. This development has to be accompanied by intense quality assessment. Within the framework of the Excellence Cluster IntCDC (Integrative Computational Design and Construction for Architecture) of the German Research Foundation (DFG) at the University of Stuttgart, a Holistic Quality Model for building systems is to be developed. This model should consider social, environmental as well as technical aspects and thus enable a holistic quality assessment of the building. For the technical part of the model quality parameters and characteristics for many different disciplines like architecture, structural engineering, engineering geodesy, mechanical engineering and system engineering should be included. This definition of the critical parameters will take place in close alignment with the co-design-based development of building systems. In addition, the construction processes are modelled in order to allow quality propagation through the construction process. This contribution will deal with a first quality concept, a first quality model as well as exemplary quality characteristics and parameters gathered from concrete and timber constructions. Exemplary quality propagation possibilities will be highlighted based on previous work at the Institute of Engineering Geodesy (IIGS). The quality will be evaluated at different decision points during the building processes in the future.

Keywords: Quality model · Quality assurance · Quality concept · Quality parameter · Quality characteristic · Tolerance

1 Introduction

Nowadays, the development of new building forms is important in order to meet the global demand for buildings. Within the framework of the Cluster of Excellence for Integrative Computational Design and Construction for Architecture (IntCDC), funded by the German Research Foundation (DFG) at the University of Stuttgart, new buildings are to be constructed to fulfill current social, environmental and technical requirements.

The aim of the IntCDC project is to develop new innovations in the building sector by using the full potential of digital technologies [1]. Here the development of a

A. Kopáčik et al. (Eds.): *Contributions to International Conferences on Engineering Surveying*,
SPEES, pp. 118–127, 2021. https://doi.org/10.1007/978-3-030-51953-7_10

comprehensive methodology of the "co-design" of methods, processes and systems is in the focus [2, 3]. For this new and innovative processes and methods also, quality assurance is important. Within the overall project, a Research Project with the name "Holistic quality model for IntCDC building systems: social, environmental and technical characteristics" was developed to rate the new buildings systems and processes. This holistic quality model will be developed within the framework of the overall project in cooperation with the Institute for Social Sciences (SOWI) and the Institute for Acoustics and Building Physics (IABP) and the Institute of Engineering Geodesy (IIGS) in order to represent the social, environmental and technical characteristics and parameters.

In Fig. 1 the planned holistic quality model is shown. Here, the subject-specific models are each represented by exemplary quality characteristics, since the final characteristics and parameters used are developed in the course of the project [4, 5]. The final quality characteristics and parameters will be developed within the project in close interdisciplinary cooperation. The quality requirements in the construction process vary greatly depending on the used material and construction method. These requirements differ also according to discipline and are implemented in a co-design-based development process.

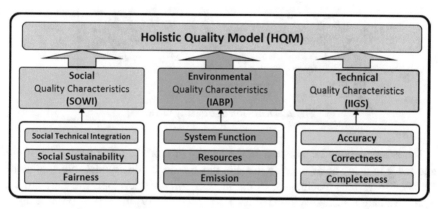

Fig. 1. Holistic quality model for IntCDC building systems [6]

Another difference compared to the previous evaluation of construction processes is that this is not a linear evaluation process. Rather, quality feedback should be provided to the disciplines involved as early as possible in order to be able to identify any problems and to initiate appropriate countermeasures, which in turn are then subject to a quality assessment.

In the context of this work, first the basic concepts of quality and their relation to previous work in the context of quality assessment will be discussed. Also general quality terms such as quality characteristics, parameters and the specific parameter tolerance correctness are defined. Then a first concept is explained regarding the identified quality characteristics and parameters. Finally, previous methods of quality propagation, which will be further developed with the project, are explained. A distinction between the propagation of the statistic quantities and the propagation of the tolerances is made.

2 General Quality Terms and Models

2.1 Overview of Technical Quality Models

The term quality is defined according to DIN EN ISO 9000 as "degree to which a set of inherent characteristics fulfils the requirements" [7]. Here the requirements are defined as "need or expectation that is stated, generally implied or obligatory" [7]. Quality assurance therefore refers to ensuring that the requirements for a product or process are fulfilled.

In general, a quality model is used to assess the quality of products and processes based on quality characteristics and parameters. Figure 2 shows the structure of a quality model.

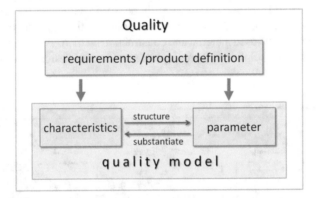

Fig. 2. Definition of a quality model [8]

The assessment and evaluation of the quality of a product or a process is of great importance in many areas of daily life. For this reason, quality models are application-orientated in various disciplines. In the field of software development, they are used for the evaluation of a developed software product. In [9] the prototype of a quality model, which evaluates a new software by means of quality characteristics such as correctness, efficiency, reliability or functionality, was developed. The qualitative aspects of software products were also considered in [10]. Here different software packages for adjustment calculation were examined. The application of quality models has also been used in the field of geodesy for a long time. As an example, quality modelling is applied to geodetic networks. For these networks, the quality characteristics accuracy and reliability are of great importance. For engineering geodetic networks, the characteristics sensitivity and separability are also used for the evaluation [11]. Quality modelling is also already being used in the construction sector. For example, within the framework of the EU project QuCon "Development of a Real Time Quality Support System for the Houses Construction Industry", a real-time quality control system has been developed for the use in housing construction processes [12]. Quality assessment also plays a role in the development of Building Information Modelling (BIM). For this purpose, a BIM based construction quality management model was developed, that does not base on characteristics and parameters, but takes into account data from the design to the construction of a building [13].

2.2 Quality Characteristics and Parameters

When setting up quality models, a differentiation must be made between quality characteristics and quality parameters. A quality characteristic describes a specific characteristic of a product or process. In DIN EN ISO 9000 the quality characteristic is therefore defined as "inherent characteristic of a product, process or system related to a requirement" [7]. These quality characteristics can now be substantiated by quality parameters. Each parameter can be quantified by a (measured) value. In addition, it should be mentioned, that a quality characteristic might have one or more quality parameters.

From [8] quality characteristics and parameters for engineering geodetic processes in building construction were extracted. As an example, these quality characteristics and parameter are presented in Table 1. Here a distinction is made between product-related and process-related quality characteristics. One of these quality parameters, the tolerance correctness, will be further described below.

Table 1. Quality characteristics and–parameters for engineering geodesy processes [8]

Quality characteristic	Parameter	Product- (pt)/process-related (ps)
Accuracy	Standard deviation	pt
Correctness	Topological correctness	pt
	Tolerance correctness	pt
Completeness	Number of missing elements	pt
	Number of odd elements	pt
Reliability	Condition density	pt
	Minimal detectable error	pt
	Vulnerability to failure	ps
Timeliness	Time delay	ps

2.3 Tolerance Correctness

The tolerance T of a building component is kept when the difference between the actual size S_{act} and the given nominal size S_{nom} is less than the difference between the maximum possible component size and the minimum possible component size. In the following, this tolerance is assumed symmetrical. This dependency is given by

$$\frac{T}{2} \geq |S_{act} - S_{nom}| = \tilde{d},$$ (1)

with $\tilde{d}$ as actual deviation. The real size is determined by measurements. However, the measured quantity S_{meas} is not identical with the actual size as they differ by the uncertainties of the measuring device. This is normally not taken into account if the keeping of the tolerances is assumed [14]. In order to be able to evaluate compliance with the tolerance under consideration of the measurement uncertainty, the term tolerance correctness is introduced [8]. For this purpose, the standard deviation σ of the measurement is first converted into the measurement tolerance T_M. The measurement tolerance T_M is given by

$$T_M = 2 \cdot k \cdot \sigma. \tag{2}$$

Here k is a factor which depends on the error probability α, e.g. $k \approx 2$ for $\alpha = 5\%$ and assumed Gaussian distribution. This measurement tolerance decreases the construction tolerance T_C, since the tolerance

$$T = \sqrt{T_M^2 + T_C^2}. \tag{3}$$

Since these two tolerances are statistically independent of each other, the quadratic propagation has to be applied. Thus, Eq. (1) results in

$$\frac{1}{2}\sqrt{T^2 - T_M^2} \geq |S_{\text{meas}} - S_{\text{nom}}| = d, \tag{4}$$

with d as random deviation. The tolerance correctness T_k now indicates whether the required tolerance was met or not. The tolerance is met if

$$T_k = \left(\frac{1}{2}\sqrt{T^2 - T_M^2}\right) - d \geq 0. \tag{5}$$

3   Technical Quality Model Within the Framework of IntCDC

3.1   Technical Quality Concept for IntCDC

The quality requirements of the different disciplines within the framework of IntCDC differ widely. Therefore, a survey was first conducted to determine the various quality characteristics as well as the standards and regulations applicable to the various disciplines. The first results of this survey concerning the general understanding of quality and the different quality requirements for the development stages of the individual construction and planning phases are presented. It shows that the relevance of technical quality modelling is of high importance for all disciplines. Furthermore, it was shown that a large part of the participants orientates their work on national and international norms and standards. In addition, urban development guidelines play a major role, especially in the planning phase. From this, a technical quality model is to be created, which will evaluate the individual requirements at various previously defined checkpoints [12]. The technical quality concept includes a first set-up of the quality model, as well as the corresponding quality characteristics and parameters.

In a first step, product related quality characteristics are extracted from the commonly used standards [15–18]. As it turns out the characteristics themselves do not differ much between the different materials used and disciplines involved, but the parameters surely differ. Here, the component properties in concrete construction and in timber construction were considered first. In addition to geometrical parameters, which must be known with a specified tolerance, the characteristics that describe the stress properties of the material are of major importance. This is for example the bearing capacity. Furthermore, characteristics such as compliance with fire protection regulations or with water permeability of walls must also be considered. The evaluation of the quality characteristics and parameters shows a multi-levelness of the parameters. Some primary parameters are combined to determine high-level condensed parameters.

3.2 Exemplary Quality Characteristics and Parameters

In the following, some quality characteristics are presented for concrete and timber building components. As already mentioned, in addition to the quality characteristics used in geodesy, characteristics such as bearing capacity, usability, strength, durability or fire protection properties are of high importance. In Table 2 some exemplary quality characteristics and parameters for the technical quality concept are shown.

Table 2. Exemplary quality characteristics and parameters for the technical quality concept for IntCDC for concrete and timber building components

Quality characteristic	Exemplary parameters
Accuracy	Standard deviation
Correctness	Tolerance correctness
Completeness	Number of missing elements Number of odd elements
Bearing capacity	Load application time Pressure, tension
Water permeability	Stress class

In the first step, special attention will be paid to the geometric parameters. These include, for example, tolerances for dimensions, angels, flatness or deviations from alignment [15]. When considering the tolerances of the dimensions, it must also be taken into account whether these are dimensions or clear dimensions. It must also be noted that the tolerances are given in relation to the size of the component.

3.3 Condensed Parameters and Primary Parameters

As already mentioned, quality characteristics are structured by several quality parameters. In this structure, a quality characteristic can have several condensed quality parameters, which in turn consist of several primary parameters. The values of the condensed parameters may be based on the values of the primary parameters. It should also be considered that a primary parameter could be assigned to several condensed parameters. This structure is shown exemplarily in Fig. 3. Here the water permeability of a building component is described by the stress class as well as the utilization class. These correspond here to the condensed parameters. The parameter stress class in turn contains the parameters granulate size and wall thickness. These correspond to the primary parameters.

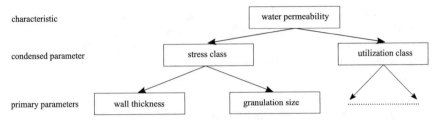

Fig. 3. Concept of use of characteristics, condensed-parameters and primary parameters

4 Quality Propagation Possibilities

4.1 Accuracy and Correctness Propagation

In order to be able to make an early statement about the quality of the final product (in this case the finished building, or an individual part of the building process), the quality characteristics and parameters must be propagated through the process. At this stage, it is not possible to discuss quality propagation for all parameters, since they are not defined up to now. However, as investigated in other projects (e.g. [19]), the propagation of quality accuracies and the propagation of tolerances plays an important role. Therefore, the propagation of the tolerance correctness should be mentioned. Regarding the tolerance correctness, it should be noted that the tolerance correctness is not propagated by the process, but the tolerance T and the measurement tolerance T_M, which are used for the calculation according to (5), have to be propagated.

4.2 Quality Propagation for Measurements

In order to be able to carry out quality propagation for statistic quantities e.g. measurements, classical variance-covariance propagation can be applied, or alternatively a Monte-Carlo simulation can be performed. This is of importance, since in many cases the size, angle and flatness cannot be measured directly, but is a function of measured values. For example, the determined component length is a function of the measured distances and angles. In classical variance-covariance propagation, the input variables are assumed to be normally distributed. Furthermore, linear dependencies between measured variable and target variable are assumed. A linearization is necessary for this [11]. In contrast, Monte Carlo simulation offers the advantage that analytical or numerical differentiation for linearization is not necessary. In addition, the input variables here do not have to be normally distributed. Instead, they can follow any distribution. However, m random numbers for the measured variable must first be generated with the assumed distribution function. The expected value, the variance or confidence intervals can then be derived by evaluating the model several times. The general procedure for both methods of propagation is shown in Fig. 4. The disadvantage of the Monte Carlo simulation compared to variance covariance propagation is that a higher computational effort is required due to the m-times calculation, exemplarily 100 000 times, and thus the computing capacity influences the duration of the solution [19]. In principle Guide to the Expression of Uncertainty in Measurement (GUM) [20, 21], that is widely applicable e.g. in mechanical engineering, purpose the use of Monte-Carlo too.

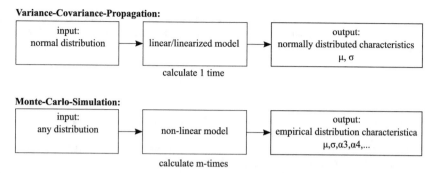

Fig. 4. Comparison of variance-covariance reproduction and Monte Carlo simulation according to [19]

4.3 Quality Propagation for Tolerances

Besides the variance-covariance propagation of the measurements, sometimes the tolerances have also to be propagated through the construction process. Different fit calculations are used for this [22]. On the one hand, values of all individual tolerances may be added up. This results in the total tolerance T_{comb} by

$$T_{comb} = \sum T_i, \tag{6}$$

where T_i are the included tolerances. This corresponds to a linear tolerance propagation, which is mainly consider in mechanical propagation processes. If a component consists of four individual building components, each has a certain tolerance. If these parts are assembled to form a common building system, the tolerances have to be summed up. On the other hand, as in geodesy, fit calculations are carried out under considering of the tolerance propagation. This results in the total construction tolerance following

$$T_{comb} = \sqrt{\sum (T_i)^2}. \tag{7}$$

This method is popularly used in civil engineering as well as engineering geodesy. The decision among (6) and (7) is also a challenge for the future.

For the tolerance correctness, the tolerance T can be propagated according to 4.3, the measurement tolerance T_M can be propagated according to 4.2 and the random deviation d can be measured.

5 Conclusion and Outlook

In conclusion, it is shown that the first quality characteristics have been identified and structured. In comparison to other quality models, condensed quality parameters and primary quality parameters are introduced. In the further progress of the project, the individual characteristics and parameters will be structured and related to each other. In addition, the quality characteristics and parameters must be extended from the component level to the manufacturing processes and the whole building systems in order to be able to carry out a quality assessment of these processes. The processes are also needed to create quality propagation methods through and for the processes.

Acknowledgements. This contribution is supported by the Deutsche Forschungsgemeinschaft (DFG, German Research Foundation) under Germany's Excellence Strategy– EXC 2120/1– 390831618. The authors cordially thank the funding agency.

References

1. Schwieger, V., Menges, A., Zhang, L., Schwinn; T.: Engineering Geodesy for Integrative Computational Design and Construction. ZfV, Heft 4/2019 (2019).
2. Menges, A.: New Cluster of Excellence: Integrative Computational Design and Construction for Architecture. [online] Available: https://icd.uni-stuttgart.de/?p=24111, last access 12/2019.
3. https://www.youtube.com/watch?v=UaegUN0XeRA, last access 2/2020.
4. 4. Kropp, C.: Controversies around Energy Landscapes in Third Modernity, Landscape Research, 43, 4 (2018), 562–573.
5. Horn, R., Dahy, H., Gantner, J., Speck, O., Leistner, P.: Bio-inspired sustainability assessment for building product development–Concept and case Study, Sustainability, 10, (2018), Art. 130. https://doi.org/10.3390/su10010130.
6. Schwieger, V., Kropp, C., Leistner, P.: Holistic quality model for IntCDC buildings systems: social, environmental and technical characteristics (2019). Unpublished.
7. DIN EN ISO 9000: Quality management systems – Fundamentals and vocabulary, Trilingual version. Normenausschuss Qualitätsmanagement, Statistik und Zertifizierungsgrundlagen (NQSZ) in DIN (Deutsches Institut für Normung e.V.), Beuth Verlag GmbH, Berlin (2005a).
8. Schweitzer, J., Schwieger, V.: Modeling of quality for engineering geodesy processes in civil engineering. Journal of Applied Geodesy, Walter de Gruyter, 5(1), 13–22 (2011). http://doi.org/10.1515/jag.2011.002
9. Ortega, M., Pérez, M., Rojas, T.: Construction of Systemic Quality Model or evaluating a Software Product. Software Quality Journal, Volume 11, Kluwer Academic Publishers, pp. 219–242 (2003). https://doi.org/10.1023/a:1025166710988.
10. Schwieger, V., Foppe, K., Neuner, H.: Qualitative Aspekte zu Softwarepaketen der Ausgleichungsrechnung. 93. DVW-Seminar: Qualitätsmanagement geodätischer Mess- und Auswertverfahren. Hannover, 10.-11.06.2010.
11. Niemeier, W.: Ausgleichungsrechnung: statistische Auswertemethoden (Ed. 2). Walter de Gruyter, Berlin, New York (2013)
12. Zhang, L., Schwieger, V.: Real Time Quality Assurance Indexes for Residential House Construction Processes. FIG Working Week, Marrakesch, Marokko, 18.-22.05.2011.
13. Chen, L., Liu, H.: A BIM-based construction quality management model and its applications. Automation in Construction, 46, 64–73 (2014). http://doi.org/10.1016/j.autcon.2014.05.009
14. Heunecke, O.: Eignung geodätischer Messverfahren zur Maßkontrolle im Hochbau. ZfV, Heft 4/2014 (2014). https://doi.org/10.12902/zfv-0021-2014.
15. DIN 18202: Toleranzen im Hochbau – Bauwerke. Normenausschuss Qualitätsmanagement, Statistik und Zertifizierungsgrundlagen (NQSZ) in DIN (Deutsches Institut für Normung e.V.), Beuth Verlag GmbH, Berlin (2019).
16. DIN 4109: Schallschutz im Hochbau - Teil 1: Mindestanforderungen. Normenausschuss Qualitätsmanagement, Statistik und Zertifizierungsgrundlagen (NQSZ) in DIN (Deutsches Institut für Normung e.V.), Beuth Verlag GmbH, Berlin (2005b).
17. DIN EN 1995-1-1: Eurocode 5: Design of timber structures – Part 1-1: General – Common rules and rules for buildings; German version EN 1995-1-1:2004 + AC:2006 + A1:2008. Normenausschuss Qualitätsmanagement, Statistik und Zertifizierungsgrundlagen (NQSZ) in DIN (Deutsches Institut für Normung e.V.), Beuth Verlag GmbH, Berlin (2010).

18. DIN EN 1995-1-1/NA: National Annex – Nationally determined parameters – Eurocode 5: Design of timber structures – Part 11: General – Common rules and rules for buildings. Normenausschuss Qualitätsmanagement, Statistik und Zertifizierungsgrundlagen (NQSZ) in DIN (Deutsches Institut für Normung e.V.), Beuth Verlag GmbH, Berlin (2013).
19. Schweitzer, J., Schwieger, V.: Modeling and Propagation of Quality Parameters in Engineering Geodesy Processes in Civil Engineering, in Kutterer, H., F. Seitz, H. Alkhatib, and M. Schmidt, The 1st International Workshop on the Quality of Geodetic Observation and Monitoring Systems [Proceedings of the 2011 IAG International Workshop, Munich, Germany April 13-15, 2011] (Heidelberg: Springer, 2015), pp. 163–168.
20. JCGM: Evaluation of measurement data — Supplement 1 to the"Guide to the expression of uncertainty in measurement" — Propagation of distributions using a Monte Carlo method. https://www.ptb.de, (2008), last access 05/2019.
21. JCGM Auswertung von Messdaten – Eine Einführung zum "Leitfaden zur Angabe der Unsicherheit beim Messen" und zu den dazugehörigen Dokumenten. https://www.ptb.de, (2009), letzter Zugriff 05/2019.
22. Steinle, A., Bachmann, H., Tillmann, M.: Bauen mit Betonfertigteilen im Hochbau. 3rd Edition, Ernst W. + Sohn Verlag (2018).

A priori vs. In-situ Terrestrial Laser Scanner Calibration in the Context of the Instability of Calibration Parameters

Tomislav Medić$^{(\boxtimes)}$, Heiner Kuhlmann, and Christoph Holst

University of Bonn, Institute of Geodesy and Geoinformation, Bonn, Germany
{t.medic,c.holst}@igg.uni-bonn.de, heiner.kuhlmann@uni-bonn.de

Abstract. Commonly, terrestrial laser scanners (TLSs) are calibrated on calibration fields so the scanner's instrumental errors are estimated a priori to a given task. Such approaches presume the stability of errors with time and external influences so that the calibration parameters are valid at future measurements. Alternatively, TLSs are calibrated in-situ during the measurement. Then, there is no need to assume the stability of the calibration parameters. However, these in-situ strategies only work out under specific conditions. Thus, it is important to know the calibration parameters' stability to optimally select or combine the calibration approaches. Herein, we investigate the parameter stability and possible causes for their changes. Our results indicate that the changes can be tracked and partially modeled based on the internal warming-up of the instrument, the ambient temperature, and the compensator bias. We reduced the parameter variability by modeling these effects and quantified the magnitudes of the parameter changes.

Keywords: TLS · Point cloud · Accuracy · Quality assurance · Temperature · Systematic errors

1 Introduction

Terrestrial laser scanners (TLSs) are used for a variety of applications, where some of them, such as deformation monitoring, require high measurement accuracy [1]. To guarantee high accuracy, TLSs are calibrated by the manufacturers [2, 3]. In the process, mechanical misalignments remaining after the instrument's factory assembly are mathematically modeled with calibration parameters (CPs). The parameters change over time due to wear and tear, suffered stress and other influences. As a result, the manufacturers recommend repeating the factory calibration at certain time intervals (typically 1–2 years). The repeated factory calibration is financially and timewise burdening for the end-users (several thousand € and several weeks at manufacturers). Hence, scientists have invested efforts in the development of user-oriented calibration approaches [4].

The existing user calibration approaches can be separated into the ones aiming at the a priori calibration, before the measurement task, and the ones aiming at the in-situ calibration, during the measurement task [5]. The most established calibration approach

© The Editor(s) (if applicable) and The Author(s), under exclusive license
to Springer Nature Switzerland AG 2021
A. Kopáčik et al. (Eds.): *Contributions to International Conferences on Engineering Surveying*,
SPEES, pp. 128–141, 2021. https://doi.org/10.1007/978-3-030-51953-7_11

is target-based self-calibration [6]. The term self-calibration denotes that no reference measurements with the instrument of higher accuracy are required. It aims at the a priori calibration in dedicated calibration fields, where the specialized targets are well distributed to allow an accurate and comprehensive calibration of all relevant instrumental errors. On the contrary, in-situ calibration approaches rely on the environment found during a measurement campaign. Thus, they do not require specialized targets and facilities. However, the calibration success strongly depends on the measurement surrounding and the comprehensiveness cannot be guaranteed [7].

One of the main disadvantages of the a priori calibration is the necessary presumption that the calibration parameters are stable over time and that they can be reused in the following measurements. Previous studies inspected the stability of the calibration parameters and noticed some significant changes over time [8, 9]. Hence, the presumption about the parameter stability is not entirely fulfilled. To further investigate this, we repeatedly calibrated two TLSs over a period of several months (multiple times within a day). Consecutively, we analyzed the variations in the parameter values over time and investigated possible causes for their changes. The main goals of this investigation are to discuss, quantify, and eventually improve the

- short-term CP instability—within one measurement campaign (1 day),
- long-term CP instability—between measurement campaigns (few months),
- and to discuss the influence of the compensator measurements on the CPs.

Based on our results, we discuss choosing between the a priori and in-situ calibration strategies, as well as the best possibilities for combining these approaches. Finally, we argue the relevance of our findings for typical engineering tasks.

2 Experiment (Materials and Methods)

We calibrated two high-end panoramic TLSs (Leica ScanStation P50 and Zoller+Fröhlich Imager 5016), both allowing two-face measurements, having high measurement accuracy and being equipped with dynamic dual-axis compensators. The instruments were calibrated using the target-based TLS self-calibration approach. They were calibrated in a dedicated calibration field with 14 BOTA8 black-and-white planar targets developed for high-accuracy [10]. The targets were placed on predefined locations that are highly sensitive towards detecting all mechanical misalignments relevant for high-end scanners (10 in total, Table 1) [11]. The targets were observed from two predefined locations using two-face measurements to estimate all parameters. The instruments were placed on heavy-duty wooden surveying tripods and secured using the stabilization. The whole calibration procedure is described in detail in [12].

The P50 was calibrated 39 times in 9 days over a period of 5 months, while the Imager was calibrated 34 times in 8 days over a period of 6 months. Each day, 2–6 consecutive calibrations were conducted (typically 5 per day), while each calibration lasted approximately one hour. During the measurements, the onboard dynamic dual-axis compensator was used to estimate the tilt of the standing axis and to aid the calibration [8]. The exception was made during 2 out of 9 days (8 for Imager) and the dynamic compensator was turned off in order to analyze if the compensator measurements influenced the

Table 1 Calibration parameters relevant for the high-end TLSs

CP	Tilts / angular errors [$''$]	CP	Offsets / metric errors [mm]
x_4	Vertical index offset	x_{1n}	Horizontal beam offset
x_{5n}	Horizontal beam tilt	x_{1z}	Vertical beam offset
x_{5z}	Vertical beam tilt	x_2	Horizontal axis offset
x_6	Mirror tilt	x_3	Mirror offset
x_7	Horizontal axis error (tilt)	x_{10}	Rangefinder offset

calibration results. During the measurements, we monitored: instrument's working time (IWT), ambient temperature, pressure, humidity, and the compensator index errors (both instruments allow a compensator calibration). Also, the Imager 5016 allows monitoring the internal temperature. All of the mentioned values are used in the analysis of the calibration parameters instability, where the special focus is placed on the temperature, as it is known to influence the calibration of total stations [13].

The target centers were estimated using the template-matching algorithm described in [10]. These observations are processed in the target-based self-calibration adjustment according to [12]. The results of the calibration adjustment are the calibration parameter values (Table 1) and their corresponding standard deviations. The CPs can be divided into tilts and offsets of the main instrument components. The tilts have a higher impact on measurements because the magnitude of the related systematic errors grows with the distance. On the contrary, the systematic errors of the offsets are constant concerning the distance and usually small in magnitude. The functional connection between each of the calibration parameters and the TLS measurements are given in Eqs. 1–3:

$$\Delta r = x_2 \sin(\theta) + x_{10} \tag{1}$$

$$\Delta \varphi = \frac{x_{1z}}{r\tan(\theta)} + \frac{x_3}{r\sin(\theta)} + \frac{x_{5z} - x_7}{\tan(\theta)} + \frac{2x_6}{sin(\theta)} + \frac{x_{1n}}{r} \tag{2}$$

$$\Delta \theta = \frac{(x_{1n} + x_2)\cos(\theta)}{r} + x_4 + x_{5n} \cos(\theta) - \frac{x_{1z}\sin(\theta)}{r} - x_{5z}\sin(\theta) \tag{3}$$

where Δr, $\Delta \varphi$, $\Delta \theta$ equal the sum of all instrument related systematic errors on range (r), horizontal angle (φ) and vertical angle (θ) measurements. These calibration parameters are analyzed in the following section.

3   Results

To demonstrate the short-term (within one day) and the long-term (within several months) parameter instability under investigation, we initially focus on one calibration parameter for one instrument. We analyze in detail the mirror tilt (x_6) in the Leica ScanStation P50, which equals the collimation axis error in total stations [5]. Figure 1—left presents the parameter values related to the instrument's working time within each day (one colored line per day). The repeatable negative drift of the CP value with time is apparent for each day.

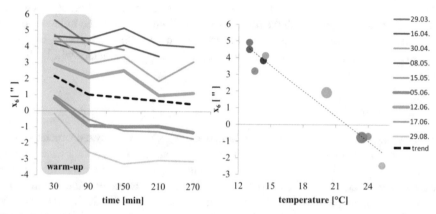

Fig. 1 Stability of the x_6 calibration parameter in Leica ScanStation P50 with the instrument's working time (left) and ambient temperature (right), 9 days of repeated calibrations

The most probable cause for this drifting effect is an internal instrument's temperature change, which causes small displacements of the instrument's mechanical components due to thermal expansion of different materials. As the users monitoring of the internal temperature is not possible for the majority of the instruments on the market, we focus on modeling this short-term drift based on the instrument's working time (IWT). From [14], we know that the instrument's temperature change can be divided into two separate stages. The first stage is the initial warming-up, with a sharp temperature change that lasts approximately one hour. The second stage is recognizable by a gradual temperature change. Therefore, we model these two stages as two separate linear trends (simplified due to a few data points per day)—one between the first two calibrations in a day and one between all calibrations after warming-up. The instrument's temperature change (warming-up) and its influence on TLS measurements is analyzed in more detail in [14].

The black dashed lines in Fig. 1—left present the average trends from all days. The slopes of both trendlines significantly differ from 0 (t-test, 99% significance), meaning that x_6 drifts over time rather constantly. Also, both trendlines are significantly different from each other (t-test, 99% significance), confirming that the warm-up period should be considered separately. The latter results indicate that it is meaningful to model the short-term parameter drift based on the IWT. Hence, this is the goal of the short-term CP stability investigation (Sect. 3.1).

Furthermore, in Fig. 1—left we can notice the clear offsets between the absolute values of x_6 between different days. This indicates that the parameter is not stable or repeatable in the long term. To investigate this parameter long-term parameter variability, Fig. 1—right presents the x_6 values against the ambient temperature (both average within each day). Fitting a linear trend (black dashed line) makes the strong correlation between the values apparent. Hence, it is possible to reduce the CP instability with temperature-based correction, which is the goal of the long-term CP instability investigation (Sect. 3.2). Again, the probable physical cause is the thermal expansion of the instrument's mechanical components.

Finally, the calibration experiments in which the dynamic compensator was turned off (thick lines Fig. 1—left and large dots Fig. 1—right, 05.06. and 12.06.) fit well within the visible trends modeled with the black dashed trendlines. Hence, the influence of the compensator can be ruled out as a possible cause of these phenomena for x_6. However, the compensator influence on the values of all calibration parameters will be further analyzed and discussed in Sect. 3.3.

3.1 Short-Term Instability

To investigate the short-term stability (within one day), we first analyzed which CPs (for both instruments) show a noticeable drift with the instrument's working time. Hence, we computed the differences of all CP values with respect to the 1st calibration within each day and we compared them with the IWT in minutes. To identify the drifting CPs, we computed the Spearman's correlation coefficients [15]—Fig. 2. We selected this correlation coefficient instead of typically used Pearson's linear correlation coefficient because one of our variables (time) is not normally distributed. Hence, it is meaningless to compute the standard deviation (necessary for computation of the Pearson's correlation coefficient). Additionally, the relationship between our variables is not necessarily linear.

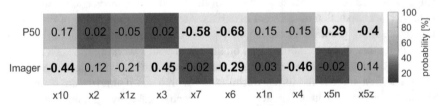

Fig. 2 Correlations between the CP change (Δx) and TLS working time (hours) for the Imager 5016 and the ScanStation P50 (color indicates probability of the null hypothesis: "Values are significantly correlated")

The short-term drift (Fig. 1—left) is only one of many causes of the CP instability, next to e.g. the measurement and the calibration uncertainty. Hence, we a priori expect rather low correlations between CP values and IWT. All parameters which were significantly correlated with IWT are bolded in Fig. 2 (nearly half CPs for both scanners). We empirically decided for the significance threshold of 90% probability (explanation follows).

We modeled the short-term drifts of these significantly correlated CPs according to Sect. 3 (linear trends during and after the warming-up) and the results are presented in Table 2. The drifts during the warming-up are few arcseconds and few tenths of a millimeter per hour, while after this period they are several times smaller. Hence, already by warming-up the scanner before the measurement task, the systematic errors can be notably reduced.

Table 2 The mean estimated calibration parameter values ($\mathbf{x}$) after 1 hour of scanning, drift of the calibration parameters ($\Delta\mathbf{x}$) with the instrument's working time (per 1 hour—1 h) during and after instrument's warm-up (w.-u.) time, standard deviation of the CPs before and after the short-term drift correction (d.c.) and the improvement due to the correction

	P50				Imager 5016			
	x_7 [″]	x_6 [″]	x_{5n} [″]	x_{5z} [″]	x_{10} [mm]	x_3 [mm]	x_4 [″]	x_{5z} [″]
x @ 1 h	−44.00	1.69	5.37	−22.84	0.27	−0.17	−3.07	2.11
Δx/ 1 h *warm-up*	−5.73	−1.15	0.63	−1.57	−0.04	0.03	−1.51	3.24
Δx / 1 h *after w.-u.*	−1.20	−0.25	0.19	−0.12	−0.02	0.01	−0.94	0.71
σ_x *before d.c.*	6.16	0.67	1.31	1.41	0.08	0.09	3.41	3.19
σ_x *after d.c.*	2.40	0.33	1.20	1.23	0.07	0.08	2.45	2.83
Impr. [%]	61.00	50.93	8.65	13.12	8.91	10.49	28.26	11.40

Furthermore, we computed the standard deviations of the calibration parameters for each day. Then, we calculated the average of these CP standard deviations (pooled standard deviation) for all days together to get a more representative value. Finally, we repeated this procedure once before and once after subtracting the short-term drift (linear trend) from the parameter values. The results (Table 2) indicate that the short-term parameter drift for these significantly correlated CPs accounts for 9–61% of the overall parameter instability. In [16], it is stated that the error in estimating the CPs smaller than 20% can be safely ignored, as they are sufficiently small to be treated as random errors. Hence, these drifts can be neglected for 5 out of 8 investigated parameters. However, for the remaining 3 parameters, functionally modeling the short-term drift and correcting the CP values concerning the instrument's working time can notably improve the CP stability. Therefore, it can reduce the systematic errors in TLS measurements. Estimating the trendlines for the remaining not-significantly-correlated calibration parameters (probability < 90%) brought no improvement to the parameter stability (explanation for the 90% probability threshold).

3.2 Long-Term Instability

A similar analysis was used to investigate the long term parameter instability (within several months). We calculated the correlations between the mean CP values and the mean ambient temperature during each day. This time we calculated the most common Pearson's linear correlation coefficient, as we expect a normal distribution and a linear relation between the CPs and temperature [15]. For the Imager 5016, we additionally estimated correlations with the mean instrument's internal temperature. The results are presented in Fig. 3. Again, all significantly correlated values are bolded (same criteria as in the Sect. 3.1).

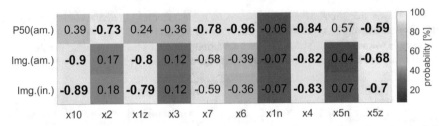

Fig. 3 Correlations between the CP value (**x**) and temperature (am.—ambient, in.—internal) for ScanStation P50 and Imager 5016 (color indicates probability of the null hypothesis: "Values are significantly correlated")

The highest correlations are found for the rangefinder offset (x_{10}) in the case of Imager (−0.90) and the mirror tilt (x_6) for P50 (−0.96), where almost the complete long-term parameter variability can be described by the temperature change. Based on Fig. 3, we estimated the temperature-related linear trends for four CPs both for P50 and Imager (x_2 for P50 was omitted due to minimal changes, < 0.003 mm / 1 °C). The results are presented in Table 3. The magnitudes of the CP changes per 1 °C are rather small, up to two arcseconds for tilt parameters and a few hundreds of millimeter for offset parameters. However, the temperature differences can easily achieve 15–20 °C over the year, making this effect much more prominent. Although small, the observed magnitudes (per 1 °C degree) are in the same order of magnitude as the precision of estimating the target centers [17]. Hence, this effect cannot be neglected.

Table 3 The estimated calibration parameter values (**x**) at standard ambient temperature of 20 °C, change of the calibration parameters (**Δx**) with ambient temperature (T) per 1 °C, standard deviation of the CPs before and after the temperature correction (t.c.) and the improvement due to the correction

	P50				Imager 5016			
	x_7 [″]	x_6 [″]	x_4 [″]	x_{5z} [″]	x_{10} [mm]	x_{1z} [mm]	x_4 [″]	x_{5z} [″]
x @ 20 °C	−46.80	0.81	1.81	−23.69	0.45	0.87	−0.79	9.32
Δx / T [1 °C]	−1.79	−0.54	−0.31	−0.76	−0.03	−0.03	−0.45	−0.84
σ_x *before t.c.*	12.05	2.94	1.91	6.77	0.17	0.23	3.25	7.47
σ_x *after t.c.*	7.60	0.79	1.02	5.47	0.07	0.12	1.93	5.65
Impr. [%]	36.94	73.09	46.40	19.17	55.54	47.83	40.49	24.33

When we compare the CP instability (standard deviation) before and after removing this linear trend, we can observe that the temperature-related changes can account for up to 73% of the overall long-term instability. This is visualized in Fig. 4, where we presented the box-plots for the calibration parameters before (blue) and after (red) removing the temperature-related trend. For the parameters that were strongly correlated with temperature the dispersion of the CP values is notably reduced, the median comes closer towards zero and the distribution of the CP values is more symmetric. The latter two indicate that subtracting the temperature-related effect almost completely removes the systematic trends in the distribution of the CP values. Hence, it can be concluded that considering the correction based on the ambient temperature can notably reduce the instability of the calibration parameters, and therefore improve the TLS measurement accuracy.

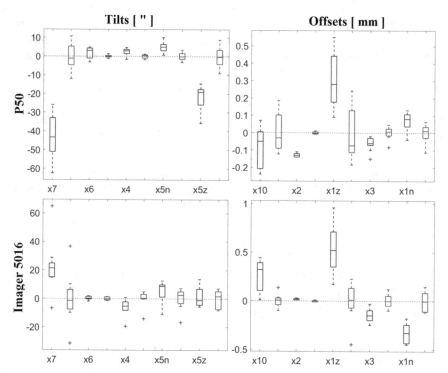

Fig. 4 Boxplots of the tilt and offset calibration parameters before (blue) and after (red) removing the temperature-related linear trend for ScanStation P50 and Imager 5016

The results for Imager are equivalent if the ambient temperature is substituted with the instrument's internal temperature. Namely, after the initial warming-up, the difference between internal and ambient temperature is a constant offset of approximately 11 °C. We additionally investigated the correlations with the pressure and humidity. However, this analysis was omitted as the correlations are expectedly lower than with the temperature and because there is no plausible physical cause that these values could directly induce the change of the instrument's mechanical components.

3.3 Dynamic Dual-Axis Compensator

The compensator index error appears when the zero index is out of the alignment with the direction of gravity. In dual-axis compensators, this error is separated into the longitudinal index error (C_L), along the line-of-sight, and the transversal index error (C_T), along the horizontal axis. If not accurately calibrated, C_L adds up with the vertical index offset (x_4) and C_T adds up with the horizontal axis error (x_7) [18]. Additionally, an incorrectly calibrated compensator can cause wrong estimates of the TLS orientation during the calibration. This further influences the estimated calibration parameters due to high correlations with the instrument's orientation in the calibration adjustment [19]. Thus, it is important to assure that the compensator measurements are unbiased as well.

As both instruments under investigation allow an in-situ calibration of C_L and C_T, we analyzed the influence of the ambient temperature (and internal for the Imager 5016) on their values (Table 4). Both errors for both instruments are significantly correlated with temperature, with the 99% probability. The linear change per 1 °C is in the same order of magnitude as the change of the CPs, so it cannot be neglected. In our datasets, the temperature changes caused 24–69% of the overall instability (standard deviation) of the C_L and C_T values. Thus, accounting for this effect can notably reduce their instability. Additionally, we observed significant mutual correlations (99% probability) of 0.67 (P50) and 0.91 (Imager) between the C_L and C_T values. Hence, incorrect calibration of only one of these errors directly influences both x_4 and x_7, as well as indirectly the majority of the remaining CPs due to their strong mutual correlations.

Table 4 Compensator index errors (C_L and C_T)—correlation with the ambient temperature—$\rho(T)$, change of the C_L and C_T (ΔC) with temperature (T) per 1 °C, standard deviation of the C_L and C_T before and after the temperature correction (t.c.) and improvement due to the correction

	P50		Imager 5016		
	C_L [″]	C_T [″]	C_L [″]	C_T [″]	
$\rho(T)$	−0.78	−0.95	−0.75 (−0.79)*	−0.52 (−0.65)*	* correlation with internal temperature
$\Delta C / T$ [1 °C]	−0.11	−0.62	−3.09	−1.81	
σ_x (before t.c.)	1.01	4.61	23.00	15.98	
σ_x (after t.c.)	0.63	1.43	14.04	12.08	
Impr. [%]	37.99	69.00	38.98	24.43	

Furtherly, we investigated the existence of the short-term drift of the C_L and C_T for the Imager 5016 within each day. In Fig. 5, we compare the C_L and C_T values (blue and red) with the internal instrument's temperature (black) for consecutive scans for three days (the most representative ones). There are two important observations. First, the reversely proportional trend corresponding to the correlation values in Table 4 is visible. Second, there is a high gradient during the initial warming-up (~90 min). Based on this data, we can conclude that there are notable variations in the compensator index error values within a day that should be accounted for (discussion follows in Sect. 4). However, further investigations should be conducted for accurate modeling of this effect.

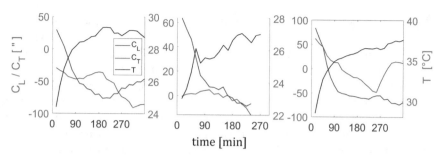

Fig. 5 Compensator index errors (C_L and C_T) and the internal instrument's temperature (T) over time for three calibration days

4   Discussion

To summarize, the results in the previous section indicate that the TLS calibration parameters are generally unstable. A large portion of this instability can be traced back to the changes in the instrument's internal temperature, which is mainly influenced by the internal heating up during the measurements and the ambient temperature. Although these calibration parameter changes are detectable, they are generally small in value when compared with the measurement accuracies given in the manufacturer's specifications (8" for P50, 14.4" for Imager 5016). Hence, for the majority of TLS applications, the manufacturer's factory calibration assures sufficient measurement accuracy. The following discussion and suggestions are primarily focused on stringent engineering tasks, such as very demanding deformation monitoring or industry quality assurance.

A general consideration for demanding engineering tasks is that the instrument should be properly warmed-up (at least 60 min) before the data acquisition. Afterward, an in-situ calibration is highly recommended for all calibration parameters that can be estimated based on the given measurement geometry. The remaining parameters should be a priori estimated on the dedicated calibration field and corrected for the temperature influence. Based on the data availability, this can be realized either by monitoring the instrument's internal temperature or by measuring the instrument's working time and the ambient temperature. Additionally, it is important to timely calibrate the dynamic compensator (in-situ, after the instrument's warming up).

When considering a measurement campaign within one day, the calibration parameters mostly change during the initial warming-up. However, the everlasting parameter drift should not be neglected. Based on the accuracy demands and the duration of the measurements, it is advisable to repeat the in-situ calibration procedure (compensator included) every few hours.

If the in-situ calibration is not possible due to the instrument or environmental limitations, relying on the a priori calibration is the best alternative. For realizing the instrument's maximal accuracy potential, the calibration should be performed directly before the measurement campaign (same day). Such an approach could be well suited for industrial applications, where a dedicated calibration hall could be situated in the proximity of the measurement subject. Unfortunately, such an approach is hardly feasible for deformation monitoring tasks.

To further investigate the stability of the calibration parameters we analyzed their standard deviations. Our calculations are explained in Fig. 6. First, we estimated the standard deviations for all parameters within each day and made an average (pooled) standard deviation as the most representative value for the 1-day instability. Then, we combined parameters of each two consecutive days together and again estimated standard deviations for each two-days-couple and averaged it. The procedure is repeated until the parameters of all nine days are combined together and only 1 representative standard deviation is calculated. This concept is similar to the Alan Variance calculation [20] and in this article it serves just as an indication of the parameter stability over time.

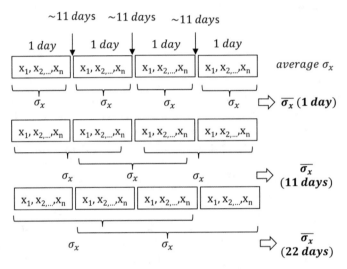

Fig. 6 A scheme of calculating the standard deviations presented in Fig. 7

The results of the analysis for the tilt parameters of the ScanStation P50 (most representative ones) are given in Fig. 7. In Fig. 7 we investigate how the standard deviations of the calibration parameters changes with time before (solid-lines) and after (dashed-lines) applying the temperature correction (Sect. 3.2). The approximate time difference between the calibrations was 11 days (median). In Fig. 7, we can observe a strong sudden increase in the standard deviation (for all CPs) if we compare standard deviations within 1 day and standard deviations when more days were grouped together. We can also observe that there is little difference when we make standard deviations of the CPs using only two consecutive days (approximately 11 days) or all days together (approximately 3 months). This especially holds true, if the temperature correction is applied (dashed lines), where the line is almost completely flat after the initial jump.

This means that we can expect a sudden increase in the uncertainty (instability) of the calibration parameters if they are used one or several days after the calibration. However, this increase in the uncertainty stops here, if we apply the temperature correction, and the parameters standard deviation is stabile over a long period of time.

Hence, our results indicate that in the long-term (>1 day), the parameter instability is strongly related to the atmospheric conditions, and not particularly related to the time

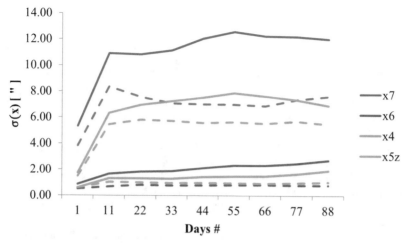

Fig. 7 Empirical standard deviations of the ScanStation P50 tilt calibration parameters with time before (solid lines) and after (dashed lines) applying temperature correction

passed since the last calibration. Therefore, the measurement at the extreme environmental conditions should be avoided, as the extrapolation of the temperature influence is expected to be inaccurate. If inevitable, it is recommended to calibrate the instrument at similar conditions, rather than in a controlled laboratory environment.

5 Conclusion

In this study, we investigated the instability of the calibration parameters estimated on the calibration field for the target-based calibration of panoramic terrestrial laser scanners. As the aim of the calibration fields is typically a posterior parameter use, the stability of the estimated parameters is of high relevance. We analyzed time series of several calibration days (2–6 calibrations per day) within half a year for the Leica ScanStation P50 and the Z+F Imager 5016. We investigated a short-term (within a day) and a long-term (between days) parameter instability. The variability of nearly half of the parameters can be reduced if the instrument's working time and the ambient temperature are measured and mathematically considered. By applying simple correction models, the short-term parameter instability can be reduced by 24% on average, while the long-term instability by 43% on average.

We have several considerations for future: investigating and quantifying further causes for the parameter instability, assessing the parameter instability within consecutive days and for multiple instruments of the same type; and in-detail investigation of the instrument's initial warming-up period. Finally, it is necessary to evaluate the impact of the observed parameter instability on the concrete engineering tasks.

References

1. Mukupa, W.; Roberts, G.W.; Hancock, C.M.; Al-Manasir, K. A review of the use of terrestrial laser scanning application for change detection and deformation monitoring of structures. *Surv. Rev.* 2017, *49*, 99–116, https://doi.org/10.1080/00396265.2015.1133039.
2. Walsh, G. Leica ScanStation P-Series – Details that matter. Leica ScanStation-White Paper Available online: http://blog.hexagongeosystems.com/wp-content/uploads/2015/12/Leica_ScanStation_P-Series_details_that_matter_white_paper_en-4.pdf (accessed on May 21, 2019).
3. Mettenleiter, M.; Härtl, F.; Kresser, S.; Fröhlich, C. *Laserscanning—Phasenbasierte Laser-messtechnik für die hochpräzise und schnelle dreidimensionale Umgebungserfassung; Die Bibliothek der Technik*; Süddeutscher Verlag onpact GmbH: Munich, Germany, 2015;
4. Holst, C.; Neuner, H.; Wieser, A.; Wunderlich, T.; Kuhlmann, H. Calibration of Terrestrial Laser Scanners / Kalibrierung terrestrischer Laserscanner. *Allg. Vermessungs-Nachrichten* 2016, *123*, 147–157.
5. Vosselman, G.; Maas, H.G. *Airborne and Terrestrial Laser Scanning*; Whittles Publishing, 2010; ISBN 9781439827987.
6. Lichti, D.D. Error modelling, calibration and analysis of an AM-CW terrestrial laser scanner system. *ISPRS J. Photogramm. Remote Sens.* 2007, *61*, 307–324, https://doi.org/10.1016/j.isprsjprs.2006.10.004.
7. Holst, C.; Medić, T.; Kuhlmann, H. Dealing with systematic laser scanner errors due to misalignment at area-based deformation analyses. *J. Appl. Geod.* 2018, *12*, 169–185.
8. Reshetyuk, Y. Self-calibration and direct georeferencing in terrestrial laser scanning, KTH Stockholm, 2009.
9. Lichti, D.D. A method to test differences between additional parameter sets with a case study in terrestrial laser scanner self-calibration stability analysis. *ISPRS J. Photogramm. Remote Sens.* 2008, *63*, 169–180, https://doi.org/10.1016/j.isprsjprs.2007.08.001.
10. Janßen, J.; Medić, T.; Kuhlmann, H.; Holst, C. Decreasing the uncertainty of the target centre estimation at terrestrial laser scanning by choosing the best algorithm and by improving the target design. *Remote Sens.* 2019, *11 (7)*.
11. Medić, T.; Kuhlmann, H.; Holst, C. Sensitivity Analysis and Minimal Measurement Geometry for the Target-Based Calibration of High-End Panoramic Terrestrial Laser Scanners. *Remote Sens.* 2019, *11*, 1519.
12. Medić, T.; Kuhlmann, H.; Holst, C. Designing and evaluating a user-oriented calibration field for the target-based self-calibration of panoramic terrestrial laser scanners. *2020*, 1–22.
13. Odziemczyk, W. Stability test of TCRP1201+ total station parameters and its setup. In *E3S Web of Conferences*; EDP Sciences, 2018; Vol. 55, p. 10.
14. Janßen, J.; Kuhlmann, H.; Holst, C. Assessing the temporal stability of terrestrial laser scanners during long-term measurements. In *Contributions to International Conferences on Engineering Surveying, INGEO & SIG 2020*; A., K., Kyrinovič, P., Erdélyi, J., Paar, R., Marendić, A., Eds.; Springer line: Dubrovnik, Croatia, 2020.
15. Chen, P.Y.; Smithson, M.; Popovich, P.M. *Correlation: Parametric and nonparametric measures*; Sage, 2002; ISBN 0761922288.
16. Ge, X. Terrestrial Laser Scanning Technology from Calibration to Registration with Respect to Deformation Monitoring, Technical University of Munich, 2016.
17. Medić, T.; Holst, C.; Janßen, J.; Kuhlmann, H. Empirical stochastic model of detected target centroids: Influence on registration and calibration of terrestrial laser scanners. *J. Appl. Geod.* 2019, *13*(3), 179–197.

18. Ogundare, J.O. *Precision Surveying: The Principles and Geomatics Practice*; 2015; ISBN 9781107671812.
19. Lichti, D.D. Terrestrial laser scanner self-calibration: Correlation sources and their mitigation. *ISPRS J. Photogramm. Remote Sens. 2010, 65*, 93–102, https://doi.org/10.1016/j.isprsjprs.2009.09.002.
20. Zhang, X.; Li, Y.; Mumford, P.; Rizos, C. Allan variance analysis on error characters of MEMS inertial sensors for an FPGA-based GPS/INS system. In *Proceedings of the International Symposium on GPS/GNNS*; 2008; pp. 127–133.

Analysis of Warm-up Effect on Laser Tracker Measurement Performance

Radoslav Choleva[✉] ⑩, Alojz Kopáčik ⑩, and Ján Erdélyi ⑩

Faculty of Civil Engineering, Department of Surveying, Slovak University of Technology,
Radlinského 11, 81005 Bratislava, Slovakia
radoslav.choleva@stuba.sk

Abstract. In the manufacture of various large-scale industrial components and machines (aerospace industry, shipping industry, automotive industry, etc.), high precision is required in determining their shape and parameters. At present, coordinate measuring systems—laser trackers—are often used to determine the dimensions of these components, with precision up to tens of micrometers. If a measurement result with such high precision is requested, factors that are specific for laser trackers need to be considered. One of these specific factors which affect the measurement is the warm-up effect. This effect is caused by warming-up the laser source, which also heats other parts of the instrument. In this paper, the tasted laser tracker was Leica AT 960-MR. The warm-up effect is most seen right after the device is turned on and decreases over time. According to the manufacturer, steady-state conditions should be achieved after approximately 30–40 min. To verify this statement and quantify how much is laser tracker affected by warming up, a series of experiments were performed. Detailed description of the tests is provided together with numerical and graphical results. The final statement of the tests is whether the instrument is fit for further use and after which time it has reached the declared precision (warm-up time).

Keywords: Leica AT960-MR · Testing · Warm-up time · Accuracy

1 Introduction

The measurement of large objects with high precision in the industry was a challenging problem in the past. Coordinate measuring machines (CMMs) were often used to solve this problem, but the biggest CMMs were only able to measure 3D coordinates of a few meters long objects. There was also a problem with transportation and placement of measured part onto CMM, because of its large size. The need for high precision measurement of large industrial components in situ starts the development of portable coordinate measuring systems. In 1987 the first laser tracker (LT) was introduced [1] with precision up to tens of micrometers, and further development is still in progress.

There are various definitions of LT, mostly depending on their usage. In the standard [2], the LT is described as a coordinate measuring system in which a cooperative target is continuously followed with a laser beam, and its location is determined in terms of

© The Editor(s) (if applicable) and The Author(s), under exclusive license
to Springer Nature Switzerland AG 2021
A. Kopáčik et al. (Eds.): *Contributions to International Conferences on Engineering Surveying*,
SPEES, pp. 142–154, 2021. https://doi.org/10.1007/978-3-030-51953-7_12

distance and two angles (azimuth and elevation angles). The principle of measurement came from total stations frequently used in geodesy (spatial polar method). After 30 years of development, new designs, components, and functions are implemented in LTs. This makes them universally used in many kinds of industries (for example, aircraft, airspace, automobile) or generally in the measurement of large objects like ship hulls or wind turbines.

With a precision that LT can achieve, various factors need to be considered. Neglections of these factors can lead to an increase in the uncertainty of results. Some aspects depend on the ambient environment, like changes in atmospheric parameters (temperature, pressure, humidity) or vibration, which is often present in the shop floor environment. Other factors depend on components or functions of LT, and one of them is a phenomenon called the warm-up effect. The warm-up effect is caused by the heating of the laser source of the distance measuring unit. Other components inside of the LT are heated together with the laser source, resulting in thermal expansion of these components. Because of thermal expansion, numerous misalignment and shifts appear, which affects the measured distance and angles.

In this paper, the theory about the warm-up effect and its impact on measurement is described. Several experiments are performed with LT Leica AT960-MR to determine this phenomenon. The final statement of the tests is whether the instrument is fit for further use and after which time it has reached the declared precision.

2 Warm-up Effect and Warm-up Time

The warm-up effect should not be confused with a common adaptation of the device with ambient temperature. Standard acclimatization of LT is a process when the tracker needs to adapt to temperature change when it is moved from different temperature environments.

The warm-up effect is caused by internal heating of the laser source used in the ranging unit. There are three technologies currently used in LT as ranging unit (or distance measuring unit): absolute distance meter (ADM), interferometer (IFM) and absolute interferometer (AIFM). Explanations of these technologies are given in [3–5]. Each technology need laser source warmed-up to perform measurement, and therefore each LT suffers from warm-up effect. In general, the warm-up effect changes among different manufacturers because of their individual design and composition of components of LT. Three main manufacturers today are Leica geosystems (Hexagon), Faro technologies and Automated Precision Incorporated (API). Each manufacturer has several models of LT with different design too, so the warm-up effect should be considered individually.

After starting the LT, the initialization process is made, during which the laser source of distance measuring unit is warming. This process lasts few minutes and measurements cannot be made. During initialization phase, other components are warming-up with laser source as well, causing thermal expansion of these components. Numerous misalignments, shifts and tilts of individual LT parts are made by thermal expansion and have a significant impact on the measured distances and angles.

The impact of the warm-up effect on the measurement is described in several ways. Some authors determined the impact of the warm-up effect on measured angles and distances [6–8], the others are evaluating changes in the center position of LT (translation, rotation) or impact on scale factor and offset of distance measurement [8, 9]. Exact values are not given in this paper because they depend on the specific model of LT. However, they always exceed the manufacturer declared precision. Vertical angles seemed to be often more affected than horizontal angles and distance. One possible explanation is that the laser source is placed near the vertical encoder circle and therefore is more susceptible to change in temperature. There is also information from the Leica geosystem given in [8] that the inclination sensor in the oil bath suffers from heating as well and changes its zero position. According to the manufacturer, this is the major contributor to the warm-up effect. However, inclination function can be often turned off in LT, and thus this effect may be reduced.

The important thing related to the warm-up effect is warm-up time. After initialization process, the temperature inside the LT is stabilizing, which gradually reduces the warm-up effect. The period, after the LT reaches a stable internal temperature level, is called the warm-up time. This value also depends on the model of LT, but research has shown that it takes around two hours to reach a stable temperature for different LTs [6–8]. According to manufacturers is this time quite shorter. For example, Faro technologies recommend on their web page one-hour warm-up time [10]. The same time recommends API in their tracker manual [11]. Leica geosystems provided information to the author of this paper that achieving specified accuracy for Leica AT960-MR should be after 30–40 min of warming-up. In addition, after 3–4 h should be the performance of the LT up to 50% better than is specified. This is not statement about how long does it take to stabilize temperature inside the LT (warm-up time) but can be also verified.

In general, the best thing about how to deal with the warm-up effect is to wait until the temperature stabilizes and no longer affect the measurement. But this process usually takes two hours or more and waiting so long can be unsatisfactory in some specific applications. In this case, manufacturers recommend doing the re-initialization process every fifteen minutes during the first two hours of measurement. This process should calibrate LT parameters to achieve declared precision. On the other hand, research [6] shows that the calibration of Faro LT was not very useful because of the instability of the system.

3   Experimental Testing of Warm-up Effect

A series of experiments were performed to evaluate the warm-up effect and warm-up time of Leica AT960-MR (Fig. 1). The tested instrument has a maximum range of 20 m with declared precision. The controller, external temperature sensor, and computer with the software are part of the system as well. The domain of the Leica LT is distance measurement technology called absolute interferometer (AIFM). It combines a heterodyne interferometer (IFM) for dynamic measurement with an absolute distance meter (ADM) to set the absolute reference distance. Angle encoders, inclination sensor, motorization, and other components are similar or same as are used in high accuracy Leica total stations.

Accuracy characteristics of Leica AT960-MR are calculated per standards [2, 12], and are shown in Table 1. These characteristics are a bit different as they are used for an instrument like a total station, which may lead to confusion. To properly understand the values shown in Table 1, a brief description is given. 3D coordinate accuracy combines

Fig. 1. Leica AT960-MR

uncertainties made by distance and angle measurement and retroreflector uncertainty. This coordinate uncertainty is defined as the deviation between a measured coordinate and the nominal coordinate of the measured point. Orient to gravity (OTG) accuracy means accuracy of 3D point coordinates with the inclination sensor turned on. Distance accuracy is divided into two parts. Dynamic lock-on is accuracy for ADM absolute distance determination (when setting reference distance for IFM). AIFM accuracy shows only IFM relative distance determination. These two values are indistinguishable because both modules measure simultaneously and contribute to the resulting accuracy of distance measurement. Angle accuracy is not interpreted in angular units as in geodesy. This specific notation comes from the methodology of the two-face test listed in the standard [2], which is used to determine the angle accuracy of the LT.

Table 1. Accuracy characteristics of Leica AT960-MR

Accuracy characteristics	Standard deviation (1σ)	Maximum permissible error (MPE) (2σ)
3D coordinates	7.5 μm + 3 μm/m	± 15 μm + 6 μm/m
Orient to gravity (OTG)	7.5 μm + 4 μm/m	± 15 μm + 8 μm/m
Distance accuracy AIFM	0.25 μm/m	± 0.5 μm/m
Dynamic lock-on	5 μm	± 10 μm
Angle accuracy	7.5 μm + 3 μm/m	± 15 μm + 6 μm/m

All experiments were performed in the Hexagon office in Bratislava. The Hexagon office is located on the ground floor with a robust foundation and away from the main traffic. Therefore, the stability of the building should be sufficient for testing such an accurate instrument. However, the temperature stability of the environment was challenging as stable conditions could not be ensured. Several temperature sensors were placed in a room where the experiment was performed to observe the temperature continuously and possibly use it for corrections. Before each experiment the LT was placed in room whole night to stabilize with ambient temperature.

3.1 Testing Warm-up Effect of Leica AT960-MR with Calibrated Steel Bar

The first experiment to calculate the warm-up effect was performed with a calibrated steel bar as a reference length. Initially, this experiment was used to test the performance of the LT, together with the usability of equipment, and more about it can be found in [13]. However, the data could be possibly used to determine the warm-up effect or estimate the warm-up time.

The calibrated steel bar (Fig. 2) used in the experiment usually serves for testing measuring arms. Three conically shaped holes are drilled into the steel bar where a retroreflector can be placed. From these three points, two reference lengths can be determined (Table 2). During the experiment, a calibrated steel bar was placed on a tripod with clamps and measured with the LT from three different distances (6, 12, 18 m) and with three measuring modes (fast, standard, precise). The steel bar was always oriented horizontally and perpendicular to the laser beam. The temperature of the steel bar was measured during the whole experiment with a contact temperature sensor for the correction of thermal expansion. The ambient temperature stability was monitored with four temperature sensors. The maximum difference in temperature in 20 meters long room during the experiment was around 1.5 °C.

Fig. 2. A calibrated steel bar

Table 2. Reference lengths of calibrated steel bar with uncertainty

Reference lengths	Length [mm]	Uncertainty [mm]
Right–left point	1015.977	0.0027
Middle–left point	514.401	0.0023

Measurement started right after the initialization phase and lasted for four hours. At first, the experiment was performed to test different measuring modes and measuring capabilities of the LT at almost its maximum range. Nevertheless, the four-hour time interval is enough to see any changes in measured data affected by the warm-up effect. At the beginning of the experiment, the data should be the most inconsistent and getting better with time.

After the measurement, 42 3D points were obtained for every distance and every measuring mode, which makes 378 points in total. Two sets of lengths were then determined from the point coordinates to compare with the reference lengths (one set for a longer length and another set for shorter length). The total number of determined lengths was 84. After the application of correction for thermal expansion, differences between measured and reference lengths can be calculated. As was mentioned, the differences between lengths should be more significant at the start of the experiment and getting smaller with time. For interpretation, only one graph of differences between measured and reference length (smaller length) is presented (Fig. 3). The graph is showing differences in lengths measured in every measuring mode and at 6 m distance from LT. Other graphs showed very similar results and can be found in [13].

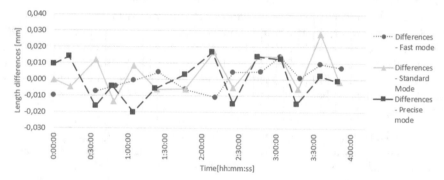

Fig. 3. Differences between reference and measured length for steel bar placed 6 m from LT

Statistical analysis was performed on all data to determine the presence of outlying (extreme) values using the Grubbs test [14]. Based on the test results, one value (in Fast mode) was removed from the sample with a probability of 95% of the right decision.

Based on the differences shown in Fig. 3, there cannot be seen any impact of the warm-up effect on the results. Differences in length are within manufacturer specifications throughout the whole experiment and not getting any better. A similar situation was observed in every case of an experiment. Why the warm-up effect cannot be seen is a combination of two things. Firstly, the measurement of two points, which were used to length determination, was performed in a short time interval (up to a few minutes). This is the reason why is the warm-up effect not changing significantly. Secondly, the warm-up effect has a systematic effect on the measurement. It means that measured points are affected by the warm-up effect in the same direction and by the same value (if the measurement is performed in a short period).

The described experiment design is not suitable for calculating the warm-up effect or warm-up time. However, there can be used some conclusions about avoiding the impact of the warm-up effect on the measurements. If the measurements are performed in a short period (a couple of minutes), the warm-up effect can be neglected. Therefore, the measurements can be performed during the warming phase. This is in congruence with the manufacturer's statement to do the re-initialization process every fifteen minutes.

3.2 Testing of Warm-up Effect of Leica AT960-MR by Measuring Single Point

The next two experiments were directly designed to calculate the warm-up effect by measurement of a single stable point during the four hours. This time interval should be sufficient to estimate the warm-up time. Individual experiments were the same in general, but have little changes in methodology to observe different behavior of the LT. In both experiments, retroreflector placed approximately 3 m away from the LT and in the same height was continuously measured. The last measured point, which should be the least affected by a possible warm-up effect, served as a reference point. The changes in the measured quantities (distances and angles) are expressed as the difference between the measured values and the last measured value. The expected behavior is that differences at the start of the experiment will be more significant and will gradually decrease over time. Warm-up time should be around two hours based on observations of other authors.

All experiments were performed in the Hexagon office in Bratislava—the same place as with the previous experiment with a calibrated steel bar. The temperature in the room was measured by non-contact temperature sensors: one placed near the LT and another one near the measured point. The reason for using only two sensors was because the distance between the LT and the measured point was only 3 m. Adding more sensors would not significantly help to describe temperature conditions in the area of performed experiments.

Testing of Warm-up Effect by Two-face Measurement. When designing the experiment, it is necessary to take into consideration the technology for measuring distances because of their different working principles. Tested Leica AT960-MR uses AIFM technology. Therefore, when aiming at target (retroreflector), the absolute distance is determined using ADM, which serves as a reference for the interferometer. Subsequently, all other distances are determined using an interferometer. However, if the laser beam is interrupted, the absolute distance is determined again by ADM. It follows that if the laser beam is not interrupted during the whole experiment, the warm-up effect on the distance measured by interferometer is tested. Nevertheless, if the laser beam before each measurement is interrupted (like with two-face measurements), the warm-up effect on distance measured by ADM is tested.

The first experiment was performed by measuring reflector in two-face mode continuously for approximately four hours. The rest of the experiment was done, as described above. The inclination sensor was turned off, so measurements were performed in the local LT coordinate system. The time interval between single measurements was around one minute to effectively determine the warm-up effect.

With this type of experiment, the impact of the warm-up effect on measured distances and angles is evaluated. Because of the two-face measurement, only the ADM performance is tested. The differences in the measured distance are shown in (Fig. 4), differences in the measured horizontal angle in (Fig. 5) and the differences in vertical angle in (Fig. 6). In the presented graphs, the temperature measured at the instrument is labeled as "sensor 1" and at the reflector as "sensor 2".

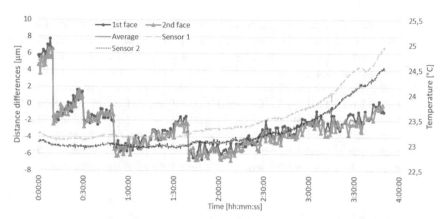

Fig. 4. Differences in measured distance (two-face measurement)

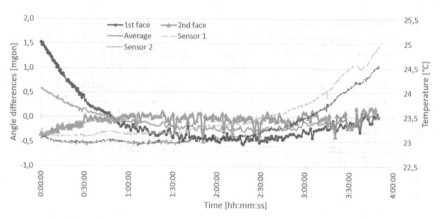

Fig. 5. Differences in measured horizontal angle (two-face measurement)

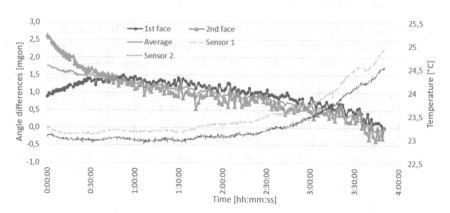

Fig. 6. Differences in measured vertical angle (two-face measurement)

Step changes in the measured distances at the level of several micrometers appeared in both instrument faces (Fig. 4). The maximum step of almost 10 μm occurred during the first 15 min, but that is still in the manufacturer's declared accuracy. A similar result happened to another author as well [8], but without any explanation of this behavior. Step changes in the measured distance ceased to occur approximately 1 h and 45 min after the start of the measurement. This period can be identified as warm-up time. After approximately another half an hour, it can be seen that the differences in the measured distance increased again. However, if these data are compared with the measured temperature, their correlation can be seen as well. After approximately 2 h and 15 min, the room temperature began to rise (by almost 2 °C until the end of the experiment), which affected the measurement.

The unusual thing is that LT should internally compensate for this rise in temperature. Device manual states that a 1 °C change in temperature can cause 1 ppm change in distance. For 3 m long distance is this value 6 μm, which is in congruence with Figure 4. The unknown fact is why the LT is not compensating the temperature changes.

Similar observations can be made based on the graph of differences in the horizontal angle (Fig. 5). When measured in the 1st face, the values stabilize as well as the distances after approximately 1 h and 45 min. The maximum difference in angle changes during this time interval is about 2.0 mgon. This value causes almost 0.1 mm deviation in a direction perpendicular to the laser beam (at 3 m distance). With a maximum instrument range of 20 m, this is approximately 0.6 mm, which is not negligible. There is also a similar phenomenon of gradual increase in differences as in the determination of distance. However, this is again caused by the rising temperature in the room. The two-face measurement can partially reduce the impact of the warm-up effect, but this type of measurement is very rarely used in practice.

When determining changes in the vertical angle (Fig. 6), the situation is not clear. Steady-state conditions cannot be estimated very well because measurements are too much influenced by temperature change in the room. If the same warm-up time as in previous graphs is considered, the maximum difference in measured vertical angles would be about 0.5 mgon for the first face and 2.0 mgon for the second face.

Testing of Warm-up Effect by Measurement in 1st Face with OTG Function Turned On. In the next experiment the reflector was measured only in the 1st face and without interrupting the laser beam during the measurement. This means that the impact of the warm-up effect on distance measured with IFM was tested. Data were collected with fifteen seconds time interval between measurements. In this case, also the OTG function was turned on. When the OTG function is turned on, the inclination sensor is measuring the deviation between the vertical axis of the LT and the plumb line (gravity vector). Measured values are then automatically corrected of this deviation. Before the start of the measurement, the LT sets the initial value for compensation to align measurement to gravity. Inclination sensor should also be affected by the warm-up effect. Therefore, after the re-initialization of the inclination sensor, the changes should be seen in the data. The variations in the measured distance are shown in (Fig. 7), differences in the measured horizontal angle in (Fig. 8) and the differences in vertical angle in (Fig. 9). As with the previous experiment, the temperature at the instrument (sensor 1) and the reflector (sensor 2) was measured too.

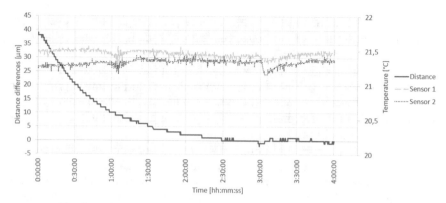

Fig. 7. Differences in measured distance (measurement in the 1st face)

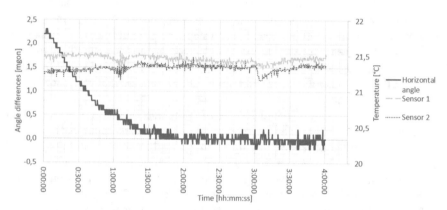

Fig. 8. Differences in measured horizontal angle (measurement in the 1st face)

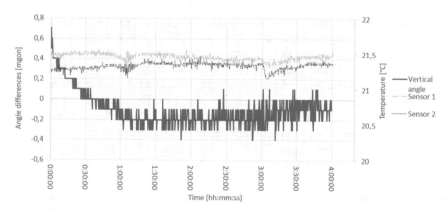

Fig. 9. Differences in measured vertical angle (measurement in the 1st face)

Differences in the distance are changed in this case (Fig. 7). There are no step changes during the measurement, and warm-up time can be set between 2 h and 2 h and 30 min.

The measured distance is affected by warming-up of about 38 μm, which is way behind declared accuracy. In this case, the temperature was stable during the whole experiment, with a maximum difference of 0.3 °C.

Horizontal angle stabilized after approximately 1 h and 40 min, with a maximum change of 2.3 mgon (Fig. 8). This value is comparable with the previous experiment and produces deviation around 0.7 mm for 20 m range of the LT. It seems that the inclination sensor has no additional effect during warming-up because similar results were obtained as in the previous experiment.

Vertical angle differences stabilized sooner than in the previous experiment (around 1 h), and the differences are also smaller (around 0.8 mgon). The question was if the OTG function causes this situation because the inclination sensor should be affected by warming-up as well. There is a possibility that the zero position of the inclination sensor drifted in such direction that it compensated the impact of the warm-up effect.

During the whole experiment, the laser beam was not interrupted. At the end of the experiment, several values were measured after breaking the line of sight on the reflector. Measured angles were not affected at all, but a step-change in the measured distance of 30 μm appeared. Such a high value can be explained by the accuracy of the dynamic lock-on of ADM, together with the impact of the warm-up effect on distance measured with ADM (Fig. 4). After the interruption of the laser beam, the IFM reference is reset, and measuring is started over.

The next step was the second initialization of the inclination sensor. After the second initialization of OTG, distance and horizontal angle were without change, but the vertical angle was affected. The difference rose by 0.5 mgon, so the maximum difference of 1.3 mgon was observed. The zero position of the inclination sensor was probably shifting during warming-up, and after the re-initialization process was this position set back to zero. This caused the jump step in the vertical angle of 0.5 mgon. This value is smaller than in the previous experiment, but it is hard to compare because of temperature influence (Fig. 6). Nevertheless, highly accurate measurement is better to perform with OTG initialization after the warming-up procedure and not sooner.

Comparison of Results with Manufacturers' Statements and Other Authors. From the presented experiments above, an approximately 2-hour warm-up time can be set for Leica AT960-MR. This value is in congruence with other authors even when they used other models of LT. However, according to manufacturers, approximately 1 h of warming-up should be enough. More specifically, around 30–40 min should be enough for tested Leica AT960-MR to reach declared accuracy. This statement can be verified from the presented graphs of the warm-up effect. For example, if differences in measured values after 40 min in the last experiment are taken, uncertainty in the position of the measured point is around 52 μm (for 3 m long distance). The maximum permissible error declared by the manufacturer for this point is 39 μm (Table 2). The manufacturer statement is a bit optimistic because the maximum permissible error of the measured point is exceeded by 13 μm. Such value is negligible in most of the applications but can be significant, for example, in the testing of the performance of the LT. A statement that after 3–4 h should be the accuracy of the LT even 50% better than declared is right. Because the warm-up time is approximately 2 h, in this way the measured distances and angles are not affected by the warm-up effect at all.

The impact of the warm-up effect on horizontal angles is around 2.3 mgon. For the maximum range of the LT (20 m), this angle deviation causes 0.7 mm in a direction perpendicular to the laser beam. Vertical angles are affected less than horizontal angles, from 1.3 to 2.0 mgon. This is a bit different when compared with other researches. In various cases, the vertical angles were affected more than horizontal angles. The different design of the LT most likely causes this situation, but the further research needs to be done.

4 Conclusion

The paper describes the impact of the warm-up effect on measured distances and angles and the evaluation of warm-up time of the LT Leica AT960-MR. The experiment with a calibrated steel bar was not suitable for the determination of the warm-up effect. However, it can be concluded that in short period measurements (a few minutes), the warm-up effect could be neglected. Overall the LT measuring performance was good and can be used for further experiments.

The other experiments were more successful and brought useful information. Two-face measurement of distance has no significant effect during warming-up. However, a two-face measurement of angles can help to reduce the impact of the warm-up effect. In general, the distance measurement with ADM is less affected then measurement with IFM. When measuring with IFM, the laser beam should be interrupted after the warming-up to get the new reference from ADM. If this condition is not fulfilled (in some rare occasion) a jump step in the data will appear.

For the maximum range of the LT (20 m), the warm-up effect is affecting horizontal angles and could cause deviation up to 0.7 mm in a direction perpendicular to the laser beam. Vertical angles are affected less and cause deviation between 0.4 to 0.6 mm. When the inclination sensor is turned on, differences in measured values seem to be minimal or any at all except for the vertical angle. If the re-initialization is made after warming-up, a jump in vertical angle by 0.5 mgon appeared. This means that the re-initialization process should be made again after warming-up or several times during warming-up.

There was also a problem with temperature compensation of measured distances when the ambient temperature changed. This indicates the wrong reaction time of external LT temperature sensor or insufficient software compensation. Also, the jumps in the data when measuring with ADM were not explained and more information is needed.

The warm-up time can be determined around 2 h based on the presented data, which is in congruence with other publications. Manufacturers, in general, declare a shorter period, around one hour. However, the warm-up time (or warm-up effect) should be considered individually for every LT model. For the LT Leica AT960-MR was a specific statement given by the manufacturer that it should achieve declared accuracy in 30–40 min. Nevertheless, this statement is a bit overrated because declared accuracy was achieved approximately in one hour. A statement that after 3–4 h should be the accuracy of the LT even 50% better was proved as right because measured values are no longer affected by the warm-up effect.

Acknowledgements. This publication was created with the support of the Scientific Grant Agency of the Ministry of Education, Science, Research and Sport of Slovak Republic and the Slovak Academy of Sciences for the project VEGA-1/0506/18. The laser tracker was lent from the Hexagon office in Bratislava, together with provided support.

References

1. NIST, https://www.nist.gov/news-events/news/2015/12/pml-expertise-and-collaborations-lead-portable-test-solution-laser-trackers, last accessed 2020/6/2.
2. EN ISO 10360-10:2016. Geometrical product specifications (GPS) - Acceptance and reverification tests for coordinate measuring systems (CMS) - Part 10: Laser trackers for measuring point-to-point distances, 2016.
3. Renishaw. Interferometry explained, https://www.renishaw.com/en/interferometry-explained -7854, last accessed 2020/2/6.
4. Muralikrishnan B., Phillips S., Sawyer D.: Laser trackers for large-scale dimensional metrology: A review. Precision Engineering 44, pp. 13–28 (2016), ISSN 0141-6359.
5. 5. Harding, K.: Handbook of Optical Dimensional Metrology. 1st edn. CRC Press, Florida (2008), ISBN 978-1-4398-5482-2.
6. Laser Tracker warm-up tips for ideal measurement conditions, https://insights.faro.com/buildit-software/5-laser-tracker-warm-up-tips-for-ideal-measurement-conditions, lase accessed 2020/2/6.
7. Sugahara R., Masuzawa M., Ahsawa Y.: Performance test of laser trackers of Faro, In: Proceedings of the 7th Annual Meeting of Particle Accelerator Society of Japan, pp. 253–256. Himeji, Japan (2010).
8. 8. Dvořáček F.: Laboratory testing of the Leica AT401 laser tracker. Acta Polytechnica 56(2), pp. 88–98 (2016).
9. Gassner G., Ruland R.: Instrument tests with the new leica AT401. International workshop on accelerator alignment, Stanford, pp. 1–5. CA, USA (2011).
10. Accurate Measurement Best Practices for the Laser Trackers, https://knowledge.faro.com/ Hardware/Laser_Tracker/Tracker/Accurate_Measurement_Best_Practices_for_the_Laser_ Tracker, last accessed 2020/2/6.
11. Automated Precision, Inc.: Tracker 3, Laser tracking system users manual v. 1.0 (2005).
12. ASME B89.4.19:2006. Performance Evaluation of Laser-Based Spherical Coordinate Measurement Systems (2006).
13. Choleva, R.: Overovanie meracieho výkonu laser trackera. Advances in Architectural, Civil and Environmental Engineering, pp. 123–129 (2019), ISBN 978-80-227-4927-5.
14. Grubbs, F.: Procedures for Detecting Outlying Observations in Samples. Technometrics 11, 1–21 (1969)

Deformation Measurement

Application of Fused Laser Scans and Image Data—RGB+D for Displacement Monitoring

Josip Peroš[1]([⊠]) [iD], Rinaldo Paar[2] [iD], and Viktor Divić[1] [iD]

[1] Faculty of Civil Engineering, Architecture and Geodesy, University of Split, Split, Croatia
{josip.peros,vladimir.divic}@gradst.hr
[2] Faculty of Geodesy, University of Zagreb, Zagreb, Croatia
rinaldo.paar@geof.unizg.hr

Abstract. A novel method for displacement monitoring by using RGB+D data has been recently proposed. RGB+D data is created by fusing image and laser scan data, where the D channel represents the distance, interpolated from laser scanner data. RGB+D image combines the advantages of the two measuring methods. Laser scans for distance changes in the line of sight and high-resolution image data for displacements perpendicular to the viewing direction of the camera. Image feature detection and matching algorithms detect and match discrete points within RGB+D images from different epochs. 3D coordinates of the points can be easily calculated from RGB+D images. In this paper, implementation of this method is proposed for measuring displacements and monitoring the behavior of structural elements under constant load in field conditions. Displacements measured with a high precision LVDT sensor will be used as a benchmark for displacement calculated from RGB+D images.

Keywords: Displacement monitoring · Image matching · Multi stations · RGB+D · Trimble SX10

1 Introduction

A. Wagner, in his research [1, 2] proposes a novel approach of fusing image and laser scan data from the same instrument, such as a multi-station, to create an RGB+D image. D channel represents the distance channel interpolated from laser scanner data. Detection and matching of identical image features from RGB+D images gathered from different measurement epoch is a use case for computer vision algorithms. 3D coordinate of the detected features can easily be calculated using RGB+D images. Matched image features can be used as substitutes for classic discrete points in displacement monitoring tasks and rigorous geodetic deformation analysis with tests of significance. Comparatively, the lack of discrete points is the biggest drawback of using laser scanner data (3D point clouds) for displacement monitoring and deformation analysis.

Proposed methodology and all the steps necessary to create RGB+D images and for their subsequent use, will be field tested in this research. The focus will be on determining the achievable accuracy of the method in field conditions.

A. Kopáčik et al. (Eds.): *Contributions to International Conferences on Engineering Surveying*,
SPEES, pp. 157–168, 2021. https://doi.org/10.1007/978-3-030-51953-7_13

The goal of the testing is to compare RGB+D detected displacement of a tested wooden beam to the benchmark values, measured by a linear variable differential transformer (LVDT) sensor.

2   RGB+D Data from Multi Stations

Combining all the sensors and functions of the Modern Multi sensor total Stations (MS) provides a new opportunity for data acquisition in the field and the subsequent data processing. The fusion of data acquired from the integrated RGB (red, green and blue) sensors and laser scanner data and its application for displacement monitoring, as proposed by A. Wagner, will be discussed and tested further in this paper.

2.1   Multi Sensor Total Stations

Multi sensor total stations (MS) offer a unique advantage for the engineering surveyors by encompassing multiple measuring sensors in one instrument. It combines the measuring abilities of the classic total station equipped with robotic functionalities with built-in RGB sensors and laser scanning function.

Total stations by themselves are multi-sensor systems that provide highly accurate angle and distance measurement to prism and less precisely prismless to nearly any other surface. Further functions are tilt correction by 2-axis inclinometers and automatic target recognition and tracking.

In recent years all significant surveying manufactures equipment have started to offer total stations with the addition of at least one RGB sensors, which predominant function is for the overview. In literature, those types of total stations are called Image Assisted Total Stations (IATS). Some manufacturers provided a second coaxial camera in their instruments built in the telescope, which uses the optics and magnification of the telescope. Coaxial cameras axis coincides with the telescope axis, and the camera center is the same as the instrument center. Using the magnification of the telescope results in a lower ground sampling distance (higher spatial resolution) of the image compared to the overview camera, albeit at a smaller field of view [2]. Most manufacturers use a 5-megapixel Complementary Metal-Oxide-Semiconductor (CMOS) sensor in their total stations.

In addition to RGB sensors in the newer generations of MS, manufacturers have started implementing laser scanner functions. The most significant disadvantage of the scanners integrated into total stations, when compared to Terrestrial Laser Scanners (TLS), is their slow rate of measuring data. Uniformly across all mayor manufactures TLS measures data at a rate of around 1 MHz, while an integrated scanner usually measures data at a rate of 1 kHz, which results in long scanning times. The newest generation of Trimble total station SX10, which is used for the testing by the author, as shown in Fig. 1, has pushed the limit of scanning rate up to 26.6 kHz [4]. Relevant specification of the Trimble SX10 for the test is shown in Table 1.

Fig. 1. Trimble multi station SX10 (*source* [3])

Table 1. Trimble SX10 specifications [4]

Trimble SX 10 specifications	
Angle measurement accuracy	1"
Prismless distance measurement accuracy	2 mm + 1.5 ppm
Scanning 3D position accuracy at 100 m	2.5 mm
Scanning measurement rate	26.6 kHz
Scanning point spacing	SUPERFINE—6.25 mm, FINE—12.5 mm, STANDARD—25 mm or COARSE—50 mm at 50 m
Camera	Telescopic, Primary or Overview
Image sensor	3 × 5 MP (2592 × 1944 px) CMOS sensor
Primary camera pixel size	4.4 mm at 50 m
Primary camera Field Of View (FOW)	13.12° × 9.86°

2.2 Fusion of Laser Scan and Image Data from Multi Stations

The resulting product from the fusion of laser scanner data an RGB image of the same scene is a generated RGB+D (Red, Green and Blue+Distance) image. The D is the 4th channel of the image that depicts distance information from the laser scanner as pixel values.

It should be noted that the RGB+D images are also used in other fields of science, especially in applications of RGB+D cameras for computer 3D reconstruction of objects [5].

The focus of this paper is the application of RGB+D images, generated from MS precise measurements, in displacement monitoring and deformation analysis.

The fusion of data from the laser scanner and camera data to RGB+D images combines the advantage of the two measuring methods. With laser scans and the resulting dense point clouds, distance changes in line of sight can be easily detected. High-resolution image data, in contrast, is most sensitive to displacements perpendicular to the viewing direction of the camera [1].

The biggest hurdle in this process is defining the precise spatial and angle relationship between the instrument, theodolite and integrated camera. Empirical tests conducted with SX10 shows that methods like direct linear transformation (DLT) [6] for determining camera offsets and parameters were not sufficiently precise for this use case. The ideal solution is to obtain the technical specification of the used MS camera directly from the manufacturer.

3   Deformation Analysis Using Point Clouds

The emphasis in classic deformation analysis is on the discrete points, which represent the object of interest. Deformation analysis compares the coordinates of the points, calculated from measurements in multiple epochs, to determine the possible displacements of the object with an appropriate statistical test of significance. Only a statistically verified result of a measurement may be used for further processing or interpretation. This is valid in particular to reliable alarm systems for the prevention of human and material damage [2].

A fundamental change in geodetic displacement monitoring is currently taking place. Nowadays, surface measurements are increasingly used instead of pointwise observations (of manual selected discrete points) as it was done in the past. Laser scanners are used, as they allow a fast, high resolution and dense acquisition of 3D information. However, it is challenging to uncover displacements and deformations and changes in multi-temporal point clouds as no discrete points are measured, and only changes in line of sight (of non-signalized areas) can be detected automatically as distance variation. Further, rigorous deformation analysis and tests of significance of the results are missing [7].

The goal of the RGB+D approach is to bypass the drawbacks of laser scanner measurement—point clouds by enhancing the data with corresponding RGB data, preferably captured with the same instrument, such as an MS, to identify corresponding (discrete) points between subsequent measuring epochs. The resulting data can then be integrated into a rigorous geodetic deformation analysis with test of significance [1].

3.1   Matching Algorithms for RGB+D Data in Displacement Monitoring

The biggest challenge of RGB+D method used for the displacement monitoring is how to identify corresponding (discrete) points between subsequent measuring epochs. Numerous matching algorithms have been developed for that purpose. The concept of feature detection and description refers to the process of identifying points in an image (interest points) that can be used to describe the image's contents such as Edges, corners, ridges and blobs [8].

Most prominent and promising feature detection algorithms, applicable to RGB+D images in deformation analysis, are:

- SIFT—Scale Invariant Feature Transformation [9],
- SURF—Speed Up Robust Features [10] and
- BRISK—Binary Robust Invariant Scalable Keypoints [11].

All of the algorithms mentioned are scale invariant, which enables them to match image features despite the scale change of an object between images (movement). David G. Lowe developed SIFT as a novel approach for image matching and feature detection. The basis of the algorithm is the transformation of image data into scale-invariant coordinates relative to the local features. An important aspect of the SIFT approach is that it generates a large number of features that densely cover the image over the full range of scales and locations [9].

SURF approximates or even outperforms previously proposed image matching algorithms with respect to repeatability, distinctiveness, and robustness, yet can be computed and compared much faster [10].

BRISK is proposed as an alternative image matching algorithm to SIFT and SURF, matching them in quality and with an emphasis on much less computation time [11]. Because of the binary nature of the algorithms descriptors, image features can be matched very efficiently, which makes the algorithm applicable for real-time use.

For postprocessing use case direct comparison of algorithms based on computation time is not necessary. Algorithms selection will be scene (surveillance area) dependent because computation time is not a limiting factor for postprocessing of data.

BRISK and SURF have a direct implementation in MATLAB [12].

MATLAB [12] also has a built-in function "matchfeatures" for matching detected features between images. In this use case, images of the same test scene taken in different epochs. There are multiple other software solutions available for matching features between images and they all function on the same or similar mathematical principles. All performed tests in this research were coded and conducted in MATLAB.

Example results of matching with the SURF algorithm are shown in Fig. 3.

4 Experimental Test Procedure

The general test procedure consists of loading a test material element sequentially with pressure. Relative movement of the element is measured with one or more LVDT.

Before every change of load is applied, a picture or multiple pictures in a panorama, and a laser scan of the element is taken.

Measured data is processed to create a fusion of a laser scan and image into RGB+D image for each load, i.e. epoch. From RGB+D images displacements are calculated for detected matched points between the epochs.

Resulting calculated displacement can then be compared to the LVDT measured data.

4.1 Testing Setup and Measurements

For the test material, wooden beams were chosen. Wood was selected because of its excellent elastic properties, and it exhibits large displacements under a small load.

The wooden beam was supported on both sides, and the pressure was applied in the middle of the beam. The pressure distribution was not measured because it has no relevance for this type of test and subsequent calculations, but the overall force is known for every step of load increment using a small load cell integrated into the loading apparatus. This setup was necessary to determine inputs for an independent numerical model of beam displacements. Small black shapes were added as additional targets on the wood beams to increase the number of detected features with the feature matching and detecting algorithms.

Displacements were measured with a Linear Variable Differential Transformer (LVDT). LVDT is a common type of electromechanical transducer that can convert the linear motion of an object to which it is coupled mechanically into a corresponding electrical signal. LVDT typical sources of uncertainty are due to nonlinear electrical responses, mechanical positioning and orientation errors, electrical transmission noises and digitalization errors. In the described measurement setup, HBM WA100 LVDT sensor [13] with a nominal measuring range 0–100 mm was used. Nonlinearity of the used sensor is in the range of 0.2–1% of its entire measurement range. For reduced measuring range 0–20 mm used in this setup, overall accuracy is ±0.02 mm (obtained by previous laboratory calibration). LVDT measuring accuracy is higher than the achievable accuracy of measurements SX10 MS by a factor of 100. The displacements obtained by LVDT were taken as a reference in this research for testing the accuracy of determined displacements from RGB+D images. LVDT was positioned in the middle of the beam where the pressure was applied to, Fig. 2 (right). The point where the LVDT touches the test beam is not visible on the images for this test setup. The beam was modeled based on the LVDT measurements to be a reference for all points on the beam, not just the one where LVDT was directly measuring.

Fig. 2. Test site (left) and test configuration (right)

MS instrument Trimble SX10 was used for the testing. The instrument was positioned roughly 5 meters from the wood element, as shown in Fig. 2 (left). The distance was chosen based on the cameras FOV, so that the whole test field is visible within one image from the integrated primary camera. For the primary camera, image size is 2592 × 1944 pixels, and at 5 m distance, the ground sampling distance (GSD) is 0.3 mm.

The instrument station was set up in a local coordinate system, axis shown in Fig. 2 (left). Control reference points (CRP) were positioned around the test site, two are shown as CRP in Fig. 2 (left). Their coordinates were determined in the local coordinate system and they were used to check the stability of the instrument between the epochs. RGB+D method does not require the instrument station to remain the same between the epochs, using the control reference points, the station stability can be checked and, if necessary, recalculated between epochs using resection.

The final test was conducted on two different wooden beams. For each element, six measuring epochs were conducted. The first and last epoch were without any pressure. For epochs two to five rising pressures was applied so that different displacements are obtained and then measured with the MS. In each epoch, a picture was taken with the MS primary camera with known horizontal and vertical angles from the MS. The test area was then scanned in the FINE resolution mode of the scanner, which resulted in a measurement density of 1 mm. Data from the LVDT was continuously logged and stored.

4.2 Data Processing

Output data from MS Trimble SX10 needed for the processing were images, point clouds and a job file. The job file contains all the metadata for a single job in an SX10; in this case, metadata about the captured images were extracted. Data processing was done in MATLAB software using the code developed by the authors.

The code was used to compare the data from two different epochs. For each epoch, an image, corresponding point cloud and image metadata were loaded. Processing of the image consists firstly of determining the location of the camera projection center in the assigned coordinate system. Secondly, calculating the exterior orientation parameters of the camera [14, 15]. Theodolite angles (Hz and V) for each pixel were calculated, with regards to the camera center, based on the interior and exterior parameters of the camera. MS primary camera is rectilinear, so it is considered to be distortion-free.

Processing of the point cloud consists of calculating theodolite angles (Hz and V) and 3D distance from the camera center. The next step is matching between the image and point cloud, based on minimal angle difference, to assigned 3D distances to some of the pixels. Because the number of pixels in the image is usually larger than the number of point cloud points, 3D distances for the empty pixels are calculated using a 2D interpolation algorithm [16]. Result of this step is an RGB+D image for both epochs. For each image pixel, RGB values, Hz and V angles and 3D distance are known. Using the SURF algorithm [10] identical points (features) were detected in both images, marked with blue circles and green x-es. Dotted red lines represent the calculated image displacement vectors between detected features in different epochs, as shown in Fig. 3.

DISPLACEMENT OF POINTS FROM EPOCH 1. TO EPOCH 2.

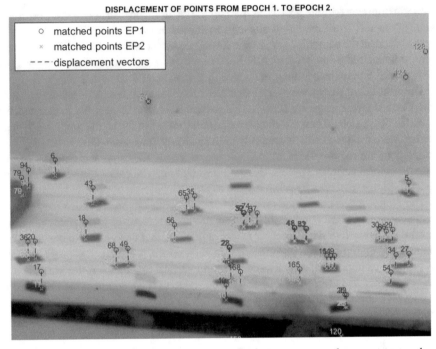

Fig. 3. Example of plotted identical points and displacement vectors between two epochs

The results of matching were filtered for outliers. In this test case, filtering was done based on the calculated projective transformation between the images and expected maximum displacements from LVDT.

Based on the data from RGB+D images and matched points 2D and 3D displacements between matching points in epochs can then be calculated.

4.3 Achieved Results and Quality Analysis

The final output of data processing are the 3D coordinates of the matched points in both epochs. Based on the coordinates, 2D and 3D displacements can be calculated.

Data was processed for all six measuring epochs for each wooden beam.

Data for the displacement measurement for epochs 210 and 214 will be discussed further as a reference example. Epochs are named based on the file name of the image taken by the instrument for that epoch. Image matching results between epochs are shown in Fig. 4.

Displacement of the center of the beam was measured with the LVDT sensor. Displacement of the center for epoch 210 was 0.00 mm, with no pressure and for epoch 214 was 12.1 mm, with maximum pressure. Based on the measurement and geometry data for the beam, measured on the test site and extracted from the point cloud and LVDT measurement data, a numerical model of the beam was created. Used numerical model is modeled in commercially available software SCIA Engineer 19.2 based on finite elements method structural analysis. The input for the model is the dimension and

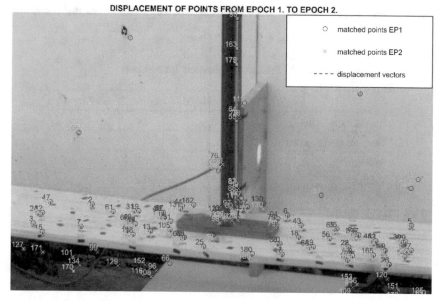

Fig. 4. Matching points between epoch 210 and 214

orientation of the wooden beam, mechanical properties of the timber material and load. The model outputs displacements and internal forces which are not used in the scope of this paper. Based on the numerical model z-axis displacement (change of height) can be calculated for every point on beam in each epoch, as shown for epoch 214 displacement in Fig. 5.

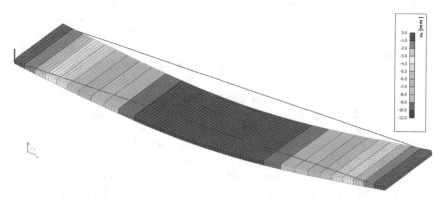

Fig. 5. Numerically calculated displacement for the beam

Matching points from each epoch that are located on the top of the beam were manually selected, with a total of 58 discrete matched points. The difference of the height displacements from the points and the model (residuals), as a reference were calculated. Statistical parameters for residuals are shown in Table 2 and the distribution is shown in Fig. 6.

Table 2. Statistical parameters for residuals

	Epoch 1.—210 (mm)	Epoch 2.—214 (mm)
Standard deviation	0.7	0.9
Standard deviation error	0.1	0.1
Average	−0.1	0.0
Median	−0.1	−0.1
Min	0.0	0.0
Max	2.6	3.4

Distributions are fitted to residuals using Maximum likelihood method. The residuals are distributed normally with some outlier elements.

5 Conclusion

A. Wagner, in his research [1, 2] proposes a methodology of fusing image and laser scan data from the same instrument, such as a multi-station, to create an RGB+D image and use the data for displacement monitoring.

Using computer vision algorithms such as SURF, BRISK and SIFT distinct features of an image can be detected and tracked through different measurement epochs. Subsequently, their 3D coordinates can be determined from RGB+D images. This acts as a substitute for discrete geodetic points for further classic deformation analysis.

The proposed methodology is objectively and numerically tested within the scope of this research. The experiment tests the achievable accuracy of the whole RGB+D method. Final accuracy is a function of the measuring accuracy of the MS Trimble SX10 and the accuracy of feature matching algorithm SURF. Accuracy of the MS and algorithm were not tested separately within the scope of this test and will be the subject of future researches. Computer code used for data processing has been developed by the authors of this paper independently of the original author's code to test the validity of the methodology objectively. After the methodology was validated through the implementation of the code on test data measured with a Trimble SX10 MS, the next step is to test the achievable precision numerically. Reference data for testing the height displacements calculated from RGB+D was from a numerical model of the beam, modelled based on LVDT measurement and beam geometry. The precision of the LVDT measurement supersedes the achievable precision of 3D point measurement with an SX10 MS (2.5 mm at 100 m), by a factor of 100. GSD for the images at 5 m distance is 0.3 mm.

The residuals (difference between RGB+D height and reference height), are distributed normally with some outlier elements, as shown in Fig. 6, which suggests that

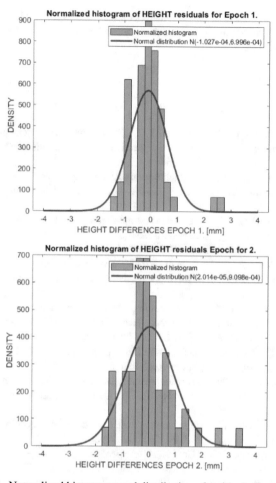

Fig. 6. Normalized histograms and distribution of residuals for epochs

there is a negligible influence of systematic errors in the measurements, and the dominant factor is measurement noise. Standard deviations for both epochs are under 1 mm with standard estimation error under 0.2 mm, Table 2.

Calculated height displacements are all within the limits of accuracy of 3D point measurement with an SX10 MS.

Conducted tests and achieved results show the applicability of the RGB+D method for construction testing in collaboration with civil engineers. RGB+D method provides the ability to track multiple discrete points on the testing element through the testing, that point clouds and relative measurement cannot provide. Thus, providing new data for testing the validity of behavior numerical models of an element being tested.

Based on the positive outcomes of the test, further research in the real-world conditions on man-made structures is needed to prove the validity of RGB+D for use in deformation analysis.

Acknowledgements. We would like to thank Geoprojekt d.d. Split, for lending us the Trimble SX10 for testing and Trimble employees C. Jorawsky and C. Grasse for their help with the technical specifications of the Trimble SX10.

References

1. Wagner A, Wiedemann W, Wunderlich T (2017) Fusion of Laser Scan and Image Data for Deformation Monitoring–Concept and Perspective. In: INGEO-7 th International Conference on Engineering Surveying. pp 157–164
2. Wagner A (2016) A new approach for geo-monitoring using modern total stations and RGB+D images. Meas J Int Meas Confed 82:64–74. https://doi.org/10.1016/j.measurement.2015.12.025
3. Trimble Geospatial (2019) Sx10 Product. https://geospatial.trimble.com/sites/default/files/2019-05/sx10-product.png. Accessed 30 Jan 2020
4. Trimble Geospatial (2017) Trimble SX10-Datasheet
5. Zollhöfer M, Stotko P, Görlitz A, et al (2018) (Survey paper) State of the Art on 3D Reconstruction with RGB-D Cameras. In: Shape Modeling and Geometry Processing-Spring 2018. pp 625–652
6. Abdel-Aziz YI, Karara HM (1971) Direct linear transformation into object space coordinates in close-range photogrammetry. In: Proc. Symp. Close-Range Photogrammetry. pp 1–18
7. Wunderlich T, Niemeier W, Wujanz D, et al (2016) Areal Deformation Analysis from TLS Point Clouds–The Challenge/Flächenhafte Deformationsanalyse aus TLS-Punktwolken–die Herausforderung. Allg. Vermessungs-Nachrichten 123:340–351
8. Salahat E, Qasaimeh M (2017) Recent advances in features extraction and description algorithms: A comprehensive survey. Proc IEEE Int Conf Ind Technol 1059–1063. https://doi.org/10.1109/ICIT.2017.7915508
9. Lowe DG (2004) Distinctive image features from scale-invariant keypoints. Int J Comput Vis 60:91–110. https://doi.org/10.1023/B:VISI.0000029664.99615.94
10. Bay H, Ess A, Tuytelaars T, Van Gool L (2008) Speeded-Up Robust Features (SURF). Comput Vis Image Underst 110:346–359. https://doi.org/10.1016/j.cviu.2007.09.014
11. Leutenegger Stefan, Margarita Chli RYS (2011) "BRISK: Binary robust invariant scalable keypoints." Computer Vision (ICCV). In: Iccv. pp 2548–2555
12. MATLAB (2016) version 9.10.0 (R2016b)
13. Hottinger Baldwin Messtechnik GmbH LVDT-B00553: Tehnical specifications. https://www.hbm.com/fileadmin/mediapool/hbmdoc/technical/B00553.pdf. Accessed 30 Jan 2020
14. Grasse C (2018) Rotation Bivector Instrument to telescope angles-Trimble SX10 [CODE]
15. Grasse C (2018) Rotation Bivector telescope to camera angles-Trimble SX10 [CODE]
16. John D'Errico (2020) inpaint_nans [CODE]

Influence of Given Parameter Errors on Accuracy of Tunnel Breakthrough by Height

Branko Milovanović[1(✉)], Slavko Vasiljević[2], Jovan Popović[1], and Petar Vranić[1]

[1] Faculty of Civil Engineering, University of Belgrade, Bulevar Kralja Aleksandra 73, Belgrade, Serbia

{milovano,popovic}@grf.bg.ac.rs, petvar994@gmail.com

[2] Faculty of Architecture, Civil Engineering and Geodesy, University of Banja Luka, Bulevar Vojvode Petra Bojovica 1A, Banja Luka 78 000, Republic of Srpska

slavko.vasiljevic@aggf.unibl.org

Abstract. There are two types of geodetic 2D and 1D networks for tunnel construction: overhead and underground. The purpose of these networks is to perform tunnel stake out with allowable deviation in transverse, longitudinal and height terms. Overhead network is developed before a tunnel is built and it connects entry and exit portal into a unique system. Underground network expands during tunnel construction. When evaluating the coordinates of newly assigned underground network points, previously established points are considered as given points. In order to realistically estimate the accuracy of underground network, it is necessary to include all previous measured quantities in the adjustment, and to define a minimum trace at all points. With this approach, estimated coordinates of network points would change each time when network is extended. To avoid this problem, when estimating new points, errors of given parameters must be taken into account. In the example of the Tunnel Vežešnik in Montenegro, when designing a 1D underground network, a case was reviewed when previously established points were taken as given and when their error was considered. To validate this procedure, an accuracy calculation was applied only when the overhead network points were defined as given. The network quality criteria were compared for all cases and it was shown that errors of the given parameters must be taken into account.

Keywords: Tunnel breakthrough · Tunnel geodetic networks · Errors of given parameters

1 Introduction

Tunnels belong to one of more complicated structures to perform. Structurally, they can, as underground objects, come up to the surface at entrances and exits (roads), only on one side (hydro-technical tunnels) or they are built underground (metros) [1]. The cost of exporting works is 4–5 times higher than the price of a road [2].

A. Kopáčik et al. (Eds.): *Contributions to International Conferences on Engineering Surveying*, SPEES, pp. 169–177, 2021. https://doi.org/10.1007/978-3-030-51953-7_14

The basic geodetic task is to provide a tunnel breakthrough in transverse, longitudinal and vertical terms within the required limits of tolerance. Tolerance depends on the terrain composition and structure itself and is expressed in the function of the tunnel length [3–5]. Geodetic works consist of establishing a geodetic overhead network that connects entrance and exit portals into a unique mathematical system in a horizontal and vertical sense. This network is established before the excavation. In order to provide tunnel breakthrough, it is necessary to successively develop an underground horizontal and vertical network, as it serves to stake out the excavation. The problem of developing underground networks due to tunnel conditions and the required accuracy has been shown in many papers with practical examples of specific tunnels [6–8].

Planning and preparing geodetic work prior to tunnel construction is extensive. It requires providing an accurate geodetic base that is negligible comparing to the allowable deviation of tunnel breakthrough, whereas the accuracy of overhead network must be negligible in regards to the accuracy of underground network. This paper describes the procedure of designing an underground levelling network, where the network is successively developing as excavation progresses. The development mode itself requires that, during the accuracy calculations, the estimated benchmark heights should not be changed, whereas the standard deviation of the previously specified benchmarks with their correlation must be taken into account. When applying this procedure and including the errors of given parameters, the standard deviations of heights will be calculated which are adequate as when the whole underground 1D network is adjusted together. In the example of Tunnel Vežešnik in Montenegro, it is shown that if the previously established benchmarks are considered as given, the information obtained about the accuracy of the network is incorrect.

2 Influence of Given Parameters on Accuracy of Geodetic Control Networks

In tunnels, as well as in other tall objects, a geodetic network develops successively by levels. In these cases, geodetic networks rely on earlier points called given parameters. As they are determined on the basis of previously taken measurements, their estimation of position or height contains errors of earlier measurements, that is, they cannot be accepted as true values.

Denote by $\mathbf{x}$ the vector of unknown parameters and by ξ the vector of given parameters. Then the equations of residuals are in matrix form

$$\mathbf{v} = \mathbf{A}\mathbf{x} + \bar{\mathbf{f}} \tag{1}$$

$$\bar{\mathbf{f}} = \mathbf{f} + \mathbf{B}\xi \tag{2}$$

where:

$$\mathbf{A} = \left[\frac{\partial l_i}{\partial x_i}\right], \mathbf{B} = \left[\frac{\partial l_i}{\partial \xi_i}\right] \tag{3}$$

l_i—the observation equation, A_{nxu}—design matrix (partial derivatives by u unknown parameters) for new established benchmarks,

$B_{n \times s}$—design matrix (of partial derivatives by s given parameters) for benchmarks that the extended network relies on, $\mathbf{f}$—vector of free terms.

Vector $\mathbf{x}$ is defined by the formula [9]

$$\mathbf{x} = -\left(\mathbf{A}^T \mathbf{K}_{\bar{f}}^{-1} \mathbf{A}\right)^{-1} \mathbf{A}^T \mathbf{K}_{\bar{f}}^{-1} \bar{\mathbf{f}} \tag{4}$$

the covariance matrix of unknown parameters is

$$\mathbf{K_x} = \left(\mathbf{A}^T \mathbf{K}_{\bar{f}}^{-1} \mathbf{A}\right)^{-1} \tag{5}$$

and the covariance matrix of measured parameters is

$$\mathbf{K}_{\bar{f}} = \mathbf{K}_f + \mathbf{B} \mathbf{K}_{\xi} \mathbf{B}^T \tag{6}$$

$$\mathbf{K}_f = \sigma_0^2 \mathbf{P}^{-1} \tag{7}$$

where

σ_0^2—is a priori dispersion factor, $\mathbf{P}$—is weight matrix of measurements. A predefined covariance matrix of given parameters $\mathbf{K}_{\xi}$ is taken from the adjustment or calculating the accuracy (which is applied when designing geodetic networks). Equations (4) and (5) can be used to estimate unknown parameters and their covariance matrix, taking into account both the error of the given parameters and their correlation dependence.

3 Underground Levelling Network

The Vežešnik tunnel is located on the Bar-Boljari highway (the south of Montenegro with the north to the border with Serbia). Its breakthrough happened in August 2018, and it was built as a part of the first section of Smoko. It is designed with two separate tunnel tubes, each for one traffic direction. The tunnel is approximately 2.45 km long. The entrance and exit of the right tunnel tube are linear. The entrance portal is at chainage 1.348 km, and the exit portal is at 3.762 km. The part of the right tunnel tube is in the curve R = 1050 m. The length of the tunnel per axis of the right tube is 2.414 km. The entrance and exit of the left tunnel tube are linear. The entrance portal is at 1.248 km and the exit portal is at 3.722 km. The part of the left tunnel tube is in the curve R = 1080 m. The length of the tunnel per axis of the left tube is 2.474 km. Given that the Vežešnik tunnel is intended as a two-tube tunnel, the tubes are connected in several places by passageways to save people and vehicles. The breakthrough comes from the south and north portals simultaneously (Fig. 1).

As the tunnel breakthrough was dug on both sides, an underground leveling network was developed from the north and south portals. In this paper, for one side of the tunnel, an accuracy calculation is done when the underground network is treated as unique, when the datum is defined on the previously established benchmarks, that is, when the errors of given parameters are and are not taken into account.

Fig. 1. The Vežešnik tunnel [10]

The calculation of the accuracy for this tunnel length [3] yielded the allowable deviation of tunnel breakthrough $\Delta_h = 2 \times \sqrt{L} = 31\text{mm}$, L–tunnel length in km, allowable deviation is given in cm.

Applying the principle of negligence, that is, stakeout the breakthrough as negligible compared to the allowable deviation, it turns out that

$$\sigma_{0B} = \frac{\Delta_h}{3 \times tp} = 4\text{mm}$$

where:

tp—quantile of normal distribution (taken for probability p = 0.99, which is 2.576). The allowable deviation Δ_h consists of the allowable deviation of construction work Δ_{civil} and the allowable deviation of geodetic stake out (the allowable deviation is half the confidence interval) $\Delta_{\text{stakeout}} = t_p.\sigma_{0B}$. Surveyors apply the principle of negligence, that is, the standard deviation error σ_{0B} is negligible with respect to the allowable deviation of tunnel breakthrough. In order for a dispersion to be negligible in relation to the total dispersion, it must not exceed the value of the first-order error in testing linear hypotheses, that is, it must not exceed 5% of the total dispersion.

Additionally, by calculating the accuracy, it was obtained that the standard deviation of the measured height difference at the station is $\sigma_0 = 1\text{mm}$, and the weights of the measurements were calculated as $p_i = 1/n_i$ (n_i—number of stations of the leveling side).

The negligence principle means that the impact of an error must not exceed 5% of the total error. This principle is applied so that the error while performing works would have the largest budget in the allowable deviation.

The underground network relies on two benchmarks of the overhead portal network (Fig. 2), which are designated as R6 and R7.

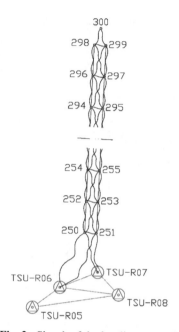

Fig. 2. Sketch of the leveling network

At the entrance to the tunnel tube, the two benchmarks (marked as 250 and 251) are supposed to stabilize. Height-differences will be observed from both overhead network benchmarks to these two underground network benchmarks. When the works are performed, the overhead network benchmarks define the datum of the underground network. As each excavation progresses, every two benchmarks are stabilized every 50 m. In this network segment, the datum is defined by two previously established underground network benchmarks. In each segment, the measured parameters are the height-differences between the given benchmarks and the newly placed ones (four height-differences) and between the newly placed ones (one height-difference). There are 25 network extensions in total. Stake out of tunnel breakthrough is performed with the last two network benchmarks. The breakthrough point is indicated by 300 (Fig. 2)

4   Accuracy Calculation When Underground Network Is Observed as a Whole

The datum is defined by the R6 and R7 benchmarks. By calculating the accuracy, a local measure of internal reliability is obtained in the interval 0.4 (measurements at breakthrough point 300) up to 0.67. A local measure of reliability is the probability of detecting a gross error in a single measurement, that is, how much it will "pollute" individual measurements due to the appearance of algebraic correlation when applying the least square method. The Table 1 and Fig. 3 show the changes in standard deviations as a function of distance.

Table 1. The changes in standard deviations as a function of distance—whole network

D (km)	0.05	0.10	0.15	0.20	0.25	0.30	0.35	0.40	0.45
STDV (mm)	0.8	1.1	1.3	1.5	1.6	1.8	1.9	2.0	2.2
D (km)	0.50	0.55	0.60	0.65	0.70	0.75	0.80	0.85	0.90
STDV (mm)	2.3	2.4	2.5	2.6	2.7	2.8	2.9	2.9	3.0
D (km)	0.95	1.00	1.05	1.10	1.15	1.20	1.25	1.30	
STDV (mm)	3.1	3.2	3.3	3.3	3.4	3.5	3.6	3.7	

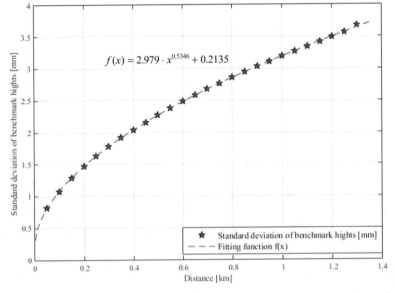

Fig. 3. Standard deviation of benchmark heights of entire underground 1D network as a function of distance

5   Accuracy Calculation When Underground Network Is Observed by Segments

The datum of a segment of network is defined by two previously established benchmarks, which are positioned 50 m away from the newly set benchmarks. The last segment consists of stake out the projected height of the point 300 from the benchmarks 298 and 299. It is obtained that the standard deviation of the point 300 is only 1 mm by height. It is clear that this approach is wrong, as the information about the breakthrough accuracy is unrealistic.

6 Accuracy Calculation When Underground Network Is Observed by Segments Together with Given Parameter Errors

When considering the given parameter errors with their covariates (formulas in Chap. 2), an accuracy calculation is obtained that is adequate for the joint adjustment of entire network. The design matrices of unknown and given parameters for the first segment have numerical values

$$
\mathbf{A} = \begin{bmatrix} 1 & 0 \\ 0 & 1 \\ 1 & 0 \\ 0 & 1 \\ -1 & 1 \end{bmatrix} \begin{matrix} \Delta H_{R\,6-R\,250} \\ \Delta H_{R\,6-R\,251} \\ \Delta H_{R\,7-R\,250} \\ \Delta H_{R\,7-R\,251} \\ \Delta H_{R\,250-R\,251} \end{matrix}, \quad \mathbf{B} = \begin{bmatrix} -1 & 0 \\ -1 & 0 \\ 0 & -1 \\ 0 & -1 \\ 0 & 0 \end{bmatrix} \begin{matrix} \Delta H_{R\,6-R\,250} \\ \Delta H_{R\,6-R\,251} \\ \Delta H_{R\,7-R\,250} \\ \Delta H_{R\,7-R\,251} \\ \Delta H_{R\,250-R\,251} \end{matrix} \tag{8}
$$

where columns of $\mathbf{A}$ are dH_{R250}, dH_{R251} and columns of $\mathbf{B}$ are dH_{R5}, dH_{R7}.

Up to the last segment, design matrices have the same dimensions and numerical values. For the last segment they take the values

$$
\mathbf{A} = \begin{bmatrix} 1 \\ 1 \end{bmatrix} \begin{matrix} \Delta H_{R\,298-R\,300} \\ \Delta H_{R\,299-R\,300} \end{matrix}, \quad \mathbf{B} = \begin{bmatrix} -1 & 0 \\ 0 & -1 \end{bmatrix} \begin{matrix} \Delta H_{R\,298-R\,300} \\ \Delta H_{R\,299-R\,300} \end{matrix} \tag{9}
$$

where the column of $\mathbf{A}$ is dH_{R300} and columns of $\mathbf{B}$ are dH_{R298}, dH_{R299}.

Covariance matrices of given parameters $\mathbf{K}_\xi$ are formed as

$$
\mathbf{K}_\xi = \begin{bmatrix} \mathbf{K}_{R_i} & \mathbf{K}_{R_i R_{i+1}} \\ \mathbf{K}_{R_{i+1} R_i} & \mathbf{K}_{R_{i+1}} \end{bmatrix} \tag{10}
$$

and for the first, penultimate and last segment, their values are

$$
\mathbf{K}_{\xi 1} = \begin{bmatrix} 0.84 & 0.34 \\ 0.34 & 0.84 \end{bmatrix}, \mathbf{K}_{\xi 24} = \begin{bmatrix} 12.34 & 11.84 \\ 11.84 & 12.34 \end{bmatrix}, \mathbf{K}_{\xi 25} = \begin{bmatrix} 12.84 & 12.34 \\ 12.34 & 12.84 \end{bmatrix}. \tag{11}
$$

The Table 2 and Fig. 3 show the changes in standard deviations as a function of distances for this approach (Fig. 4).

7 Conclusion

Measurements of height-differences in the underground 1D network are loaded with the following errors: instrument (the most dominant is the non-horizontal view), the non-verticality of the level staff, the error of conditions during the measurement. Errors of condition are caused by problems of illumination and refraction. Low illumination results

Table 2. The changes in standard deviations as a function of distance—segments

D (km)	0.00	0.05	0.10	0.15	0.20	0.25	0.30	0.35	0.40
STDV (mm)	0.4	0.9	1.2	1.4	1.5	1.7	1.8	2.0	2.1
D (km)	0.45	0.50	0.55	0.60	0.65	0.70	0.75	0.80	0.85
STDV (mm)	2.2	2.3	2.4	2.5	2.6	2.7	2.8	2.9	3.0
D (km)	0.90	0.95	1.00	1.05	1.10	1.15	1.20	1.25	1.30
STDV (mm)	3.1	3.1	3.2	3.3	3.4	3.4	3.5	3.6	3.7

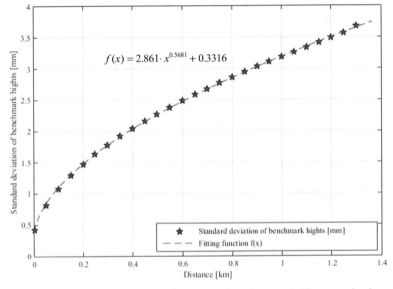

$$f(x) = 2.861 \cdot x^{0.5681} + 0.3316$$

Fig. 4. Standard deviation of benchmark heights of underground 1D network observed by segments as a function of distance

in higher reading errors for both digital levels and higher level staff non-verticality, as the bubble of the spirit level in the dark is more difficult to see. Additionally, the refraction is higher because the view extends along the walls of the tunnel, which are cold and damp. To decrease of the impact of these measurement errors, measurements must be well planned.

The project also defines the procedure for estimating unknown parameters. Based on the presented standard deviation calculation of point heights, errors increase in a function of the distance from the benchmarks, which define datum, by the exponential function. The problem with this network is that these benchmarks are not properly placed but are at the beginning of the network. If we take the pre-determined benchmarks, it is clear we will get an unrealistic picture of the precision of the network.

It is most reliable to adjust the entire network together, but this is impossible as the network is expanded as excavation progresses. In order to obtain objective information about the precision of the network, the errors of the given parameters with their covariances must also be taken into account, since the same precision estimates are obtained as in case of joint adjustment. The additional advantage of this procedure is that even the benchmark height estimates are not changed.

References

1. Begović, A.: Engineering geodesy 2 (in Serbian). 1st edn. Naučna knjiga, Belgrade (1990).
2. Zrinjski, M., Džapo, M., Redovniković, L.: Underground Geodetic Basis of the Tunnel "Mala Kapela". Shaping the Change, XXIII FIG Congress, pp. 1–17. Munich, Germany, October 8–13 (2006).
3. Cvetković, Č.: Application of geodesy in engineering (in Serbian). 1st edn. Građevinska knjiga, Belgrade (1970).
4. Krüger, J.: Geodätische Netze in Landes- und Ingenieurvermessung II, Absteckungsnetze, speziell für Tunnelabsteckungen. In: Vorträge des Kontaktstudiums Februar 1985 in Hannover, Konrad Wittwer Verlag, Stuttgart (1985).
5. Shults, R.: Development and research of the methods for analysis of geodetic monitoring results for the subway tunnels. In: 4th Joint International Symposium on Deformation Monitoring (JISDM), pp. 112. Eugenides Foundation, Athens, Greece (2019).
6. Chrzanowski, A., Trevor Greening, W., Grodecki, J., Robbins, J.: Design of geodetic control for large tunneling projects. In: Annual Publication of the Tunneling Association of Canada, pp. 1–11. Tunnelling Association of Canada, Richmond (1993).
7. Korritke, N., Wunderlich, T.: Le gyromat, un gyrotheodolite de grande precision et son emploi dans l'Eurotunnel. XYZ 40, 25–28 (1989).
8. Stiros C., S.: Alignment and breakthrough errors in tunneling. Tunneling and Underground Space Technology 24, 236–244 (2009).
9. Mihailović, K.: Geodesy 2 (in Serbian). 1st edn. Naučna knjiga, Belgrade (1987).
10. BEMAX Homepage, https://bemax.me/project/autoput-bar-boljare/, last accessed 2020/01/21.

Design and Implementation of Special-Purpose Geodetic Networks for the Infrastructure Facilities of the Highway

Radovan Đurović[1](✉), Zoran Sušić[2] ⓘ, Mehmed Batilović[2] ⓘ, Gojko Nikolić[3],
Miloš Pejaković[4], and Željko Kanović[2]

[1] Faculty of Civil Engineering, University of Montenegro, Podgorica, Montenegro
radovandj@ac.me
[2] Faculty of Technical Sciences, University of Novi Sad, Novi Sad, Serbia
[3] Institute of Geography, University of Montenegro, Nikšić, Montenegro
[4] Faculty of Polytechnic, University of Donja Gorica, Podgorica, Montenegro

Abstract. Special-purpose geodetic networks are established for various stages of geodetic works, starting from an idea, geodetic setting-out, inspection of the construction of facilities, up to monitoring the behavior of the exploited facility. In case of large infrastructure facilities such as the highway, the design of the independent micro geodetic networks for all separate highway facilities (bridges, tunnels, viaducts, retaining walls, overpasses, underpasses, etc.) and fitting the referred networks into a single highway coordinate system represents a challenge. Regarding certain stages of works, such as geodetic setting-out, surveying the installations and lines to be surveyed, a state coordinate system is used, whilst for purposes of conducting the deformation measurements and the analysis of certain facilities on the highway, a local coordinate system is used, relieved from any impact of given quantities, which may impede the measurement results within the monitoring process. The paper will provide an overview and an analysis of the current design of monitoring the bridge Gornje Mrke at the part of the Bar—Boljare highway, which is a fundamental project currently being implemented in the territory of Montenegro, taking into account a very unfavorable configuration of the terrain at which the construction of the highway is underway. In the optimization process, Particle Swarm Optimization, a global optimization algorithm from the group of evolutionary metaheuristics, will be applied, and the results will be compared to the traditional network optimization concept. The criterion of optimality will be determined based on the parameters of sensitivity of geodetic monitoring network.

Keywords: Geodetic network · Deformation · Monitoring · Bridge · Highway

A. Kopáčik et al. (Eds.): *Contributions to International Conferences on Engineering Surveying*,
SPEES, pp. 178–186, 2021. https://doi.org/10.1007/978-3-030-51953-7_15

1 Introduction

The purpose of optimization of geodetic network design for engineering objects, such as bridges, is to determine the optimal solution in accordance to selected objective function. When dealing with the geodetic monitoring of bridges, the network is usually designed in such a way to satisfy desired aspects of accuracy, precision and reliability, according to final goal, which is the least detectable deformation that can be "surely" measured, having in mind the chosen probability $(1-\alpha)$ and the power of test $(1-\beta)$. Also, the equally important aspect that needs to be considered during design is of the economic nature, in order to reduce the number of measurements and to satisfy all the above-mentioned criteria at the same time. The network can be designed as a combination, to include all measured values: horizontal directions, zenith angles, slope distances, GNSS vectors etc. In case of geodetic monitoring of bridges, the points of the referent network from which potential displacements and deformations are observed are being stabilized by concrete pillars with plate for enforced centering, in order to eliminate the influence of errors caused by instrument and signal centering.

In accordance to classic concept of optimization [1], the division into orders has been defined as follows, depending on constant and unknown parameters of functional and stochastic model. Zero-order design (ZOD) problem seeks for optimal coordinate system, first order design (FOD) problem deals with finding optimal network design, second order design (SOD) problem implies determining the optimal weights or accuracies of planned measurements in the network, and third order design (TOD) problem enables the improvement of the network in the sense of design and accuracy.

Global optimization techniques are nowadays widely used in numerous disciplines, including optimal geodetic network design problem. Namely, genetic algorithm (GA) and particle swarm optimization (PSO) are successfully applied in geodetic network design [2, 3] and in improvement of deformation analysis methods efficiency [4]. Besides the accuracy, precision and reliability, objective function can include also the robustness of the geodetic network. The reliability is expressed as the ability to detect outliers within the observations (internal reliability) or the influence of undetected gross errors to the vector of unknown parameters (external reliability). Since the vector of unknown parameters depends on the datum of the network, classic optimization concept is combined with strain analysis that enables the network to be independent of datum, but only of geometry and observation accuracy, which results in robustness [3]. The methods of multicriteria decision making for optimization of trilateration networks, including the sensitivity analysis of geodetic network, are also present in the literature [5].

2 Motivation

The subject of the paper is the optimization of geodetic network for monitoring of the bridge Gornje Mrke, which is located at the route of the currently built highway, in very demanding section Smokovac–Mateševo, in Montenegro.

The bridge goes over a dry valley, which is about 300 m long and 30–35 m deep with respect to the roadway. Under the bridge there are two roads to the nearby village of Gornje Mrke. As presented in the elaborate plan of geo-engineering studies, the geological characteristics of the locality where the bridge was built indicate a karstic depression formed in the framework of the frontal part of Kučka Kraljušt, near the settlement of the same name Gornje Mrke. The bridge is entered from an open track about 1100 m long preceded by the tunnel of Vežišnik. On the object locality the karstic depression is built by dolomites and dolomitic limestones. The karstic depression bottom is covered by terra rossa. The steep limestone slope, northeast of the karstic depression formed from dolomites and dolomitic limestones, is partly covered by a deluvium-eluvium clayey crumbly material [6].

The bridge in the left-hand track occupies from the survey mark km 4 + 843.115–km 5 + 140.21, in the right-hand one from the survey mark km 4 + 830.54–km 5 + 125.49. The interval width of the roadway axes of bridges in the basis and at the beginning is 16.38, 25.64 m at the end. The total length with the shore piers for both bridges is 311 m. The roadway width is 8.70 m, and the total width of both bridges is 9.62 m. The horizontal curvature of the left bridge is 1050, 1080 m on the right-hand side. The vertical curvature is 10000 m for both. The transversal fall is constant for both bridges, equal to 4%, whereas the average longitudinal one is 2%. The bridges have frame constructions. Piers S3–S9 are tightly connected to the construction, whereas in the case of A01, S1, S2, S10 and A02 this is done through a bearing. The ratio of the end—and middle spans is 22/28 = 0.78, it is suitable to a continual frame constructive system where the bridge construction is carried out by using the "field by field" method [6] (Fig. 1).

Fig. 1. Example of stabilization of a reference/control geodetic network point (left) and the way of installing the adapter for the purpose of setting of prisms/marks (right)

The concept of bridge monitoring is created in such a way that, using conventional geodetic measurements (horizontal directions, zenith angles and slope distances), one can observe points located on bridge piers (one at the top and at the bottom, and one at each span from the points of the referent geodetic network.

The deformation network model is designed in such a way that it can provide, through its stabilization, geometric shape, kind of measured quantities and treatment manner in the deformation analysis, proper quality indicators for the geodetic network as to the accuracy, precision and reliability, so that significant shifts in the space sense at all points which geometrically interpret the object can be discovered. The rules followed by the 3D network are:

- to set the points on a geologically stable terrain,
- to set the points to be beyond the influence of deformations affecting the objects under study,
- to secure the permanence of the 3D points,
- the observation of the objects under study to be possible from the 3D network points and
- the disposition of the 3D network points should make it possible applying the conventional measuring method.

3 Geodetic Network and Design Criteria

As mentioned above, for the purposes of monitoring the displacements and deformations of the structure of the Gornje Mrke Bridge, a geodetic network was established consisting of 11 reference points and 52 points on the object (Fig. 2). Within the project, the geometry of the network with the observation plan is defined. The observation plan provides for the measurement of 175 horizontal directions, 183 slope distances and 175 zenith angles. Also, within the design is envisaged that the slope distances will be measured with an accuracy of 1 mm + 1 ppm, and the horizontal directions and zenith angles with an accuracy of 0.5". Thus, a total of 533 measurements are surveyed within the observation plan.

In design of geodetic monitoring of displacements and deformations of complex construction objects, it is often necessary to consider the proposed design solution from the point of view of economy. In this sense, it is very often necessary to reduce the number of measurements due to limited economic resources. Accordingly, this paper proposes a solution for SOD problem of geodetic network of the Gornje Mrke bridge by application of the PSO algorithm in order to reduce the number of measurements, according to the methodology proposed in [7]. In order to apply the PSO algorithm in the process of geodetic network optimization, it is necessary to define variables, the feasible region of search and optimality criterion. Since the SOD problems deal with finding optimal values of planned measurements accuracy, the weights of planned measurements are taken as variables in optimization problem. Thus, the feasible region directly depends on planned measurement accuracy. Therefore, the following weight constraints are defined:

$$0 \le p_{\alpha_i} \le 4, \ 0 \le p_{d_i} \le 1, \ 0 \le p_{z_i} \le 4 \tag{1}$$

It is well known that the point positioning error is the largest in the direction of the semi-major axis of the confidence ellipsoid (ellipse), so the mentioned direction represents the most unfavorable case of testing the stability of the point in the deformation

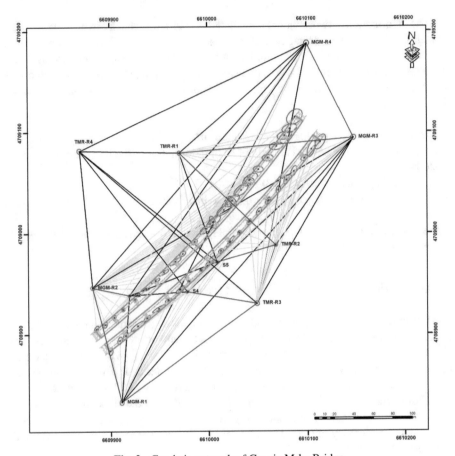

Fig. 2. Geodetic network of Gornje Mrke Bridge

analysis. Accordingly, the optimality criterion is formulated based on the smallest intensities of displacement vectors that can be detected in the direction of the semi-major axis of confidence ellipse, as

$$\rho = \|\mathbf{d}_A - \mathbf{d}_A^c\|_1 = \min. \tag{2}$$

where $\mathbf{d}_A$ is the vector of the smallest intensities of the vector of displacing points of the network and $\mathbf{d}_A^{cT} = [5\text{mm}, \ldots, 5\text{mm}]$ is the criterion vector of the smallest intensities of the vector of displacing points of the network. The constraints for measurement weights defined by expression (1) are integrated into the optimality criterion by the penalty function method, as follows

$$\rho = \|\mathbf{d}_A - \mathbf{d}_A^c\|_1 + \sum_i \sum_j \beta \cdot g_{ij} = \min \tag{3}$$

where g_{ij} is the corresponding penalty, β is the penalty factor that is determined empirically, and j is the constraint index. Therefore, for each discrepancy of the defined limits, the corresponding penalty value is formed as

$$g_{ij} = \begin{cases} |p_i - p_{max}|, p_i > p_{max} \\ 0, p_{min} \leq p_i \leq p_{max} \\ |p_{min} - p_i|, p_i < p_{min} \end{cases} \tag{4}$$

where p_i is the weight value of the corresponding measurement and p_{min} and p_{max} are the minimum and maximum permissible weight values of the measurement. The previously defined optimization problem can be successfully solved by applying the PSO algorithm.

4 Particle Swarm Optimization

Particle swarm optimization (PSO) algorithm is a swarm-based global optimization technique, inspired by the social behavior of animals moving in large groups (particularly birds) [8]. Although very simple and easy to implement, this algorithm is quite robust and solves very complicated optimization problems. It uses a set of particles called swarm to search for optimal solution. Each particle is described by its position ($\mathbf{x}$) and velocity ($\mathbf{v}$). The position of each particle is a set of variables' values that represents the potential solution to the optimization problem. The best position (in the sense of objective function value) that each particle achieved during the entire optimization process is memorized ($\mathbf{p}$). The swarm as a whole memorizes the best position ever achieved by any of its particles ($\mathbf{g}$). The position and the velocity of each particle in the kth iteration are updated as

$$\mathbf{v}^{(k+1)} = w \cdot \mathbf{v}^{(k)} + cp \cdot rp^{(k)} \cdot \left(\mathbf{p}^{(k)} - \mathbf{x}^{(k)}\right) + cg \cdot rg^{(k)} \cdot \left(\mathbf{g}^{(k)} - \mathbf{x}^{(k)}\right), \tag{5}$$

$$\mathbf{x}^{(k+1)} = \mathbf{x}^{(k)} + \mathbf{v}^{(k+1)}.$$

Acceleration factors cp and cg control the relative impact of the personal (local) and common (global) knowledge on the movement of each particle. Inertia factor w keeps the swarm together and prevents it from diversifying excessively. Random numbers rp and rg are mutually independent and uniformly distributed in the range [0, 1].

There are many modifications of PSO algorithm presented in literature so far. In this paper, the early PSO variant with variable inertia weight and constant acceleration factors has been used. This version of PSO is implemented in Matlab's Global Optimization Toolbox. Inertia factor w is decreased from 1.1 to 0.1, enabling thorough space search. Cognitive factors cp and social factor cg are set to 1.49, as suggested in literature.

5 Results and Discussion

The SOD optimization problem of the geodetic network of the Gornje Mrke bridge was solved by applying the PSO algorithm within the Matlab software package. The search process was conducted through 1000 iterations, using a swarm of 200 particles. The tolerance for premature stopping the search is set to the default value of 10^{-6}. Figure 3 shows the best values of the optimality criterion in the entire swarm during the search process.

In this way, the optimal values of the weights,, i.e. the optimal accuracy of the planned measurements were obtained. The analysis of the optimal values of the weights of the planned measurements revealed that there are many measurements whose weights have very small values. The very low value of the weight of the corresponding measurement indicates that its influence in the optimization process is negligible. Therefore, measurements that have very low values of weights should be eliminated from the observation plan. In this regard, 70 horizontal directions, 65 slope distances and 43 zenith angles, i.e. totally 178 measurements, were excluded from the observation plan. However, in order to obey some basic principles of geodetic network design and to preserve the geometry of the network, 17 measurements were selectively restored, namely 9 horizontal directions, 6 slope distances and 2 zenith angles. Finally, an optimal observation plan was obtained in which 114 horizontal directions, 124 slope distances and 134 zenith angles are being measured, with a total of 372 measurements. It is important to note that in this way the number of measurements is reduced by 30%.

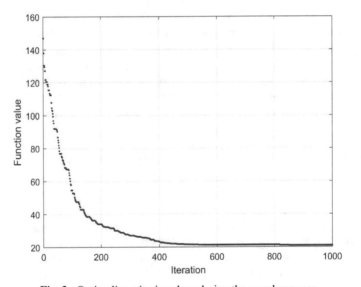

Fig. 3. Optimality criteria values during the search process

Based on the optimal observation plan and the adopted accuracy of the planned measurements, the procedure of preliminary analysis of the quality of the geodetic network was carried out, with the date defined by a minimal trace to the reference points of the network. The geodetic network quality analysis includes the analysis of the accuracy, reliability and sensitivity of the geodetic network. As part of the sensitivity analysis of the geodetic network, the most unfavorable and the most favorable case of point matching is considered. Consequently, the smallest intensities of the 2D displacement vectors in the direction of the semi-major and semi-minor axes of error ellipses for each network point are calculated. The statistical values of the smallest intensities of the displacement vectors are shown in Table 1. Based on the results in Table 1, it can be concluded that the obtained solution completely satisfies the adopted optimization criterion.

Table 1. Characteristic value of the smallest intensities of the displacement vectors that can be detected in the semi-major and semi-minor of the confidence ellipse

	d_A (mm)	σ_{d_A} (mm)	d_B (mm)	σ_{d_B} (mm)
Minimum	0.99	0.32	0.64	0.21
Average	2.80	0.90	1.31	0.42
Median	2.83	0.91	1.30	0.42
Maximum	4.90	1.58	2.68	0.86

The previous estimation accuracy calculated the standard deviations of the coordinates and position of the points, as well as the parameters of the error ellipsoid for a confidence level of 95%. The statistical values of the accuracy parameters of the geodetic network are shown in Tables 2 and 3.

Table 2. Characteristic values of standard deviations of coordinates and position of points

	σ_Y (mm)	σ_X (mm)	σ_H (mm)	σ_P (mm)
Minimum	0.20	0.15	0.08	0.28
Average	0.51	0.47	0.21	0.75
Median	0.49	0.45	0.19	0.72
Maximum	1.03	0.88	0.56	1.16

Table 3. Characteristic values of the confidence ellipsoid parameters

	A (mm)	B (mm)	C (mm)
Minimum	0.63	0.22	0.41
Average	1.80	0.51	0.86
Median	1.86	0.43	0.84
Maximum	3.19	1.29	1.71

In the procedure of the previous reliability analysis, the parameters of the internal and external reliability of the geodetic network were calculated. Table 4 shows the statistical values of the internal reliability coefficients r_i and the external reliability coefficients u_i. It should be emphasized that the very small values of the internal reliability coefficients r_i relate to measurements towards the points on the object, since it is not possible to perform bidirectional measurements.

Table 4. Characteristic values of internal and external reliability coefficients

	r_i	u_i
Minimum	0.00	0.01
Average	0.47	0.53
Median	0.51	0.49
Maximum	0.99	1.00

6 Conclusion

This paper presents an application of PSO algorithm in optimization of geodetic network. Second order design problem has been solved, with objective function defined in such a way to be able to guarantee detection of displacements greater or equal to 5 mm. At the same time, the optimization process resulted in reduction of measurement number of 30%. After the analysis of network quality, it has been shown that the suggested measurement plan satisfies the demanded network quality. Thus, it has been confirmed that global optimization methods, such as PSO, can be efficiently applied in geodetic network optimization problems.

References

1. Grafarend, E. W.: Optimization of geodetic networks. The Canadian Surveyor 28(5), 716–722 (1974).
2. Mevlut, Y., Cevat, I., Cemal, O. Y.: Use of the particle swarm optimization algorithm for second order design of levelling. Journal of Applied Geodesy 3(3), 171–178 (2009).
3. Yetkin, M.: Metaheuristic optimisation approach for designing reliable and robust geodetic networks. Survey Review 45(329), 136–140 (2013).
4. Batilović, M., Sušić, Z., Kanović, Ž., Marković, Z. M., Vasić, D., Bulatović, V.: Increasing efficiency of the robust deformation analysis methods using genetic algorithm and generalised particle swarm optimization. Survey Review, 1–13 (2020).
5. Kobryn, A.: Multicriteria decision making in geodetic network design. Journal of Surveying Engineering 146(1), 040190181–0401901810 (2020).
6. Geotechnical final report for Main design for Bridge Mrke and open route part, China Road & Bridge Corporation (Montenegro Branch), Ministry of Transport and Maritime Affairs, Montenegro (2016).
7. Maksimović, J., Kuzmić, T., Batilović, M., Sušić, Z., Bulatović, V., Kanović, Ž.: Design of geodetic networks by using global optimization methods. In: 7th International conference Contemporary achievements in civil engineering, pp. 975–984, Faculty of Civil Engineering, Subotica, Serbia (2019).
8. Kennedy, J., Eberhart, R.: Particle swarm optimization. In: Proceeding of IEEE international conference on neural networks, pp. 1942–1948, Perth, Australia (1995).

The Models of Structural Mechanics for Geodetic Accuracy Assignment: A Case Study of the Finite Element Method

Roman Shults$^{(\boxtimes)}$ (iD)

Kyiv National University of Construction and Architecture, Kiev 03037, Ukraine
`shults.rv@knuba.edu.ua`

Abstract. It is well-known that two of the third parts of geodetic works comprise staking-out works and geodetic monitoring. Both of these tasks of applied geodesy as a scientific discipline face with the question of the measurement accuracy assignment. There are many ways how to assign an appropriate accuracy of measurements. However, all of them lack a rigorous approach. There is only one right way that consists of using the structural mechanics' models for the calculation of probable construction displacements to assign the appropriate accuracy as a part of total construction's displacements. The measurements' accuracy is being treated as a function of those displacements. Geodesists have to have special skills in the structures' design and simulation to find such displacements, which, in turn, is a complicated task. The solution is in using the finite element method (FEM) that allows modeling large structures with straightforward computations. The paper considers the general idea of the FEM and its models with the application to the geodetic accuracy assignment. Based on these models, the general approach for the accuracy assignment was developed and implemented. As a hands-on example, the different structures were analyzed, and its finite element models were constructed. For the case of staking-out works, the problem of high-rise building erection was considered. The FEM simulation allowed us to determine the influence of the structure displacements on the accuracy of staking-out works. The geodetic monitoring problem was studied on the example of underground structures. For this structure, the FEM simulation was carried out. By the simulation results, an appropriate accuracy of monitoring was assigned.

Keywords: Structural mechanics · Displacement · Finite element method · Measurements accuracy

1 Introduction

The problem of geodetic accuracy assignment is one of the points among geodetic tasks. During the last century, coarse methods of accuracy calculation were developed. Most of those methods were based on typical standard values that were got under some approximative conditions and had a restricted application. Different regulations have different classifications for structures that, in turn, leads to incorrect accuracy assignment.

A. Kopáčik et al. (Eds.): *Contributions to International Conferences on Engineering Surveying*,
SPEES, pp. 187–197, 2021. https://doi.org/10.1007/978-3-030-51953-7_16

In such a case, the accuracy for geodetic measurements sets from the observer experience or by regulations without any scientific approach. In these circumstances, there is a threat to set up inappropriate accuracy. The main lack of existing methods of the accuracy assignment is treating of a structure as a static and uniform object. Such an assumption does not account for the structure's deformation during assembling and exploitation. Besides, for the last decades, revolutionary changes in construction technologies and material manufacturing have occurred. It is essential to point out two crucial changes: geometric parameters (length, height, diameter, etc.), and the construction environment (to date, no constraints for a ground, seismic conditions, etc.). Both these changes lead to new, unstudied loads and hence for more complicated deformations of modern structures (shells, high-rise buildings, tunnels, etc.). Deformations are dimensionless measures of displacements. Obviously that these deformations have to be accounted for carrying out of geodetic measurements. On the other hand, the values of the displacements may be used as a reference for measurement correction and, what is more valuable, for the accuracy assignment. The question is: how to determine the values of the displacements and find out the places of their maximum? The answer is in using the models and methods of structural mechanics for structures simulation.

Cutting-edge technologies allow architects, designers, and civil engineers to use strict and mathematically correct models for construction simulation. During the simulation of anticipated deformations, one considering the relationship "ground—structure—environment". All the simulations are conducted using two approaches: differential equations or finite element method (FEM). The designers and civil engineers have a deal with precise calculations. However, for the accuracy assignment, there is no need to carry out precise calculations. Surveyors can do such calculations with the use of the FEM.

Over the years, an enormous amount of research has been devoted to the application of the FEM in geodesy [19, 22]. Conventionally, FEM helps us for the interpretation of our results during geodetic monitoring, as can we find in [9]. A number of scholars have researched the FEM application for global geodynamic monitoring [5, 8]. Numerous studies have investigated the FEM application for geodetic monitoring analysis of miscellaneous engineering structures, for example, large dams and their facilities [4, 6, 16, 18], bridges [10, 17, 20], tunnels [11, 23], wind turbines [7], high-rise buildings [13], and local studies of constructions behavior [21, 24]. In recent years, researchers have become increasingly interested in FEM-BIM integration, especially for historical building monitoring [1–3]. This list by no means exhaustive. However, analysis of monitoring is not the only application of the FEM. To date, no study has explicitly looked at the accuracy assignment using the basic principles of structural mechanics, particularly the FEM. Surveyors may use this method to resolve an inverse task; namely, use preliminary FEM simulation results to find the necessary accuracy of geodetic measurements for stacking-out and monitoring. This study is an attempt to address the issue of geodetic accuracy assignment for the modern structures and modern geodetic equipment using the models of the FEM. Given the centrality of this issue, as the first step, let us consider the key features of the FEM and present its advantages and disadvantages.

2 A Brief Theory of FEM

Before considering the models of FEM, let us study the simple example from structural mechanics as an excellent case of the FEM advantages. As a simple case, we may consider the beam laid on the ground. The differential equations describe the relationship between the ground and beam. Solving the equations, one finds a general closed-form solution for the relationships of the moment, shear force, and sag (vertical displacement). This model is one of the simplest in structural mechanics, but even so, it needs special skills. These skills, by far, are not a subject of training surveyors. At the same time, many software systems implement numerical simulation methods that allow finding stress and deformation fields numerically [12]. One of the most popular and effective numerical modeling methods is the finite element method. The idea of the finite element method is based on the principle according to which any structure, field, or body can be represented as a partition into small pieces (finite elements). The FEM is premised on the assumption that the law of deformation for each small piece (element) can be approximated by simple mathematical dependencies, such as, for example, linear or polynomial of low degrees. Since the deformation process can be described by the system of differential equations and boundary conditions, the FEM approximates these equations by the system of algebraic equations that has a general form [15]:

$$\mathbf{Ku} = \mathbf{f} \tag{1}$$

where $\mathbf{K}$—stiffness matrix, $\mathbf{f}$—loads, $\mathbf{u}$—displacements.

The displacements $\mathbf{u}$ being approximated at the nodes of joining of the separate finite elements. The structure of the matrix depends on the type of finite element. The current literature abounds with examples of different finite elements with a detailed description of the stiffness matrix structure [14, 15].

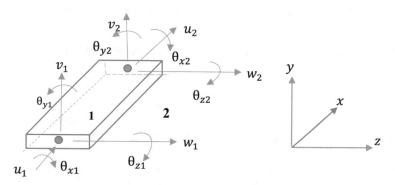

Fig. 1. Scheme of displacements for a spatial linear finite element [14]

The simulation of displacements consists of four essential steps with appropriate subsections. The first step is the structure simulation. During this step, it is necessary to choose a type of finite element. This choice depends on structural elements (frames, trusses, beams, arches, cables, shells) used to make up the structure. The most general type of the finite element for different structures' description is a spatial linear finite element (Fig. 1).

The element has twelve degrees of freedom. The displacements vector looks:

$$\mathbf{u}^T = \left(u_1 \; v_1 \; w_1 \; \theta_{x1} \; \theta_{y1} \; \theta_{z1} \; u_2 \; v_2 \; w_2 \; \theta_{x2} \; \theta_{y2} \; \theta_{z2}\right)$$

The elements of a stiffness matrix $\mathbf{K}$ depend on the geometrical and physical parameters of the finite element. In order to construct a stiffness matrix for each element, the material properties being assigned. The next important issue is the partitioning of the structure into finite elements. The simplest way to split can be done using the structural features of the structure, namely the geometry of the structure created by the designer. Finally, after partitioning, the boundary conditions and loads being applied.

The second step is a ground simulation. For this, materials of geological surveys are used. The simulation of the ground is also carried out using the finite element method. The most common way for ground simulation is using the planar finite elements, the most popular of which are triangular elements (Fig. 2).

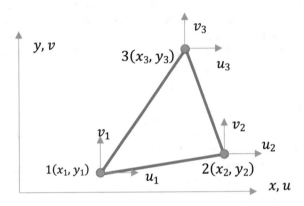

Fig. 2. Scheme of displacements for a triangular planar finite element [14]

The vector of displacements of the triangular finite element has the following view:

$$\mathbf{u}(x)^T = (u_1 \; v_1 \; u_2 \; v_2 \; u_3 \; v_3).$$

The third step is joint modeling of the structure and the ground according to the principle "ground—structure—environment". As a result, the moments, axial forces, shear forces, and displacements of the structure are obtained. The final step is the analysis of the results.

Comparing the solution in closed-form and the FEM, we may specify the following FEM features:

- FEM works for structures with any complex geometry;
- FEM works with different complex loads and constraints;
- FEM generates an approximate solution (the level of accuracy can be increased through the elements' density);
- closed-form solution allows examining system reaction to changes in different parameters (without new simulation, as for the case of the FEM).

A feature of the FEM is a relatively simple computational scheme, which is easily automated and can be grasped by surveyors to perform the necessary calculations. The next sections provide arguments supporting our decision to use the FEM and its opportunities for accuracy determination.

3 Accuracy Assignment for Staking-Out Works: A Case Study of High-Rise Building Assembling

3.1 Accuracy Calculation Approach

In applied geodesy, the accuracy calculation is a root for the choice of method of measurements and equipment to be used. To carry out the accuracy calculation, it is necessary to identify the parameters that affect the accuracy of the resulting parameter. As parameters can be considered: the dimensions of the elements, their configuration, the distance between the axes, etc. As the main parameter for the assembling of concrete high-rise buildings, designers suggest taking the requirement for verticality. For high-rise buildings, there is a typical requirement on an allowable deviation from vertical δ equals 50 mm. Considering root mean square error (RMS) $m = \delta/2.5$, we get $m = 20$ mm.

The accuracy equation has the following form:

$$m = \sqrt{m_{\mathrm{man}}^2 + m_{asm}^2 + m_{geo}^2 + \left(\Delta_{def}\right)^2} \le 20\mathrm{mm}, \tag{2}$$

where m_{man}—RMS of the construction's manufacturing; m_{asm}—RMS of the construction's assembling; m_{geo}—RMS of geodetic works; Δ_{def}—allowable displacement of construction's deformation. The parameter Δ_{def} is systematic, which in turn means that it is necessary to use a mixed error model. The systematic errors of other parameters can be neglected.

Let us use the Eq. (2) to find the allowable range of constructions' displacement. Values for the main processes typically set to the following figures: $m_{\mathrm{man}} = 2$ mm; $m_{asm} = 10$ mm; $m_{geo} = 10$ mm. Then having the expression (2), we get $\Delta_{def} = 13.7$ mm. Under adopted conditions, the accuracy of separate processes fits the primary requirement. Yet, the model (2) works just for the case when the displacements' range is known. In the opposite case, surveyors get fallen in difficult circumstances. They have to keep in mind that it is not possible to change the accuracy of the construction's manufacturing and assembling, and of course, no one can change the construction's deformation. Actually, the surveyors do not know the range of displacements as deformations' function, whether displacements are in an appropriate range or not. Moreover, the construction's displacement is growing with time and has a different meaning for different parts of the structure. Consequently, surveyors have to adapt their accuracy depending on the value of construction's displacement. The best way to manage this problem is by using the results of preliminary structure simulation by the FEM. The next subsection deals with accuracy adjustment based on the FEM simulation.

3.2 Accuracy Simulation by FEM for High-Rise Building Assembling

Let us consider a high-rise building with a height of 100 m to demonstrate the effectiveness of the FEM simulation and its effect on geodetic accuracy. The structure of twenty floors has been divided into 48,586 finite elements. The total number of nodes equals 44,111 (Fig. 3).

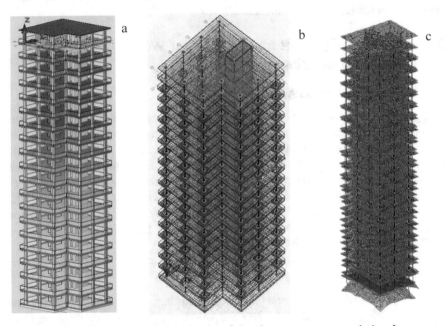

Fig. 3. Scheme of the structure partitioning into finite elements: **a** structure design; **b** computational model with finite elements; **c** superimposed initial and deformed FEM models

The types of finite elements that have been chosen are linear elements for columns and rectangular elements for slabs. The structure simulation has been performed. By the simulation results have been created the displacement fields (Fig. 4).

The data generated by the FEM simulation are summarized in Table 1.

The data yielded by this study provide convincing evidence that deformation affects the result of Eq. (2). Under the conditions from Sect. 3.1, using Eq. (2), we may substitute $\Delta_{def} = 6.8$ mm, and find the new value for accuracy of geodetic measurements $m_{geo} = 15.5$ mm. So, in order to satisfy the requirement (2) surveyors may decrease the necessary accuracy 1.5 times. On the contrary, for vertical displacements, the range equals almost 50 mm. It means that before staking-out the constructions by height, surveyors have to know the values of the displacements and account them in advance. Therefore the results of the FEM simulation considerably change the distribution between particular elements of Eq. (2) that, in turn, lead to technology rearrangement. The results prove the effectiveness of the FEM simulation that assists surveyors to assign and adjust the correct accuracy for measurements.

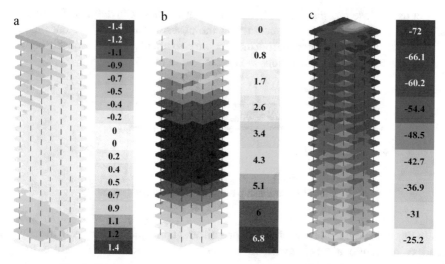

Fig. 4. Displacement fields (mm) along coordinate axis: **a** X; **b** Y; **c** Z

Table 1. Primary displacements characteristics

Descriptor	X, mm	Y, mm	XY, mm	Z, mm
Mean	0.1	4.3	4.3	−52.4
Minimum	−1.4	0.0	0.0	−71.9
Maximum	1.4	6.8	6.8	−25.2

4 Accuracy Assignment for Geodetic Monitoring: A Case Study of Subway Tunnel Exploitation

4.1 Accuracy Calculation Approach

During the geodetic monitoring, the basic criteria are the accuracy of the observations and the number of cycles for the observation period. To establish the accuracy of observations, it is necessary to know the value of the allowable displacement ΔS of the structure. In fact, the allowable displacements are the sum of two displacements: the displacement of the structure under its weight, temperature and other external loads Δe_x, Δe_y, Δe_z, and the displacements of the base of the structure as a function of the pressure of the weight of the structure (w) on the ground $\Delta g_x(w,t)$, $\Delta g_y(w,t)$, $\Delta g_z(w,t)$. Thus, to find the allowable displacement of the structure, it is necessary to determine the displacement of the structure and the base. The displacement of the base leads to the emergence of additional stresses in the structures, which in turn lead to additional displacements. Such a displacement model is very complicated. Moreover, the closed- form solution of this issue does not exist. That is why the present study employed the FEM to find the allowable displacement ΔS.

Let us suppose that the allowable displacement ΔS has been defined. The question is: how to transform ΔS to the accuracy of the observations? To figure out this, it is recommended using an approach from the random error theory. The relationship between ΔS and accuracy of the observations m is presented in Fig. 5.

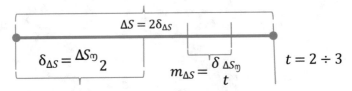

Fig. 5. The relationship between structure and ground

In Fig. 5, t is the Laplace coefficient for a different probability density. In the applied geodesy, conventionally being used values in between two to three. Therefore, for $t = 2.5$ we obtain monitoring accuracy

$$m_{\Delta S} = 0.2|\Delta S|. \tag{3}$$

The next subsection deals with accuracy assignment for geodetic monitoring based on the FEM simulation.

4.2   Accuracy Simulation by FEM for Subway Tunnel Geodetic Monitoring

As a case study of the geodetic monitoring, the two tunnels construction has been considered—these two tunnels in 17 m apart constructed in the middle depth 20 m. The tunnels' radius is 5.6 m. In order to reduce the boundary effects, the width of the simulation area was accepted 70 m. By the results of geological study, were identified four layers (sand, a mix of sand and gravel, a mix of sand and clay, and clay) with an approximate thickness of 5–6 m each.

In the first step, the finite element mesh with a step equals to 2 m has been generated. The partitioning has been done into triangular elements. In the places of tunnels excavation, the triangles density was intentionally exaggerated using a step equals to 0.5 m. Totally, 4581 elements were generated. At the next stage, the ground deformation has been calculated (Fig. 6).

The deformation field was constructed, and the forces, moments, and stresses were estimated. The largest displacements were detected underneath the tunnels. The maximum vertical displacement reached almost 10 mm. It is essential to point out that under accepted conditions, there is no effect of the construction of the tunnels to each other.

At the final step, the displacements of the tunnel's shell have been estimated (Fig. 7). The shell surface was partitioned into spatial linear elements with a step 0.5 m.

The data gathered in the simulation suggest that maximum displacement is in the bottom part of the shell. Then, using the condition (3), it is easy to assign the necessary accuracy of monitoring $m_{\Delta S} = 2.8$ mm. Unlike of closed-form solution, owing to the FEM solution, it is possible to determine the points for the monitoring. Under such

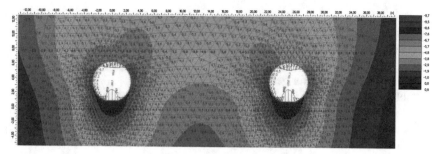

Fig. 6. The diagram of the ground displacements (mm) with partitioning into triangular elements

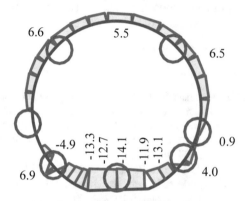

Fig. 7. The scheme of the distribution of the displacement for the tunnel shell (mm)

points, one considers the places with maximum positive and negative displacements and the places where displacements are changing their signs (red circles). Therefore it is recommended to conduct geodetic monitoring in seven points. It should be noted that the suggested scheme of observations is entirely different from a traditional scheme, in which the measurements were carried out for two mutually perpendicular tunnel radii.

5 Conclusion

The article has presented an approach to the accuracy assignment of geodetic measurements. It was pointed out that the up-to-date methods of preliminary accuracy calculation do not satisfy the requirements and conditions of state-of-the-art building technologies. Unlike the obsolete classical approaches, a new calculation method is proposed, which is based on the use of the models of structural mechanics to calculate structures' allowable displacements. Two models for displacements calculation have been compared: differential equations and the finite element models. It was proposed to use the finite element method as the primary method for structures modeling. The models of finite element and suggestions about discretization have been presented. The suggested approach uses the simulated values of the allowable displacements to assign the accuracy of geodetic measurements.

There are three main arguments that can be advanced to support the using of the FEM in the applied geodesy, namely:

assigning and adjust the correct accuracy for measurements during staking-out works;

assigning the accuracy for geodetic monitoring;

establishing the location of observation points in the case of geodetic monitoring.

These conclusions are borne out by simulation for geodetic support of high-rise building erection and geodetic monitoring of a subway tunnel. The underlying argument in favor of the FEM is that ordinary surveyors may comprehend this method. The results yielded by this study provide convincing evidence that the models of structural mechanics may play a prominent role in the applied geodesy. Further research in this area may include a complex study of constructions' assembling and geodetic works that influence the structure durability, stiffness, and stability.

References

1. Barazzetti, L., Banfi, F., Brumana, R., Gusmeroli, G., Oreni, D., Previtali, M., Roncoroni, F., Schiantarelli, G. (2015) BIM from laser clouds and finite element analysis: combining structural analysis and geometric complexity. The International Archives of the Photogrammetry, Remote Sensing and Spatial Information Sciences. XL-5/W4: 8103–8120. http://dx.doi.org/10.5194/isprsarchives-XL-5-W4-345-2015.
2. Castagnetti, C., Cosentini, R.M., Lancellotta, R., Capra, A. (2017) Geodetic monitoring and geotechnical analyses of subsidence induced settlements of historic structures. Struct Control Health Monit.: 24, 1–15. https://doi.org/10.1002/stc.2030.
3. Crespi, P., Franchi, A., Ronca, P., Giordano, N., Scamardo, M., Gusmeroli, G., Schiantarelli, G. (2016) From BIM to FEM: the analysis of an historical masonry building. WIT Transactions on the Built Environment. 149: 581–592. https://doi.org/10.2495/BIM150471.
4. Dardanelli, G., Pipitone, C. (2017) Hydraulic models and finite elements for monitoring of an earth dam, by using GNSS techniques. Periodica Polytechnica Civil Engineering. 61(3): 421–433. https://doi.org/10.3311/PPci.8217.
5. Galgana, G.A., Newman, A.V., Hamburger, M.W., Solidumd, R.U. (2014) Geodetic observations and modeling of time-varying deformation at Taal Volcano, Philippines. Journal of Volcanology and Geothermal Research: 271, 11–23. https://doi.org/10.3390/s141121889.
6. Gikas, V., Sakellariou, M. (2008) Horizontal deflection analysis of a large earthen dam by means of geodetic and geotechnical methods. 13th FIG Symposium on Deformation Measurement and Analysis, 4th IAG Symposium on Geodesy for Geotechnical and Structural Engineering, 12–18 May 2008 Lisbon, Portugal, pp 1–9.
7. Hesse, C., Heer, R., Horst, S., Neuner, H. (2006) A concept for monitoring wind energy turbines with geodetic techniques. In Proc. 3rd IAG / 12th FIG Symposium, 22–24 May, 2006, Baden, Germany, pp 1–10.
8. Hickey, J., Gottsmann, J., Mothes, P., Odbert, H., Prutkin, I., Vajda, P. (2019) The ups and downs of volcanic unrest: insights from integrated geodesy and numerical modelling. Advs in Volcanology: 203–219. http://dx.doi.org/10.1007/11157_2017_13.
9. Jäger, R., González, F. (2005) GNSS/LPS based online control and alarm system (GOCA)—mathematical models and technical realization of a system for natural and geotechnical deformation monitoring and hazard prevention. In Proc. IAG Symposium, 17–19 March, 2005, Jaén, Spain, pp 293–303.

10. Kapović, Z., Novaković, G., Paar, R. (2005) Deformation monitoring of the bridges by conventional and GPS methods. 5th International Multidisciplinary Scientific GeoConference - SGEM2005, Albena, Bulgaria, pp 1–8.
11. Kontogianni, V.A., Stiros, S.C. (2005) Induced deformation during tunnel excavation: Evidence from geodetic monitoring. Engineering Geology. 79: 115–126. https://doi.org/10.1016/j.enggeo.2004.10.012.
12. Lee, H.-H. (2015) Finite Element Simulations with ANSYS Workbench 16: Theory, Applications, Case Studies. SDC Publications.
13. Li, H.-N., Yi, T.-H., Yi, X.-D., Wang, G.-X. (2007) Measurement and analysis of wind-induced response of tall building based on GPS technology. Advances in Structural Engineering. 10(1): 83–93.
14. Liu, G.R., Quek, S.S. (2003) The Finite Element Method: A Practical Course. Butterworth-Heinemann.
15. Logan, D.L. (2012) A First Course in the Finite Element Method. 5th ed. CENGAGE Learning, University of Wisconsin-Platteville.
16. Pantazis, G., Skarlatos, D., Pelecanos, L. (2019) Long-term geodetic monitoring of seasonal deformations of earth dams and relevant finite element verification. In Proc. 4th Joint International Symposium on Deformation Monitoring (JISDM), 15–17 May 2019, Athens, Greece, pp 1–6.
17. Roberts, G., Meng, X., Meo, M., Dodson, A., Cosser, E., Iuliano, E., Morris, A. (2003) A remote bridge health monitoring system using computational simulation and GPS sensor data. In Proc. 11th FIG Symposium on Deformation Measurements, 2003, Santorini, Greece, pp 1–7.
18. Shamshiri, R., Motagh, M., Baes, M., Sharifi, M.A. (2014) Deformation analysis of the Lake Urmia Causeway (LUC) embankments in northwest Iran: insights from multi-sensor interferometry synthetic aperture radar (InSAR) data and finite element modeling (FEM). Journal of Geodesy. 88: 1171–1185. https://doi.org/10.1007/s00190-014-0752-6.
19. Szostak-Chrzanowski, A., Chrzanowski, A., Massiéra, M., Bazanowski, M., Whitaker, C. (2008) Study of a long-term behavior of large earth dam combining monitoring and finite element analysis results. 13th FIG Symposium on Deformation Measurement and Analysis, 4th IAG Symposium on Geodesy for Geotechnical and Structural Engineering, 12–18 May 2008 Lisbon, Portugal, pp 1–10.
20. Taşçi, L. (2015) Deformation monitoring in steel arch bridges through close-range photogrammetry and the finite element method. Experimental Techniques. 39: 3–10. https://doi.org/10.1111/ext.12022.
21. Tsakiri, M., Ioannidis, C., Papanikos, P., Kattis, M. (2004) Load testing measurements for structural assessment using geodetic and photogrammetric techniques. 1st FIG International Symposium on Engineering Surveys for Construction Works and Structural Engineering, 28 June–1 July, 2004, Nottingham, United Kingdom, pp 1–14.
22. Welsch, W.M., Heunecke, O. (2001) Models and terminology for the analysis of geodetic monitoring observations - Official Report of the Ad-Hoc Committee of FIG Working Group6.1. The 10th FIG International Symposium on Deformation Measurements, 19–22 March 2001 Orange, California, USA, pp 390–412.
23. Yalçinkaya, M., Satir, B., Akköse, M. (2006) Determining the displacement occurred in the tunnels using different measurement and finite elements methods: a case study for Trabzon-2 tunnel, in Turkey. In Proc. 3rd IAG / 12th FIG Symposium, 22–24 May, 2006, Baden, Germany, pp 1–11.
24. Yang, H., Xu, X., Neumann, I. (2018) Optimal finite element model with response surface methodology for concrete structures based on terrestrial laser scanning technology. Composite Structures. 183: 2–6. http://dx.doi.org/10.1016/j.compstruct.2016.11.012.

Geodetic Monitoring of Bridge Structures in Operation

Ladislav Bárta[✉], Jiří Bureš, and Otakar Švábenský

Faculty of Civil Engineering, Brno University of Technology, Institute of Geodesy, Brno, Czech Republic
{barta.l1,bures.j,svabensky.o}@fce.vutbr.cz

Abstract. This paper deals with the monitoring of vertical displacements and deformation of large concrete bridge objects. The subject of geodetic monitoring was the substructure and bearing structure of two parallel bridges. The length of each bridge is 700 m. The substructure consists of 15 supports. The heights of the pillars are up to 30 m. Six stages of measurements covering 6 years were processed. The measurement technology was designed with respect to the specific influence of the external environment and traffic on the measuring system itself when measuring the points of the reference system, the points at the heels of the bridge pillars, the points at the supports and the point at the ledges of the load-bearing. The aim is to analyze vertical displacements and deformation of the monitored bridge object in relation to the estimation of temperature states of the object in individual phases of measurement.

Keywords: Bridge · Monitoring · Vertical displacement · Deformation

1 Introduction

This paper deals with the determination of vertical displacements and deformations of bridge structures SO 203 on the R1 expressway on the section Selenec - Beladica in Slovakia. The main subject of geodetic monitoring is settlement of the substructure and determination of deflections of bridge deck sections. For this purpose, a precise levelling was used to completely measure the points of the reference grid, the points on the bridge substructure and the points on the bridge deck structure. The input for the evaluation was 6 stages covering the annual periods between year 2012 and 2017.

The conversion of the evaluated vertical displacements of the load-bearing structure to the reference temperature state of the bridge structure based on direct temperature measurement can be quite complicated. The problem is both in the accuracy of the mean temperature estimation of the bridge structure and in the knowledge of the exact value of the coefficient of thermal expansion of the concrete used. Therefore, this article tests an alternative approach based on determining the direct relationship between the detected height changes of the load-bearing structure and its height above terrain.

The technology of precision geometric levelling is the basic method of determining the height components of points of geodetic networks [1] and for determining the

A. Kopáčik et al. (Eds.): *Contributions to International Conferences on Engineering Surveying*,
SPEES, pp. 198–210, 2021. https://doi.org/10.1007/978-3-030-51953-7_17

settlement and deformation of the building objects using suitable discrete points. Exact levelling can be used, for example, to determine the displacement and deformation of dams [2] and determining the settlement of foundations of industrial objects or industrial technologies [3]. In case of greater height differences, the trigonometric height measurement or trigonometric levelling may be used [4]. Precision levelling technology can also complement very well with methods of determining horizontal deformations of objects too. An example could be the monitoring of a railway bridge [5]. Monitoring of buildings is now also carried out using terrestrial laser scanners. This technology is advantageous when fast and detailed documentation of the monitored object is needed [6]. Special application of terrestrial laser scanners for static load tests of bridges is also described. See [7, 8].

2 Subject of Measurement

The subject of further interest will be long-term monitoring of the bridge structure with the designation SO 203 consisting of two parallel bridges, each of length 760 m (Fig. 1).

Fig. 1. View of the bridge SO 203

The maximum height of the bridge deck above the terrain is 30 m. The supporting structure of each of the bridges consists of 14 bridge spans.

The bridge deck has a width of 13 m and a thickness of 3 m. The abutments and pillars have a pile foundation. The cross-section of each pillar is 3.8 m × 3.8 m. The supporting structure dilates from a pair of pillars P7 and P8 with fixed bearings. Longitudinal section of the bridge SO 203 is in Fig. 2.

3 Scope of Measurements

Long-term geodetic monitoring of bridge structures usually follows the requirements of a geodetic survey project from the time of construction or a long-term survey project.

The reference network for geodetic monitoring of displacements and deformations became the local survey network of the bridge, consisting mainly of survey pillars. The

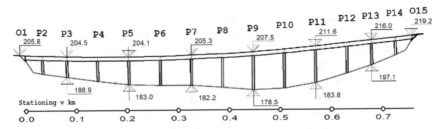

Fig. 2. Longitudinal section of the bridge SO 203

project to supplement the reference network with points with deep stabilization was completed and implemented in the course of long-term monitoring of bridges. The aim was to strengthen the stability of the reference system. The reference system of the bridge SO 203 currently consists of 10 control points (Fig. 3).

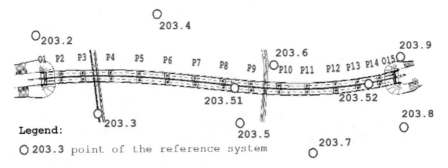

Fig. 3. Reference system for geodetic monitoring of the bridge SO 203

The monitored points on the bridge substructure consist of benchmarks placed in pairs at the footings of the pillars and in the front side of the abutments. The monitored points on the bearing structure are fitted with nail marks in the ledges of the bridge above the supports, in the middle of the first and last bridge spans and in thirds of the other bridge spans (Fig. 4). The total number of installed points is 56 on the substructure and 168 on the supporting structure.

4 Measuring Methodology

The aim of the geodetic epoch-wise measurement, that is required in long-term survey project, was to achieve a conclusive value of vertical displacement with a maximum risk of 5% error at any of the monitored points up to ±2 mm. This corresponds to the accuracy of the determination of the point heights in the individual epochs of the measurement with a maximum standard deviation of ±0.70 mm computed according to Eq. (1). Symbol δ represent the conclusive value, symbol σ represent the standard deviation and symbol t represent the reliability factor. In this case is the reliability factor equal to 2.

$$\sigma = \frac{\delta}{t\sqrt{2}} \tag{1}$$

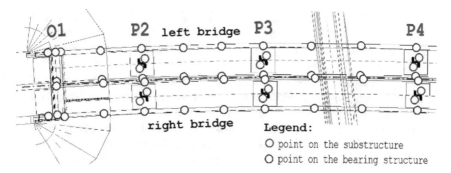

Fig. 4. Monitored points on the bridge SO 203

The stability of the reference network over time is assumed. The precision level-ling technology was used according to aforementioned facts. The individual epochs of the monitoring were measured with the Trimble DiNi levelling instrument with the accuracy given by the mean error of the two-way levelling of 0.3 mm/km using invar code staffs. The substructure was measured by one core two-way levelling line. Two lines were simultaneously measured on the supporting structure with one levelling instrument simultaneously in one direction and two lines simultaneously in the direction backwards (Fig. 5). On the outer ledges were created one-way measured lines and on one of the central ledges was created a two-way levelling line. The interconnection of measurements in the transverse direction ensured the individual positions of the levelling instrument.

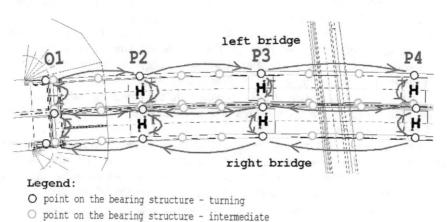

Fig. 5. Scheme of measurements on the load-bearing constructions of the bridge SO 203

As standard deviation of one levelling section for realizing measurements between two adjacent points, typically with 1–3 assemblies, was considered value ±0.11 mm for substructure and ±0.17 mm for load-bearing structure. For a given measurement method, a priori accuracy of primary variables, the specified number of measurements and net-work configuration was through an a priori analysis using a LSE method defined the accuracy of the point height determination on the substructure with a standard deviation up to ±0.35 mm and on the load-bearing structure up to ±0.45 mm. Corresponding values of vertical displacements are from ±1.0 mm on the substructure and from ±1.3 mm on the load-bearing structure.

In the process of data collection and constructing the geodetic network model of the each measured phase the following were assessed: (a) differences of back and forth measurements in levelling sections and sub-sections, (b) differences in total elevation determined by measurements on the substructure and load-bearing structure, and (c) differences in the adjustment results determined from the measurement back and forth separately. The final assessment of the measurement quality was performed on the basis of estimates of a posteriori characteristic of the accuracy of the final adjustment of all partial measurements in the network by the LSE method.

5   Evaluation of Substructure Measurements

The aim of monitoring the substructure is to capture the settlement of the bridge object in terms of overall vertical displacements and uneven settlement of the object. The different temperature conditions of the bridge structure caused by the influence of the external environment will not have a significant effect on the height of the levelling marks placed within 0.2 m above the terrain. In this case, the external environment will only affect the measurement technology.

Monitoring of bridge substructures is characterized by air flow under the bridges (drafts) which disturbs the local temperature microclimate. Positioning of measurements in the shade of load-bearing bridge structures is only possible for bridge structures with low pillars. Usually there is no vegetation under the bridges. Complications with vegetation occur only during measurements from bridge to points of reference network. It is also unpleasant to overcome considerable elevations between (a) the footings of the filling body, (b) marks on the front side of the abutments, and (c) points on the ledges above the abutments. Increased attention is paid to the stability of the staff changeover points, usually placed on access staircases, in water drainage gutters, and in stone facings of slopes at abutments.

The independent indicator of the relative accuracy of the measurements are the differences in vertical displacements of pairs of points on individual pillars (Fig. 6). The same values of vertical displacements can be expected at points placed on the same part of the structure, see Graph 1. Realized epochs are named after the date of measurement. The first number is a month and the second number is a year.

The slight variation in vertical displacement values may also be affected by changes in longitudinal and transverse tilt of the pillars. The agreement of vertical displacements of point pairs on the same pillars for the bridge object (Fig. 6) in question can be described by a standard deviation of ±0.22 mm.

Good matching of vertical displacement values is also typical for side-by-side bridge pillars, regardless of whether they support the same or different bridge structures. The compliance of vertical displacements of side-by-side supports in the case of the bridge object in question corresponds to a standard deviation of ±0.29 mm.

Settlement differences on substructure of the left bridge
pairs of points on individual pillars

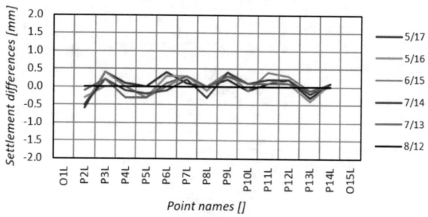

Graph 1. Differences of the vertical displacements on the points at the heels of the pillars

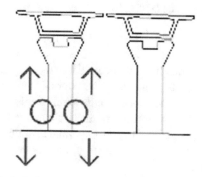

Fig. 6. Vertical displacements on the points at the heels of the pillars

Uneven settlement of substructure of bridges is typical for longitudinal direction of the structure. The detected deformation usually corresponds to continuous changes in the character of the subsoil. The most pronounced uneven settlement usually occurs between the bridge supports and adjacent pillars. It is caused by the added load of the subsoil by adjacent road embankments.

Graph 2 shows the evolution of vertical displacements and deformations within the individual measurement epochs. There is an increasing uneven settlement of the P5, P6 and P11 pillars.

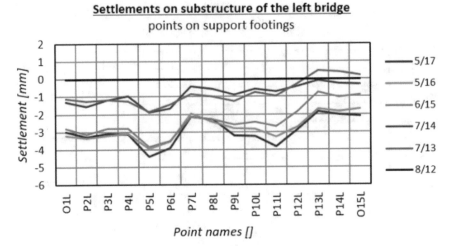

Graph 2. Vertical displacements on the substructure of the bridge

6 Evaluation of Bearing Structure Measurements

The aim of monitoring the bearing structure of a bridge is to determine its deformation in time. The evaluated vertical displacements of points on the ledges of the supporting structure have two components. It is a vertical shift of the substructure of the bridge and the deformation of the bridge structure due to the external environment. The uncertainties of the determined vertical displacements can be significantly influenced by the current atmospheric conditions and the vibrations of the structure caused by traffic.

Deflections of bridge spans (Fig. 7) are an important resulting parameter, see Graph 3. The deformation of the load-bearing structure in the transverse direction (Fig. 8) can also be evaluated, see Graph 4.

The most acceptable atmospheric conditions for measuring load-bearing structures can be considered the morning hours. Considering the dynamics of gradual changes of deflection of bridge spans and theoretically of the whole structure, it is necessary to realize individual measurements quickly, efficiently and in a coherent manner. Especially in the case of soft constructions, where it takes longer time to stabilize the structure after the passage of vehicles, it is advisable to carry out the measurements in times and days with minimal traffic. Restriction or exclusion of traffic is practically justifiable for the purposes of geodetic monitoring only with respect to the safety of the surveying team in case when it is necessary to move on the road without barrier protection.

From the vertical displacements (Fig. 9) determined at points in the ledges of the load-bearing structure, the component caused by the settlement of the bridge object can be subtracted (Graph 5). The vertical displacement values detected at the points at support footings are used for this purpose. Black dots in Graph 5 represents height of the load-bearing structure above terrain divided by value 10.

This way calculated vertical displacements describe the deformation of the bridge object in vertical direction (Fig. 9). The deformation may have a reversible and irreversible component. The reversible deformation is usually related to the object current

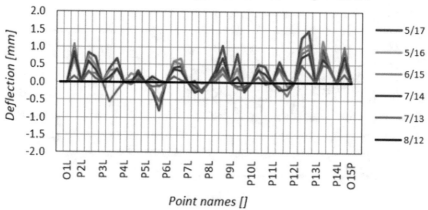

Graph 3. Deflection of bridge spans

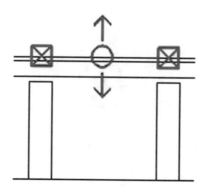

Fig. 7. Bridge span deflection

temperature state. Especially for high-pillar bridges, the part of the vertical displacement due to the temperature change of the structure to the height of the bridge may exceed the conclusive values of vertical displacements or become dominant compared to the component of the vertical displacement due to the settlement of the object. An example of the irreversible deformation of a bridge can be a permanent change of its bearing structure, e.g. in the transverse direction. Due to (a) the current temperature state of the load-bearing structure, (b) the influence of the direction of insolation and also (c) permanent changes in the load-bearing structure different height changes can occur at points on the bridge deck points with respect to the points in the bridge axis, see Fig. 10.

One of the ways to convert the evaluated vertical displacements into a uniform thermal state of the structure is to use physical measurement of the structure temperatures. However, uncertainties in the determination of the actual temperature of the load-bearing structure and substructure as well as uncertainties in the longitudinal thermal expansion

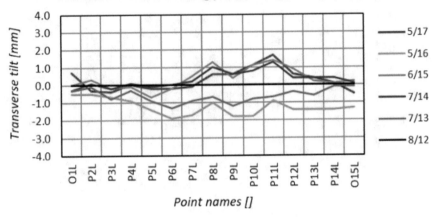

Graph 4. Transverse tilts of the supporting structure

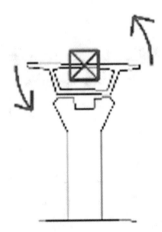

Fig. 8. Transverse tilt of the load-bearing constructions

of the material can have a significant effect on the values of the calculated corrections. Especially in the case of large dimensions of the structure, the uncertainties of the calculated temperature corrections may exceed the uncertainties of the realized geodetic measurements.

Another way is to use regression analysis to find the direct mathematical relationship between the heights of the measured points on the ledges of the bridge above the terrain and the calculated deformations of the bridge object, see Graph 5 and Fig. 9. The advantage of this approach is its independence from measuring the structure temperature. The thermal deformation of the bridge structure can theoretically be eliminated by a linear function, where (a) the functional value will be the correction of deformation of the bridge structure, (b) the variable value will be the elevation of the bridge deck

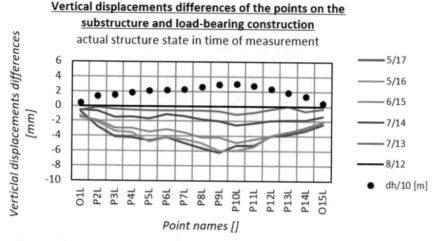

Graph 5. Vertical displacements differences of the points on the bridge substructure and bridge load-bearing construction—measured values

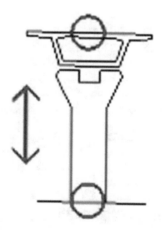

Fig. 9. Deformation of the bridge object in vertical direction

above terrain, (c) the multiple member shall be given by the coefficient of linear thermal expansion of the bridge construction material used and by the deviation of the structure temperature from the reference state, and (d) the absolute member will be given by the value of transverse deformation or creep of the supporting structure. First, the parameters of the function are determined by regression analysis. Then the vertical displacement values of the points on the ledges of the bearing structure are corrected.

Subsequently, Graph 6 shows the values of vertical displacements on the load-bearing structure after filtering out the settlement of the substructure of the bridge and after converting the measurements to a uniform temperature state of the structure. When the

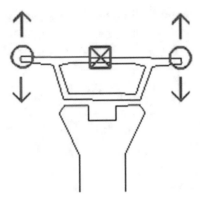

Fig. 10. Deformation of the load-bearing structure in the transverse direction

measurement was converted to the reference state of the structure, the vertical displacements on the substructure and on the bearing structure of the bridges were harmonized with a standard deviation of ±0.40 mm.

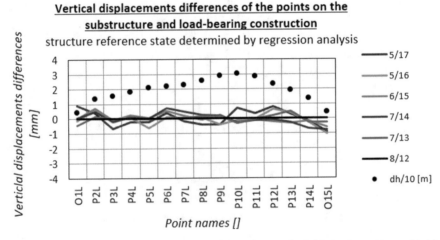

Graph 6. Vertical displacements differences of the points on the bridge substructure and bridge load-bearing construction—after application of the parameters of the regression model

Obtained parameters of linear function determined by regression analysis at selected coefficient of thermal expansion of concrete were used to calculate changes in the structure temperature from the reference state. Results of the direct temperate measurements are calculated as the difference of construction temperature in reference and actual epoch. The values determined from the parameters of the regression model and determined as differences of physically measured temperatures of the construction are compared in Table 1.

Table 1. Changes in structure temperatures from the reference state

Temperature changes (°C)	E6	E5	E4	E3	E2	E1
	5/17	5/16	6/15	7/14	7/13	8/12
Regression model	−16	−17	−11	−6	−3	0
Direct measurement	−10	−13	−10	−5	−3	0

The difference between the detected temperatures of 6 °C for the dimension of 30 m causes a length change of about 2 mm in question.

7 Conclusion

This paper describes the technological measurement procedure applied during long-term geodetic monitoring of bridge structures on the R1 highway in the section Selenec - Baladice in Slovakia. Increased attention was paid to monitoring the load-bearing structure. Measured here were doubled levelling lines in the direction back and forth, and the individual lines were interconnected in the transverse direction.

Relatively complicated is the problem of converting the measurement results to a uniform temperature state of the structure. For this purpose, point measurement of the structure's temperature by contact measurement seems to be relatively uncertain. One way to circumvent the direct measurement of the structure's temperature is to use the knowledge of certain dimensions in the reference and measured states. In this paper, the dependence between vertical displacements on the bearing structure and the heights of the supports including the deck was searched by regression analysis.

The described analyses correlate the determined vertical displacements on different parts of the bridge object in order to detect non-standard results in the evaluation. One of the causes of unexpected results may be a problem with the measuring marks caused, for example, by their damage. Another cause of non-standard evaluation results may be confusion of measuring marks during measurement or a gross measurement error or processing error. These procedures can therefore be understood as the last stage of control of geodetic data processing. The identification and explanation of non-standard behaviour of the bridge object is primarily the task of the statics engineer. For this purpose, it is necessary to use geodetically measured states of the structure and exact modelling of the bridge object behaviour in time and in dependence on the temperature. This article has been prepared with the support of project FAST S-18-5324 "Research of real behavior of building structures in operational conditions by geodetic methods".

References

1. Zrinjski, M., Barković, D., Milat, A.: Monitoring and analysis of vertical movements at the test geodetic base network. 10th International Multidisciplinary Scientific, (2010).
2. De Lacy, M.C., Ramos, M.I., Gil, A.J., Franco, O.D., Herrera, A.M., Avilés, M., Domínguez, A., Chica, J.C.: Monitoring of vertical deformations by means high-precision geodetic levelling. Test case: The Arenoso dam (South of Spain). Journal of Applied Geodesy, 11 (1), pp. 31–41, (2016).
3. Ehigiator, R.I., Ehiorobo, J.O., Ehigiator, M.O., Beshir, A.A.: Determining the subsidence of oil storage tank walls from Geodetic levelling. Advanced Materials Research, 367, pp. 467–474, (2012).
4. Chirilă, C., Albu-Budusanu, R.M.: Applying trigonometric levelling for monitoring the vertical deformations of engineering structures. Environmental Engineering and Management Journal, 18 (9), pp. 1859–1866, (2019).
5. Zhang, L., Zha, X.: Monitoring and result analysis of temporary railway bridge construction. Journal of Geomatics, 43 (6), pp. 113–116, (2018).
6. Erdélyi, J., Kopáčik, A., Lipták, I., Kyrinovič, P.: Pedestrian bridge monitoring using terrestrial laser scanning. Advances and Trends in Engineering Sciences and Technologies - Proceedings of the International Conference on Engineering Sciences and Technologies, ESaT 2015, pp. 51–56, (2016).
7. Gawronek, P., Makuch, M.: TLS measurement during static load testing of a railway bridge. ISPRS International Journal of Geo-Information, 8 (1), art. no. 44, (2019).
8. Mill, T., Ellmann, A., Kiisa, M., Idnurm, J., Idnurm, S., Horemuz, M., Aavik, A.: Geodetic monitoring of bridge deformations occurring during static load testing. Baltic Journal of Road and Bridge Engineering, 10 (1), pp. 17–27, (2015).

Non-contact Monitoring Execution with the Purpose to Determine Potential Bridge Damages

Boštjan Kovačič[(✉)], Sebastian Toplak, and Samo Lubej

Faculty of Civil Engineering, Transportation Engineering and Architecture, University of Maribor, 2000 Maribor, Slovenia
bostjan.kovacic@um.si

Abstract. Monitoring of the objects is executed primarily to determine outer and inner influences on the object and mostly for the sake of the safety of people, animals, and material assets. The monitoring can be executed with various methods which depend on the object, conditions for the execution and the purpose of the monitoring. Our focus lies in the monitoring of the footbridge over the Lobnica creek in the town of Ruše in Slovenia. The monitoring of the vibrations and displacements was carried out as the loading test under the influence of the dynamic load with the purpose to determine the elastic response. The elastic response of the very decrepit footbridge was one of the conditions for decision about the meaningfulness of the renovation of an extremely corrosive damaged bridging object. For this purpose, the physical methods were used with the application of the seismograph, the micrometre and geodetic method with the robotic total station (RTS), which was controlled in two ways: by the built-in Leica GeoCom protocol and additional software GeoCom/Zg. The emphasis of the experiment was on the application of the non-contact geodetic methods, which are nowadays in use for the measurements of the dynamic response and enable the measurements with the RTS up to 30 Hz.

The article presents the application of various procedures of non-contact data including the bridge, comparison, and the analysis of the obtained values in the dynamic response monitoring of the construction.

Keywords: Geodesy · Dynamic response · Monitoring

1 Introduction

Nowadays, the monitoring of the bridging objects is based mostly on measuring the displacements, deformations, and vibrations of the bridges. The data about the static and dynamic response of the construction under the load is usually obtained with the loading test. Generally, the data on the long-term behaviour of the objects is obtained from the loading test and continuation of the monitoring. This is significant for highly loaded objects on highways or on railways. Technological development of the measuring

© The Editor(s) (if applicable) and The Author(s), under exclusive license
to Springer Nature Switzerland AG 2021
A. Kopáčik et al. (Eds.): *Contributions to International Conferences on Engineering Surveying*,
SPEES, pp. 211–222, 2021. https://doi.org/10.1007/978-3-030-51953-7_18

equipment has influenced the area of geodetic instruments as well. Until recently, the geodetic measurements were able to determine only vertical and horizontal displacements. However, the development of the geodetic equipment, an increase of the resolution, added sensor for faster reading and automatic monitoring of the prism, together with the geodetic measurements and the very precise static measurements, enable to perform the measurements of the dynamic response on various monitoring as well [1, 2]. It is necessary to know that geodetic methods allow for high-quality data about the behaviour of the construction on sites in a very simple way, where other methods cannot be employed due to the inaccessibility or difficulty.

The monitoring of the dynamic response in the class of larger frequencies was primarily executed with the GNSS system. It was concluded that only equal and larger oscillations can be noted by using this equipment since it could contain the data with the speed up to 10 Hz [3–7]. Currently, the data with the speed up to 100 Hz can be employed with the development of the GNSS technology [8].

The RTS system is used nowadays, enabling the automatic monitoring of the prism up to the speed of 30 Hz in dynamic monitoring of the displacements of the size category of a few centimetres, specifically due to the large technological development of the geodetic equipment [9–13]. The additional software, which allows the control of the RTS over the laptop, was developed at the Faculty of Geodesy in Zagreb. The software enables recording of the angles, length, and time or only of the angles and time, depending on the needs. Smaller and unequal oscillations up to 20 Hz can now be measured with this additional protocol, now [12] and the amplitude of a few mm can be measured as well. The authors [14] also describe the possibility of measuring the vibrations up to 5 Hz.

If the construction, conditions and of course, time, allow it, the best approach to execute the measurements employs various methods (GPS, RTS and accelerometers) and compare the results as presented in the work [15].

2   The Description of the Measuring Equipment

The article presents the application of geodetic and physical methods in monitoring the dynamic behaviour of the smaller bridging object with the range between the riparian supports of length 8.6 m and width 4.8 m. The dynamic load of the construction was evoked when the three-axle fire truck with a mass of 14.2t drove across the object constantly in the same direction with 30 km/h.

Geodetic part of the monitoring was executed with the instrument Leica TS50 in two ways. The oscillation was simultaneously measured by the built-in protocol Geocom with the speed of reading 6 Hz and with additional software/protocol GeoCOM/ZG, which enables up to 30 measurements per second. The software GeoCOM/Zg for monitoring the target with the speed of reading per millisecond was developed at the University of Zagreb in Croatia at the Faculty of Geodesy. The instrument was controlled by a computer using the GeoCOM protocol. The connection to the computer was established with a serial cable GEV269, using a baud rate setting of 115, 200. The software was developed at the Faculty of Civil Engineering in Zagreb and enables the measuring of the length, Hz and V of the angle per millisecond or the measuring of only the Hz and V of the angle, where the exclusion of the distance meter results in even more equivalent measurements per second. The accuracy of measuring angles for geodetic instruments Leica TS50 is 0.5" and for measuring distances is 0.6 mm + 1 ppm.

The seismograph by the manufacturer Instantel Canada—MinimatePlus is the equipment used to measure vibrations. Vibration velocity measurement accuracy in the frequency range between 4 and 125 kHz is ±0.5 mm/s or 5%. The hardware by the same manufacturer Blastware enables the analysis of the results.

Vibration measurements were also performed with a POLYTEC PDV-100 portable digital vibrometer that enables surface vibration velocity measurements without contact in the frequency range from 0.05 to 22 kHz and cover a velocity range of 0.05 μm/s to 0.5 m/s. Variable working distances of the vibrometer are from 0.1 m up to 30 m.

The micrometres by the manufacturer Mahr enable the monitoring of the displacements with the accuracy of 0.0025 mm at the sampling of 1 Hz. The hardware Mahr Professional was applied for data containing. The characteristics of the applied equipment are shown in Table 1.

Table 1. The characteristics of the applied measuring equipment

Sensors	Name	Frequency (Hz)
Total Station Leica	TS50 GeoCOM protocol	6
	TS50 GeoCOM/ZG protocol	26
Triaxial seismograph	Instantel MinimatePlus	2–250
Vibrometer	Polytec PDV-100	0.05 Hz–22 kHz
Micrometre	Mahr Cator 1087 Ri	1

3 The Execution of the Monitoring

The measurement of the dynamic response of the construction was executed for four equal dynamic loads—the ride of the truck across the footbridge as shown in Fig. 1.

The measurement of the own oscillation of the construction without load was measured for 260 s. Four measurements of the dynamic response of the structure were performed after the measuring of the own oscillation. All measurements lasted for 25 s.

3.1 The Description of Bridging Construction

The bridging construction contains four steel rods and a reinforced-concrete deck of the bridge. The geometry of the construction and schematic overview of the load with the truck is shown in Fig. 2.

Fig. 1. The bridging object under the load

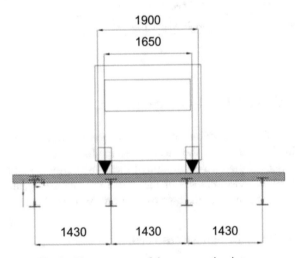

Fig. 2. The geometry of the construction in mm

3.2 The Placement of the Equipment

Non-contact measuring equipment, RTS and vibrometer were placed on a stable and non-deformable area directly at the footbridge. The distance from the observe point was 4.5 m. The sensors of the stated measuring equipment were placed—three-axle geophone, micrometre, and geodetic prism GPR121 Leica on the site of the expected largest displacements in the middle of the range on the outer steel rod. All equipment was held on a longitudinal rod with a clamp. The location of the installation of the measuring equipment and the sensors used is evident from the geodetic plan in Fig. 3. The scaffold was placed under the object for easier access to the measuring point and for the placement of the micrometre, as shown in Fig. 1.

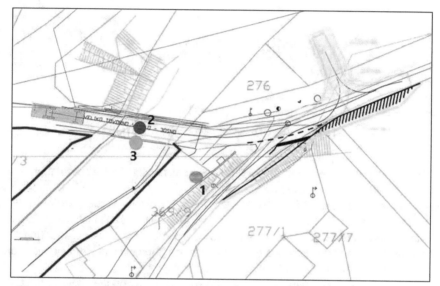

(Note: 1 – Location of RTS and vibrometer, 2– Location of micrometer and seismograph, 3 – Location of geodetic prism GPR121)

Fig. 3. The location of the installation of the measuring equipment and the sensors

4 The Measuring Results and the Discussion

The result analysis of the monitoring of the bridging object, obtained with the geodetic equipment, is shown graphically for all loading examples. The maximal displacement in dynamic load was determined from the geodetic measurements. Then, the results were compared with the displacements, measured with a micrometre and a vibrometer. However, the comparison of this quantity with the measured results by the seismograph is not possible since the relative displacements are obtained with the seismograph. The software tool by the manufacturer Instantel–Blastware Advanced Module, was used for graphical and analytical evaluation of the speed of vibration, and for the analysis of the results, which were obtained with the seismograph.

Later, the results of the own oscillation and four examples of the dynamic response, measured with all three methods, are presented.

All loading examples were measured continuously with RTS 6, 26 Hz, seismograph Instantel and vibrometer Polytec. The results of both methods are presented in the following figures. Figure 4 shows the measurement of the own oscillation with geodetic measuring equipment.

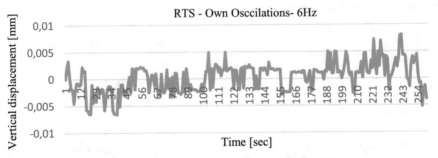

Fig. 4. Measured own oscillation with RTS at 6 Hz

Hence, it is seen from the results of the measurements that the own oscillation of the bridging construction, which is evoked by the water flow, hitting both ending supports, is very small, only 0.008 mm. The PPV V (Peak Particle Velocity Vertical) 0.127 mm/s at the frequency larger than 100 Hz was measured with the seismograph Instantel. Similarly, the vibrations and displacements are due to the own oscillation of the construction very small—on the limit of the detection of the measuring equipment and they can, therefore, be excluded from the foregoing analysis of the results.

Figure 5 shows the dynamic response of the construction on the effect of the truck riding across the footbridge for all four loading examples, which were measured with the micrometres.

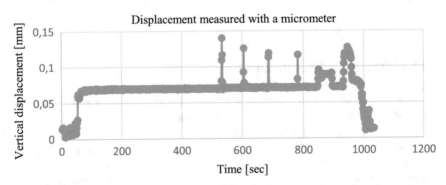

Fig. 5. Measured dynamic response of all four loading examples with a micrometre

The values of measured vertical displacements are very unequal and are 0.08 mm maximal at the first ride of the truck and 0.055 minimal at the last fourth ride of the truck.

Figure 6 shows the measurements of the displacements, measured with the geodetic equipment. The measured vertical displacements were very equal and are in all four examples 0.15 ± 0.05 mm.

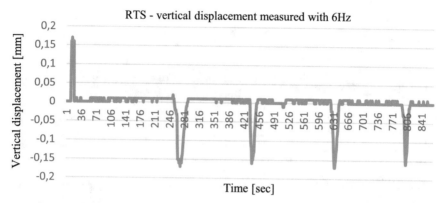

Fig. 6. Measured displacements of all four loading examples with RTS at 6 Hz

Figure 7 shows the measurements of PPV V with the measuring equipment seismograph Instantel and vibrometer Polytec for all four loading examples. The results of the maximal values of PPV V are evaluated in Table 2.

Figure 8 shows the measurements of the vertical displacements with geodetic equipment Leica TS50 and vibrometer Polytec, the results of the measurements—vertical displacements are evaluated in Table 3. The graphs in Fig. 8 show that the measurements were performed for about 25 s. The problem is that the instruments are triggered at the same time, so the recorded maximum displacements are delayed. In the first case, the maximum displacement at 8.1 s was recorded by the geodetic method, while the maximum displacement at 11.9 s was measured with a Vibrometer Polytec. The closest time was in the fourth measurement case. The value of the measured displacements, however, coincides with all measured cases, as can be seen in Table 3.

The value of the measured displacements, however, coincides with all measured cases, as can be seen in Table 3.

Table 2 and the shapes of the graphs in Fig. 7 show that measured speed values of the vibrations (PPV V) with the physical measuring equipment are comparable. It is seen from Table 3, i.e. evaluation of the values of maximal measured vertical displacements, that measured values obtained with physical equipment—vibrometer Polytec and geodetic equipment Leica TS50 are very comparable, the difference of measured values is lower than ±0.01 mm.

The first loading example

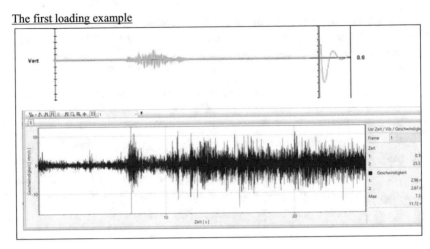

The second loading example

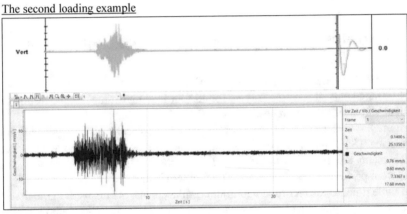

The third loading example

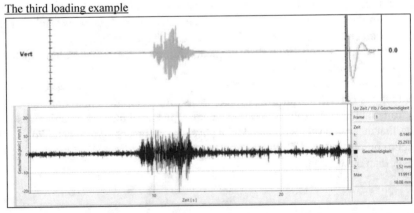

Fig. 7. Measured PPV V with the measuring equipment seismograph Instantel and vibrometer Polytec for all four loading examples

The fourth loading example

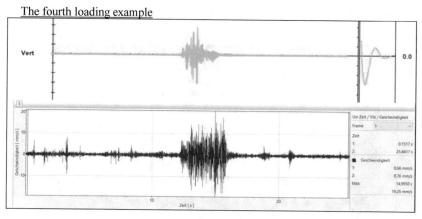

Fig. 7. (*continued*)

Table 2. The evaluation of the results of PPVmax V

Loading example	PPV$_{max}$ V measured with MINIMATE PLUS (mm/s)	PPV$_{max}$ V measured with POLYTEC PDV-100 (mm/s)
First	3,302	10
Second	8,636	5
Third	8,890	7
Fourth	15.11	10

5 Conclusion

The measurements on the object over the Lobnica creek have shown that geodetic methods with fast sampling are highly appropriate for the execution of the monitoring of the dynamically loaded objects. Comparable results of the displacements of the construction with the measuring equipment, the vibrometer Polytec, which is used for the monitoring of the physical quantities of the dynamic response, such as vibration speed, acceleration and displacement, can be obtained with fast sampling, which follows the oscillation of the construction and with the sampling frequency of approximately 30 Hz.

The monitoring of the object, which was employed to obtain representative and engineeringly correct data during the experiment, is definitely reasonable to execute with the application of geodetic as well as physical methods since only all of them can provide the whole insight into the behaviour and response of the construction in its exploitation period.

The equipment with low data containing is inappropriate for the monitoring of the dynamic effects which was proved with the measurement. The comparison of the graphs shows that measurements performed with micrometres with the containing frequency of only 1 Hz did not follow the oscillation of the construction. As a result, the equipment did not note the extreme.

The first loading example

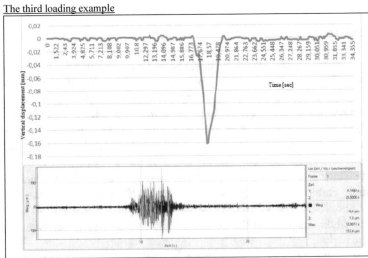

Measurement with the geodetic equipment Leica TS50 in the second loading example was not noted.

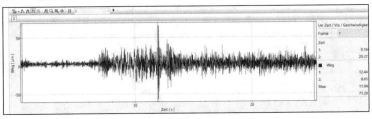

The third loading example

Fig. 8. Measurements of the vertical displacements with the geodetic equipment Leica TS50 and vibrometer Polytec

The fourth loading example

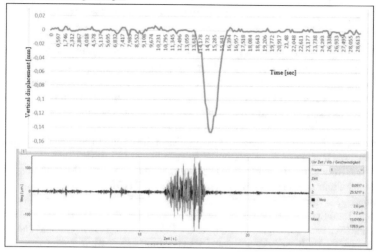

Fig. 8. (*continued*)

Table 3. Evaluation of the results of maximall vertical displacements

Loading example	Vertical displacement measured with Leica TS50–26 Hz (mm)	Vertical displacement measured with Vibrometer Polytec (mm)
First	0.161	0.025
Second	–	0.1101
Third	0.161	0.1574
Fourth	0.143	0.1399

Note Measured values of the displacements and PPV V for the example of the first ride of the truck across the footbridge, which were obtained with the measuring vibrometer Polytec, were excluded from the analysis since they do not correspond to the parallel obtained measurements with the measuring geodetic equipment Leica TS50 and also not to the physical measuring seismograph Instantel MinimatePlus. The cause of this phenomenon can be ascribed due to the unsuccessful prequel calibration of the instrument—Vibrometer Polytec

References

1. Lienhard, W., Ehrhart, M., Grick, M.: High Frequent Total Station Measurements for the Monitoring of Bridge Vibrations. Proceedings of 3rd Joint International Symposium on Deformation Monitoring (JISDM), Vienna, Austria (2016).
2. Lienhard, W., Ehrhart, M.: State of the Art of Geodetic Bridge Monitoring, Proceedings of International Workshop of Structural Health Monitoring (IWSHM), Stanford, USA (2015).
3. Celebi, M., Sanli, A., GPS is Pioneering Dynamic Monitoring of Long-Period Structures. Earthq Spectra;, 18(1), 47–61 (2002)

4. Chen, Q., Huang, D.F., Ding, X.L., Xu, Y.L., Ko, J.M.: Measurement of Vibrations of Tall Buildings with GPS, Proceedings of Health monitoring and management of civil infrastructure systems, Bellenham (WA), SPIE (2001)
5. Roberts, G.W., Meng, X., Dodson, A.H.: The use of Kinematic GPS and Triaxial Accelerometers to Monitor the Deflections of Large Bridges, 10TH FIG INTERNATIONAL SYMPOSIUM ON DEFORMATION MEASUREMENT, Orange, California (2001).
6. Ogaja, C., Wang, J., Rizos, C.: Detection of Wind-induced Response by Wavelet Transformed GPS Solutions, Journal of Surveying Engineering, 129(3), 99–104 (2003)
7. Meng, X., Dodson, A.H., Roberts, G.W.: Detecting Bridge Dynamic with GPS and Triaxial Accelerometers. Engineering Structures, 29, 3178–3184 (2007)
8. Marendič, A., Kapović, Z., Paar, R.: Possibilities of Surveying Instruments in Determination of Structures Dynamic Displacements, Geodetski list, 3, 175–190 (2013)
9. Koo, K.Y. Brownjohn, J. M. W.: Structural health monitoring of the Tamar suspension bridge. Structural Control and Health Monitoring, 20, 609–625 (2013)
10. Psimoulis, S., Stiros, S.: Measuring deflections of a Short-span Railway Bridges Using Robotic Total Station, Journal of Bridge Engineering, 18, 182–185 (2013)
11. Marendić, A., Paar, R., Grgac, I., Damjanović, D.: Monitoring of Oscillations and Frequency Analysis of the Railway Bridge "Sava" Using Robotic Total Station. Proceedings of 3rd Joint International Symposium on Deformation Monitoring (JISDM), Vienna, Austria (2016).
12. Marendić, A., Paar, R., Duvnjak, I., Buterin, A.: Determination of Dynamic Displacements of the Roof of Sports Hall Arena Zagreb, FIG, 6TH INTERNATIONAL CONFERECE ON ENGINEERING SURVEYING, Prague, Czech Republic (2014).
13. Psimoulius, P., Stiros, S.: Measurement of Deflections and Oscillation Frequencies of Engineering Structures Using Robotic Theodolites (RTS), Engineering Structures, 29, 3312–3324 (2007)
14. Kopáčik, A., Kyrinovič, P., Kadlečíková, V.: Laboratory Tests of Robotic Stations. Proceedings of FIG working week, Cairo (2005).
15. Lekidis. V., Tsakiri, M., Makra, K., Karakostas, C., Klimis, N., Sous, I.: Evaluation of Dynamic Response and Local Soil Effects of the Evripos Cablestayed Bridge Using Multi-Sensor Monitoring Systems, Engineering Geology, 179, 7–17 (2005).

Influence of External Conditions in Monitoring of Building Structures

Jiří Bureš(⊠), Ladislav Bárta, and Otakar Švábenský

Faculty of Civil Engineering, Institute of Geodesy, Brno University of Technology, Brno, Czech Republic
{bures.j,barta.l1,svabensky.o}@fce.vutbr.cz

Abstract. Building constructions are influenced by various external factors, especially by changes in temperature, insolation, humidity etc., and their state is changed, which is manifested, among other things, by geometric deformation. The article shows examples of the influence of external conditions during the monitoring of the shaft of localization of the accidents in the Dukovany Nuclear Power Plant near Brno, Czech Republic, during the verification overpressure test, and during the monitoring of the footbridge over Morava River in Kroměříž within its static securing.

Keywords: Monitoring · External conditions · Bridge · Constructions

1 Introduction

Current commercially available robotic total stations (RTS) enable the measurement of angular and linear quantities with high accuracy, which allows the determination of spatial relationships at sightline lengths up to 100 m, with sub-millimetre accuracy in the 3D position. In the area of measurement of displacements and deformations of building structures, it is often necessary to work at the very limit of the accuracy of the measurement methods used. Building structures are simultaneously exposed to many external influences, in particular to changes in temperature, insolation, humidity, etc. Expansion effects of structures are a typical manifestation.

When high measurement accuracy is achieved, even small deformations of building structures due to external conditions can be detected. The degree of influence due to external conditions is essential in the context of the magnitude of the measured displacements and deformations purposefully induced by the loading of the structure, e.g. during their loading or verification tests. The influence of external conditions can be negligible or insignificant, but it can also exceed the values induced by the intended load of the building structure [1].

External conditions also affect the measuring instruments themselves or measuring systems and it is therefore important to know the extent to which they can be affected and to minimize their adverse impact on the measuring instruments and thus on the measurement results.

A. Kopáčik et al. (Eds.): *Contributions to International Conferences on Engineering Surveying*, SPEES, pp. 223–235, 2021. https://doi.org/10.1007/978-3-030-51953-7_19

The following are two examples of determining the influence of external conditions on structures of different kinds, which were not negligible. The first example deals with the monitoring of the accident localization shaft in the Dukovany Nuclear Power Plant, which was implemented in connection with the process of extending the life of individual units of the nuclear power plant. The second example deals with monitoring of the static securing of the footbridge over the Morava river in Kroměříž by its additional prestressing.

2 Related Works

The influence of external conditions on building structures is solved in many applications in the context of structural health monitoring (SHM). Multi-sensor systems for continuous data monitoring are implemented on the building structure. The knowledge can be obtained by analyzing long-term data collected by methods of regression analysis. E.g. in the framework of structural health monitoring of the Westend Bridge, which is located on the A100 Highway in Berlin, a cluster analysis (k-mean clustering analysis method) was used for the measurement of cracks and deformations [2]. It was thus possible to identify the "turning points" of the structure's behavior, which indicate its annual periodic change. The different bridge states in each year can be explained by temperature changes. The paper analyzes long-term stress changes in prestressed cable and concrete, in longitudinal and transverse slopes in the load-bearing column and in cracks between concrete decks, based on continuous dynamic monitoring of the prestressed bridge over 13 years. Significant annual fluctuations in these variables are observed due to temperature.

The analysis of the time delay between the response of the bridge structure and its thermal load is solved by the phase shift method based on the Fourier series method [3]. Almost every day there is a time lag of varying magnitude, the hysteresis loop area is larger in summer (July) than in winter (November).

Significant influence of external conditions was also observed during long-term monitoring of The Slovak National Uprising Bridge [4]. Verification of strain gauge and geodetic measurements during long-term monitoring of Gagarin bridge in working conditions is solved in [5]. Influence of temperature and humidity during long-term vibration monitoring of RC reinforced concrete structural part (RC slab) was investigated in [6]. The paper investigated the relation between dynamic properties and the environmental factors temperature and humidity. It is found that temperature deteriorates the modulus of elasticity of concrete significantly. Consequently the natural frequencies decrease with temperature increase.

The Sutong Cable-Stayed Bridge (SCB) with a main span of 1088 m over the Yangtze River in Jiangsu province of China, was the longest cable-stayed bridge in the world when it was opened to public traffic. Modeling and forecasting of temperature-induced strain of along-span bridge using an improved Bayesian dynamic linear model was applied here [7].

Direct monitoring provides an important source of information about the actual structural loading and response. This article [8] presents an integral approach to identify fatigue damage of a reinforced-concrete deck as a function of the relevant actions for fatigue using monitoring data. This includes a long-term monitoring system to measure strain and temperature in the most loaded parts, an inverse method using monitoring data to reconstruct traffic actions from the structural response, and a simulation of traffic loading and its effects using a compiler and a finite element model to estimate fatigue damage. The presented approach can be used as a base on how to monitor and analyze recorded data to evaluate the fatigue safety of existing reinforced-concrete slabs in road bridges. For assessment of existing reinforced-concrete bridges, only few rules and recommendations are available, and engineers meanwhile apply design codes for new bridges to evaluate the fatigue safety of existing bridges leading to non-realistic approaches and conclusions. Design codes for new structures are often based on the worst scenarios, and they are not made to assess existing structures with specific loadings and material properties.

3 Monitoring of Temperature Field of the Accident Localization Shaft in the Dukovany Nuclear Power Plant

A part of the process of extending the lifetime of the individual production units of the Dukovany Nuclear Power Plant also included the verification of the integrity of the so-called containment, i.e. the protective envelope. Put simply, the protective envelope is a hermetically enclosed space formed in part by a reinforced concrete accident locating shaft (ALS) that protects the reactor and eliminates the effects in the event of an accident. Dimensions of the ALS building: length 40 m, width 22 m and height 50 m. The subject of the complex monitoring was the accident location shaft construction during the overpressure test, in course of which the object was gradually pressurized up to the pressure of 130 kPa and then back-depressurised. The whole pressure test process lasted 65 h and was supposed to prove the expected strength behavior of the shaft construction, i.e. to confirm the fact of further possible safe operation of the building. During the pressure test, the deformation of selected parts of the accident location shaft (roof and walls) was measured and the expert inspections also visually monitored the condition of the whole building from inside and outside for failure in terms of tightness and mechanical resistance.

The critical value of deformation at the centre of the roof structure of the ALS building was, according to the static model, less than 9 mm and should not be exceeded. The expected deformation value corresponding to the highest pressurisation in course of the pressure verification test should be maximally 4 mm [9].

With low deformation values and with respect to the length of the overpressure verification test including daytime and nighttime external conditions, it is also necessary to determine to what extent the measurement results would be affected only by the external conditions. For this reason, the ALS object was monitored by geodetic polar method with use of robotic measuring stations 48 h before the verification overpressure test. Along with geodetic monitoring, the parameters of external conditions were measured by sensors of air temperature, surface temperature of the monitored structure, relative air humidity, wind direction and speed and amount of precipitation. The contact temperatures of the surface of the monitored structure were determined in its lower and upper part. The quantities listed were continuously registered at 10 min intervals. The temperature field of the outer walls surface of the monitored ALS structure was documented in detail at intervals of 0.5 to 1 h, at 10 height levels, using thermovision technology. The thermovision monitoring lasted a total of 3 × 24 h.

Figure 1 shows a diagram of the monitoring of the temperature field of the ALS object measured by temperature sensors and thermovision in the context of monitoring control points with more types of high precision robotized total stations, e.g. Trimble S8, Trimble S9, Leica MS60, Topcon PS 101. The marked directions of the Cartesian coordinate system axes in which the object was monitored represent the longitudinal, transverse and vertical directions.

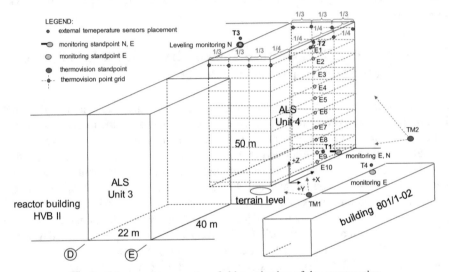

Fig. 1. Scheme of temperature field monitoring of the construction

The graph of Fig. 2 shows the evolution of the measured external air temperatures at different levels and the contact-wise measured surface temperatures of the ALS object. Figure 3 shows the evolution of the corresponding values of relative humidity, wind speed and precipitation.

In Fig. 4 is an example of infrared camera imaging of the monitored object. From the thermovision monitoring, extreme temperature states of the surface of the ALS walls were analyzed at 6:00, 12:00, 16:00 and 23:00 h, with each wall having an extreme temperature state at a different time (Figs. 5, 6).

For an illustrative idea, in the interval 5:30–6:30 on the outer surface of the ALS was the difference in extreme temperatures max–min = 3.4 °C, in the interval 11:30–12:30 max–min = 9.3 °C, in 15:30–16:30 max–min = 11.0 °C and in 22:30–23:30 max–min = 5.3 °C. During the thermovision monitoring period, the extreme surface temperature difference was max–min = 19.1 °C. The main objective of detailed temperature data acquisition was the possibility of subsequent deeper model analysis of the temperature behavior of the monitored structure.

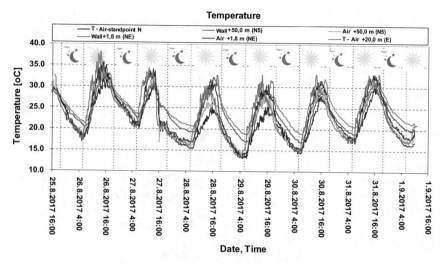

Fig. 2. External temperature evolution

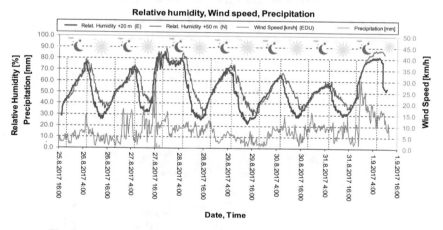

Fig. 3. Course of external relative humidity, wind speed and precipitation

Figure 7 shows the time evolution of geodetically measured deformations in the longitudinal, transverse and vertical directions at the point E1 (Fig. 1) located in the upper part of the front wall of the ALS. Simultaneously, the development of contact-measured wall surface temperature at that point is shown in red. In Table 1, the deformation at E1, E2 (at the top of the wall) and E10 (at the bottom of the wall—Fig. 1) points is numerically indicated.

Geodetically measured values given in Table 1 represent the overall influence of the external conditions reflected in the measurement results, including the component of the influence acting on the meters, the refractive state of the environment and the influence of the external conditions acting on the ALS object. Average values did not exceed 0.5 mm,

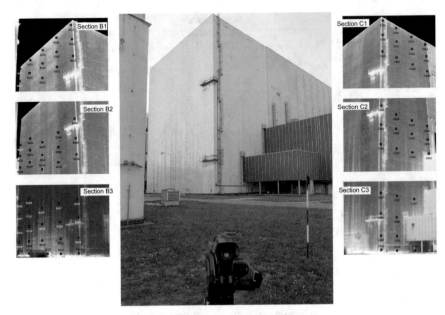

Fig. 4. Imaging of the monitored object by thermovision camera

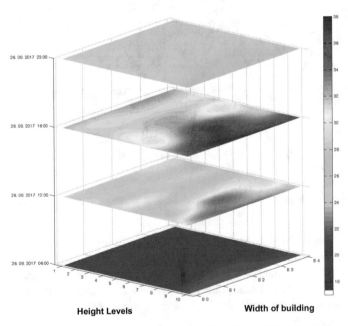

Fig. 5. Temperature hypsometric map of the front wall of the ALS

partial extreme values did not exceed 1.6 mm. The sub-millimetre measurement accuracy is given by the use of robotized total stations with an angular accuracy of 1 " and a length

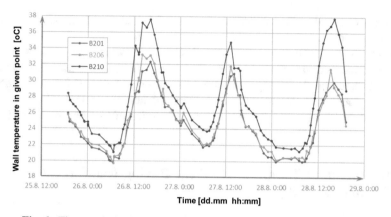

Fig. 6. Time evolution of front wall temperatures at different height levels

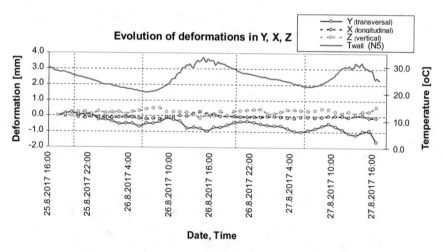

Fig. 7. Evolution of deformations in Y, X, Z at E1 point

Table 1. Average and <Min. Max> values of deformations in 48 h interval before pressure test

Point	ØdY <min;max> (mm)	ØdX <min;max> (mm)	ØdZ <min;max> (mm)
E1	−0,5 <−1,6;+0,2>	0,0 <−0,2;+0,2>	+0,3 <−0,2;+0,6>
E2	−0,3 <−1,2;+0,2>	0,0 <−0,1;+0,2>	+0,1 <−0,2;+0,4>
E10	0,0 <0,0;+0,1>	0,2 <−0,2;+0,5>	0,0 <−0,1;+0,1>

accuracy of $1.5 + 2$ mm with a registration of values at 0.1 mm with a repeatability better than 0.3 mm. High internal accuracy of measured changes is given by short sightlines (max. 50 m), repeatability characteristics of the same measurement of a particular RTS, evaluation of the epoch value from multiple measured values obtained by repeating measurements and the effect of reducing systematic effects in the difference of values from partial epochs.

During the verification test, when the ALS was pressurized at 130 kPa, the measured deviation from the vertical in the upper part of the wall was max. 3.4 mm in the transverse direction dY. Taking into account the average value of 0.5 mm in dY due to external conditions, the proportion of external conditions is 15%. Considering the extreme measured value of 1.6 mm in dY, the proportion of external conditions is up to 47%.

4   Monitoring of Temperature Field During Reinforcement of Concrete Bridge

Bridge structures are also affected by external influences. It can be said that the thinner the bridge deck and the greater the span of the bridge, the more the geometric changes induced by temperature changes and insolation occur. Load tests of large bridge structures are therefore often carried out under night conditions. Nevertheless, during load tests or installation during construction, daily conditions cannot be avoided and the influence of external conditions on the structure is a very unfavorable factor.

In connection with the aging of bridge structures, geodetic methods for measuring the settlement of the substructure and deflection changes are a suitable tool for their complex geometric diagnostics. It turns out that reinforced concrete structures can be unpredictable in terms of their condition, as evidenced by the media coverage of recent years, when, for example, the footbridge in Prague Troja collapsed. IDNES Troja [online] [10], or the tragic event of the collapse of Morandi Bridge in Genoa [11].

Fig. 8. Pedestrian bridge over the Morava river in Kroměříž

One of the possibilities of extending the service life of reinforced concrete bridges is their additional reinforcement by prestressing technologies. This technology was also used in the reinforcement of the footbridge over the Morava River in Kroměříž (Fig. 8) put into operation in 1984, which, as proved by its diagnostics, was damaged by corrosion of supporting ropes. The length of the reinforced concrete deck is 62.8 m, width 3.8 m, thickness 0.3 m. In principle, the reinforcement technology consisted of lining the pre-tensioning ropes under the bridge deck and pre-tensioning them and anchoring them to the bridge supports. Certain risks to this technology consisted essentially in the unknown strength of existing bridge abutments, resp. impossibility to determine the current state of strength of anchoring of abutments to the ground. Therefore, the convergence geodetic monitoring of the mutual length between the bridge supports was part of the realization of prestressing. The subject of the monitoring was also measurement of changes of deflection of the deck in the middle of its span in order to prove the effectiveness of prestressing, which should be manifested by a change of deflection of the deck by 29 mm upwards (see Fig. 9).

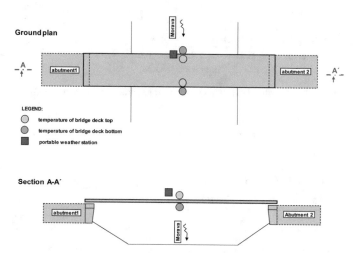

Fig. 9. Scheme of measuring system of external conditions

Because the bridge deck is thin, it is strongly influenced by changes in external conditions, in particular changes in temperature and insolation. Experimental continuous geodetic monitoring of the abutments and deflection of the deck during 24 h, performed before the static securing step by additional prestressing, proved that the changes of the deflection of the deck only due to external influences (temperature, sunlight) reach up to 48 mm (Fig. 10). which was twice the expected change in deflection caused by the post-preload. The main objective of the experimental measurement was to "calibrate" the behavior of the deck when changing external conditions, applicable in the static securing process.

The gradient of the day deflection increase in the heating phase of the bridge deck from solar radiation proved to be up to 6.8 mm at a temperature change of 1° C. Overnight, during the cooling phase of the structure, the deck returned to its original state with a different gradient of 6.3 mm at a temperature change of 1° C (Table 2).

A measuring system of external conditions was installed on the bridge deck in the middle of the span (Fig. 9). The subject of measurement were contact temperatures of concrete surface measured at the top and bottom of the deck. Temperatures were

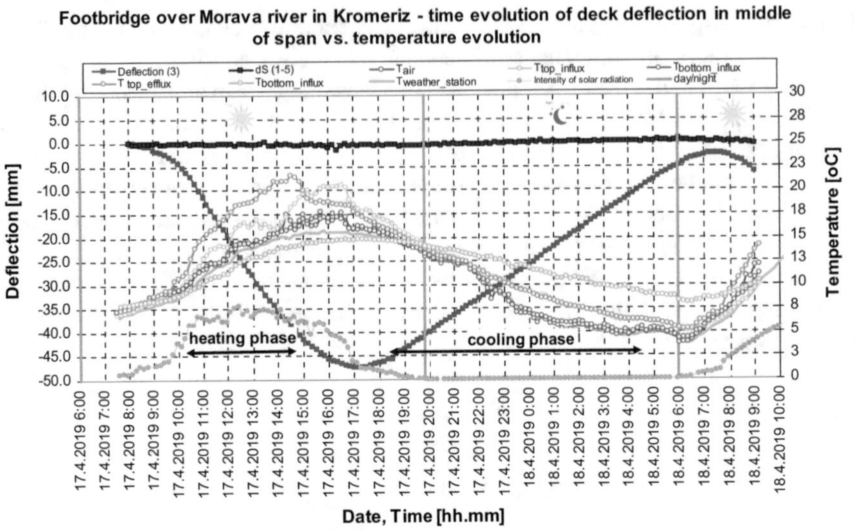

Fig. 10. Time evolution of the deck deflection and measured parameters of external conditions

Table 2. Average, maximum and minimum surface temperatures of concrete deck measured over 24 h

Spring conditions 4/2019 (time 8.30–8.30)			
T_{bet}24 h (average)	T_{bet} Max 24 h	T_{bet} Min 24h	Max–Min
11,5 °C	17.9 °C	5.7 °C	12.2 °C
Deflection gradient in the heating phase dp/dT	6.8 mm/1 °C	Deflection gradient in the cooling phase dp/dT	6.3 mm/1 °C

measured at 10 min intervals with Comet registration thermometers with an accuracy of 0.1 to 0.3 °C. The monitoring of external conditions included a portable weather station installed on the railing of the deck, which recorded air temperature, atmospheric pressure, relative humidity, solar radiation intensity and total rainfall. Geodetically by polar method using robotized total station Topcon GT1001 (angular accuracy 1", distance accuracy 1 mm + 2 ppm with resolution 0,1 mm, repeatability 0,2 mm) were measured the changes in length between supports and changes in deflection of the deck in the middle of the span at intervals of 10 min. The length of sightline did not exceed 60 m. The time evolution of the change in lengths between abutments (dS), deflection of the deck (deflection) and measured parameters of external conditions is shown in Fig. 10.

Figure 11 shows the development of the stability monitoring of the abutments (dS) and the deflection of the deck (deflection) during the additional prestressing in the context of external conditions. During the day conditions changed quite significantly.

The preloading process started in the morning and ended in the evening, with a temperature difference of only +0.7 °C at the end and at the beginning. The change

Time evolution of monitoring of abutments and bridge deck deflection in the context of external conditions

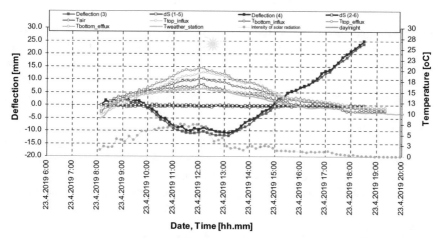

Fig. 11. Time evolution of monitoring of abutments and bridge deck deflection during additional prestressing in the context of external conditions

in deflection after additional prestressing was slightly different for the bridge deck edges (23.8 and 24.9 mm).

After the introduction of the 4.6 mm deflection correction from the temperature change, after taking into account the construction state, the average gradient from the previous calibration measurement was finally used (Table 2) and the resulting temperature corrected deflection was 29 mm (Table 3), which corresponded to the model calculation [12].

Table 3. Result of monitoring after additional prestressing

Day, time	T(bet) average (°C)	dS(5-1) (mm)	dS(6-2) (mm)	dp (3) (mm)	dp (4) (mm)
23.4.2019 8.10	+10.6	0.0	0.0	0.0	0.0
23.4.2019 18.30	+11.3	**−0.5**	**−0.2**	+23.8	+24.9
Temperature change	+0.7	Change in deflection due to temperature change (mm)		4.6	4.6
Corrected deflections due to temperature change (mm)				+28.4	+29.5
Mean value of change of deflection after prestressing (mm)				**+29.0**	

Remark: Sign ± of dS value indicates the distance enlargement/shortening of the bridge abutments, sign ± of dp value indicates a deflection change upward/downward

5   Conclusion

The influence of external conditions is often a less or more complicating to limiting factor in the implementation of testing or securing structures, or in their construction and assembly. The influence of external conditions should be investigated and, if necessary, taken into account by introducing additional correction or by appropriately adding to the uncertainty of the measurement results. The problem is that the temperature gradients of heating and cooling of the structure differ. Corresponding temperature gradients also differ between different days due to the delay (time-lag) of the actual thermal state of the structure. Therefore, the calculated gradients are only applicable in the short term under similar conditions. Taking into account the influence of external influences in the measurement results or their conversion to a uniform temperature state of the structure is a very complicated issue. It is necessary to consider the possibility of applying an inverse problem where, on the basis of geometric changes representing the influence of external conditions, it is possible to judge the possible temperature state of the structure.

Acknowledgements. This article has been prepared with the support of a specific research project FAST-S-18-5324 "Research of real behavior of building structures in operational conditions by geodetic methods".

References

1. WANG, Y.-B., LIAO, P., JIA, Y., ZHAO, R.-D.: Effects of Cyclic Temperature on Time-Dependent Deformation Behavior of Long-Span Concrete Arch Bridge, Bridge Construction, 49 (3), pp. 57–62, (2019).
2. HUL, W. H., SAID S., ROHRMANN, R., CUNHA, Á., TENG, J.: Continuous dynamic monitoring of a prestressed concrete bridge based on strain, inclination and crack measurements over a 14-year span. Structural Health Monitoring 2018, Vol. 17(5) 1073–1094. https://doi.org/10.1177/1475921717735505, (2018).
3. KANG, Y., YOULIANG, D., PENG, S., HANWEI, Z., FANGFANG, G.: Modelling of Temperature Time-Lag Effect for Concrete Box-Girder Bridges. APPLIED SCIENCES-BASEL Volume: 9, Issue: 16, Article Number: 3255, https://doi.org/10.3390/app9163255, (2019).
4. ERDÉLYI, J., KOPÁČIK, A., KYRINOVIČ, P., SOKOL, M., VENGLÁR, M.: Long-term monitoring of the Slovak National Uprising bridge. International Multidisciplinary Scientific GeoConference Surveying Geology and Mining Ecology Management, SGEM, 17 (22), pp. 389–396. https://doi.org/10.5593/sgem2017/22/s09.049, (2017).
5. BURES, J., KLUSACEK, L., NECAS, R., FIXEL, J.: Verification of strain gauge and geodetic measurements during long-term monitoring of gagarin bridge in working conditions. Applied System Innovation - Proceedings of the International Conference on Applied System Innovation, ICASI 2015, pp. 33–37, (2016).
6. XIA, Y., HAO, H., ZANARDO, G., DEEKS, A.: Long term vibration monitoring of an RC slab: temperature and humidity effect. Engineering Structures, 28 (3), pp. 441–452, (2006).
7. WANG, H., ZHANG, Y.M., MAO, J. X., WAN, H. P., TAO, T. Y., ZHU, Q. X.: Modelling and forecasting of temperature-induced strain of a long-span bridge using an improved Bayesian dynamic linear model. Engineering Structures, Volume 192, Pages 220–232, ISSN 0141-0296, https://doi.org/10.1016/j.engstruct.2019.05.006, (2019).

8. BAYANE, I., MANKAR, A., BRÜHWILER, E., SORENSEN, J.D.: Quantification of traffic and temperature effects on the fatigue safety of a reinforced-concrete bridge deck based on monitoring data. Engineering Structures, 196, art. no. 109357, (2019).

9. BURES, J.: Measurement of the accident locating shaft of Unit 4, technical report. Geodetic monitoring and monitoring of external temperature field of the structure. Brno University of Technology. In Czech, (2017).

10. IDNES Troja [on-line], available on-line: https://www.idnes.cz/praha/zpravy/ctyri-zraneni-zamestnanci-prahy-6-praha-trojska-lavka-troja.A171208_145720_praha-zpravy_nuc. In Czech, last accessed 2020/02/18.

11. IDNES Janov [on-line], dostuné on-line: https://www.idnes.cz/zpravy/zahranicni/italie-most-zriceni-janov.A180814_124236_zahranicni_dtt, last accessed 2020/02/18.

12. KLUSACEK, L., NECAS, R.: Construction work for pedestrian bridge no. L07 across the Morava River in Kroměříž. Project documentation, Brno University of Technology. In Czech, (2018).

Periodic Monitoring of the Kostanjek Landslide Using UAV

Ivan Jakopec[1]([✉]) [iD], Ante Marendić[1] [iD], Rinaldo Paar[1] [iD], Igor Grgac[1] [iD],
Hrvoje Tomić[1] [iD], Martin Krkač[2] [iD], and Tomislav Letunić[3] [iD]

[1] Faculty of Geodesy, University of Zagreb, Zagreb, Croatia
{ijakopec,amarendic,rpaar,igrgac,htomic}@geof.unizg.hr
[2] Faculty of Mining, Geology and Petroleum Engineering, University of Zagreb, Zagreb, Croatia
martin.krkac@oblak.rgn.hr
[3] Habitat Geo Ltd., Dubrovnik, Croatia
habitatgeo@yahoo.com

Abstract. The Kostanjek landslide is located in the western residential area of the City of Zagreb. The landslide was activated in 1963 and the main cause of sliding was excavation of the marl at the foot of slope. Since 1976 a several different projects of landslide displacement measurements were done. In the period from 1976 to 1994, geotechnical investigations, geodetic surveys and aerial photogrammetric surveys were carried out in order to determine an engineering-geological model of the Kostanjek landslide. In 2012, real time monitoring system of the landslide was established which consists of 15 GNSS sensors for displacements monitoring. In order to obtain more detailed information about landslide displacements, two UAV surveys (April 2017 and May 2019) of the landslide were carried out. Since landslide velocities have been changing over the last 50 years, ranging from extremely slow to very slow, in this paper we wanted to analyze suitability of UAV for monitoring of slow-moving landslide. The accuracy of UAV survey was examined by comparison of the point coordinates and displacements determined by UAV survey with coordinates and displacements determined by GNSS RTK method. Also, determined displacements were compared with displacements determined by GNSS sensors from monitoring system.

Keywords: Monitoring · Landslide · Displacements · UAV survey

1 Introduction

The Kostanjek landslide with the area of approximately 1 km^2 is the largest landslide in the Republic of Croatia. It is located in the western residential area of the City of Zagreb. According to [1], landslide velocities have been changing over the last 50 years, since the landslide activation until the present day, ranging from extremely slow to very slow according to the classification of [2].

Determination of landslide dynamics is one of the top interests to scientist that study landslides. Since 1976 several different projects of landslide displacement measurements

A. Kopáčik et al. (Eds.): *Contributions to International Conferences on Engineering Surveying*,
SPEES, pp. 236–245, 2021. https://doi.org/10.1007/978-3-030-51953-7_20

were done. In the period from 1976 to 1994, geotechnical investigations, classical geodetic surveys and aerial photogrammetric surveys were carried out in order to determine an engineering-geological model of the Kostanjek landslide [3]. In September 2009, permanent geodetic points were installed at the landslide, and GNSS measurements were carried out in 2009, 2010 and 2012 [4]. Real time monitoring system of the landslide was established in the period from 2011 to 2014 [5].

In order to obtain more detailed information about Kostanjek landslide displacements, in April 2017 and May 2019, UAV surveys of the landslide were carried out. Measurements were made by fixed wing UAV with integrated GNSS RTK receiver. From UAV surveys from different epoch, movements of the points on the landslide surface can be detected by the comparison of orthophotos as well as digital surface models (DSMs). Results and suitability of UAV surveys for landslide monitoring was presented in several studies [6–9].

In this research, we examined the accuracy of UAV survey by comparison of the point coordinates determined by UAV survey with coordinates measured by GNSS RTK method. For that purpose, we used 20 control points (CP) signalized on the landslide that weren't used for georeferencing as a ground control points (GCP). Also, we tested the precision of detected point movements from orthophotos obtained from different dates comparing them with displacements determined by GNSS RTK method. Due to the reason that GCPs in many cases cannot be placed easily or safely [10, 11], georeferencing was done using the camera (without the use of GCPs).

2   History of Displacement Measurements on Kostanjek Landslide

Kostanjek landslide was activated in 1963 and the main cause of sliding were mining activities, i.e., undercutting of slope toe and uncontrolled massive blasting at cement factory "Sloboda". Immediately after beginning of the blasting, the first damage on the factory buildings and other structures, including settlements and cracks has occurred. Excavation at the quarry stopped in 1988 after it was concluded that the marl exploitation was the main trigger of sliding.

Since 1966, many of displacement determination projects have been conducted at the Kostanjek area using various surveying methods. Between 1966 and 1976 [3], vertical displacements were determined at 28 benchmarks installed on industrial buildings of the old factory. In the following years, horizontal and vertical displacements in the factory area (1973–1976) and vertical displacements at 105 points that were set up on private and industrial buildings north of the cement factory (1978–1979) were determined by classical geodetic surveys. The maximum amount of horizontal displacement was 1559 m over a period of 3.5 years and a maximum vertical displacement was 0.56 m.

The horizontal displacements of the ground surface in the period from 1963 to 1988 were determined from the photo interpretation of aerial stereo pairs from 1963 to 1988. Displacements were in a range from 2 to 6 m (average 12–24 cm per year according to [12]).

For the purpose of determining displacements of the landslide Kostanjek, 35 new geodetic points in the landslide area were stabilized and measurements by GPS relative static method were performed in October 2009, March 2010 and February 2012. [4]

compared the displacements determined by the satellite method from 2009 to 2012 with the displacements determined by the photogrammetric method (1963–1988), and conclude that the landslide is still active today and that the displacement directions coincide with the displacement directions from the period from 1963 to 1988. Terrestrial laser scan of part of the Kostanjek landslide was conducted in 2011 and 2013. Analyzing data from two measurement epochs [13], concluded that there were displacements from 13 to 20 cm in the central part of the landslide over a two-year period.

In the period from 2011 to 2014 a real-time monitoring system was installed on Kostanjek landslide. The sensor network encompasses approximately 40 sensors for the monitoring of landslide movement and landslide causal factors [1, 14]. Monitoring system is operational and measurements from installed sensors are analyzed in near–real time at the data acquisition-processing center located at the Faculty of Mining, Geology and Petroleum Engineering, University of Zagreb. Within the monitoring system, 15 GNSS sensors are installed for the displacement measurements (see Fig. 1). The displacements in different parts of landslide, measured by GNSS sensors in period from 2013 to 2014, were from 3.8 to 42.6 cm [15]. The GNSS measurements also revealed that the landslide movement can be divided into the periods of faster movements and periods of slower movements, depending on external triggers, i.e. rainfall [5].

Fig. 1. Observatory Kostanjek (left) GNSS sensor (right)

3   UAV Survey of the Kostanjek Landslide

UAV surveys of the Kostanjek landslide were performed in April 2017 and May 2019. Both flights were performed using a fixed wing senseFly eBee RTK (see Fig. 2 left) URL [16]. SenseFly eBee has integrated RTK GNSS receiver for precise determination of image capturing coordinates in real time (RTK GNSS receiver taking RTK corrections from CROatian POsitioning System—CROPOS). In the first flight, 20 MP images were taken using camera model senseFly S.O.D.A., with RGB 1 inch sensor (fixed focal length 10.6 mm, aperture F/2.8–11, global shutter). In second flight, 24 MP images were taken using camera model senseFly Aeria X, with RGB APS-C sensor (fixed focal length

18.5 mm, aperture F/2.8–16, global shutter). RTK corrected position and orientation of the camera has been automatically recorded for each image.

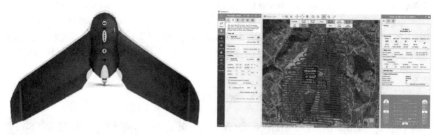

Fig. 2. SenseFly eBee RTK (left), the flight plan in SenseFly eMotion3 software (right)

The flights were planned in senseFly eMotion3 software (see Fig. 2 right). Flight parameters in the first flight were; average height of 160 meters above ground level, forward overlap of 70% and a side overlap of 75%, a total of 577 acquired images with ground sampling distance (GSD) 3.95 cm. Due to changes in legal regulations in the Republic of Croatia average height in the second flight was 70 m above ground level, with forward overlap of 80% and a side overlap of 70%, that resulted in a triple number of collected photos (1530 photos) with 2.61 cm GSD. In the second flight, a higher resolution camera was used, which, in addition to the larger number of photographs collected, resulted in four times denser point cloud. Since the height difference between the highest (north part of the landslide) and lowest terrain point (south part of the landslide) was 115 m (see Fig. 3 right), digital elevation model was used to adjust the UAV altitude during the flight in order to take images from a constant height relative to the ground.

All the flights were performed between 11:30 and 12:30 with good weather conditions (the sky was sunny and clear and wind speed was up to maximal 8 m/s).

Fig. 3. Point cloud (left) and digital elevation model (right)

At the landslide area, 20 CPs were signalized by white circles marked on the asphalt (the darker background). Those points were used for testing accuracy of UAV surveys and their coordinates were determined by GPS RTK method (Trimble R10 GNSS) using CROPOS High Precision Positioning Service—VPPS with registration interval of 30 s in three cycles (achieved 3D precision of 1–2.5 cm).

Processing of the images was done using Pix4D software. Image orientation was performed on direct georeferencing. Point cloud (see Fig. 3 left) was generated with average point density of 45 points per m^3 (first flight) and 142 points per m^3 (second flight). Ortho-mosaic with pixel size of 1 GSD of the whole area are shown on Fig. 4.

4   Achieved Results and Accuracy Analysis

In this paper, results from two UAV surveys (April 2017 and May 2019) are shown. In previous research [17] authors analyzed the accuracy of UAV survey of Kostanjek landslide when the georeferencing was done using only the GNSS RTK corrected camera positions and when the georeferencing was done using the camera positions and 10 GCPs. Although in previous research better accuracy was achieved for approach with the use of GCPs for georeferencing and in this UAV survey we had 20 CPs on the landslide area with measured coordinates, we didn't use those points for georeferencing which was done using the camera positions obtained by GNSS RTK receiver in UAV. Evaluation of the accuracy of UAV survey for landslide displacement measurements, was done by comparing detected point movements from orthophotos created from two UAV surveys and point movements of the same points measured by GNSS RTK method. For the analysis we used 20 signalized control points. Movements of the CPs determined by GNSS RTK method we used as a reference since the achieved 3D precision of point coordinates was 1–2.5 cm. Table 1 shows coordinate differences between coordinates of CPs determined by GNSS RTK method (RTK2017, RTK2019) and from orthophotos (DOF2017, DOF2019) from both UAV surveys.

CP 19 in UAV survey April 2017 wasn't visible on orthophoto. From the results shown in Table 1 we can see that standard deviation of CP coordinate differences are up to 4 cm for both surveys. Also, lower values of coordinate differences along the coordinate axes for the CPs from UAV survey May 2019 were obtained as a result of orthophoto 2019 higher resolution and lower flight height.

We analysed accuracy of detected point movements of 20 CPs from orthophotos by comparing them with displacements determined by the GNSS RTK method since this method provides better accuracy (Table 2).

Displacements of CPs 12, 13 and 17 could not be determined because points were destroyed between two UAV surveys and we had to stabilized new points. Since we could not stabilize the new points in exactly the same place, displacements of those points were not analyzed. Displacements determined by GNSS RTK methods are from 1 to 14 cm which corresponds to the displacements determined by the monitoring system (15 GNSS points—displacements from 0.5 to 16 cm). Displacements determined from orthophotos are in a range from 1 to 20 cm. From Table 2 we can see that differences between displacements determined by GNSS RTK and orthophotos on the few CPs are more than 10 cm. Those differences are results of lower resolution orthophoto from first flight and hardly recognizable image details on orthophotos. Also, displacements of a few points located on the lower part of the landslide (CP 11, 16, 18 and 20) have

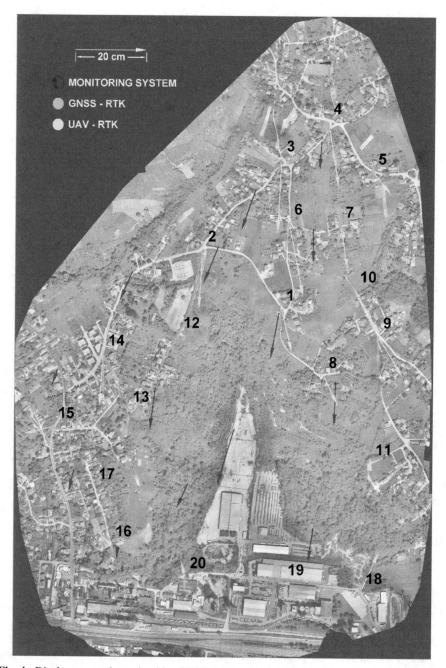

Fig. 4. Displacements determined by GNSS RTK method, GNSS RTK monitoring system and from orthophotos

Table 1 Coordinate differences of CPs for both UAV surveys

CP	RTK2017–DOF2017		RTK2019–DOF2019	
	ΔE [m]	ΔN [m]	ΔE [m]	ΔN [m]
1	0.08	0.02	0.02	0.04
2	0.04	−0.06	0.04	0.02
3	0.09	−0.02	0.04	0.04
4	0.06	−0.06	0.01	0.01
5	0.06	−0.02	0.00	0.02
6	0.09	−0.04	0.05	0.07
7	0.05	−0.01	0.02	0.05
8	0.05	0.01	0.01	−0.05
9	0.07	−0.02	0.03	0.01
10	0.06	−0.02	0.03	0.00
11	0.05	0.06	0.03	−0.02
12	−0.08	0.04	0.01	−0.01
13	0.06	0.03	0.03	−0.02
14	0.00	−0.06	0.03	−0.01
15	−0.02	−0.02	0.04	−0.04
16	0.06	0.02	0.04	−0.05
17	0.02	0.04	0.03	−0.03
18	0.05	0.06	0.02	−0.06
19	–	–	0.02	−0.05
20	0.05	0.03	0.04	−0.06
Min	−0.08	−0.06	0.00	−0.06
Max	0.09	0.06	0.05	0.07
Mean	0.04	0.00	0.03	−0.01
StD	0.04	0.04	0.01	0.04
RMSE	0.06	0.04	0.03	0.04

different directions which can be seen on Fig. 4. The main reason for that is the fact that the orthophoto map is an image obtained by vertical parallel projection and its quality is conditioned by the quality of a digital surface model (DSM) and there may be slight displacements of the projected orthophoto map in areas with steep terrain and big elevation differences.

Figure 4 shows displacements determined by GNSS RTK method and from orthophotos. The blue color shows the displacement vectors determined by GNSS RTK ("true value"), and the yellow color shows displacement vectors determined from orthophotos.

Table 2 Displacements determined by GNSS RTK and from orthophotos

CP	Displacement					
	RTK 2019–2017			DOF 2019–2017		
	ΔE [m]	ΔN [m]	ΔHz [m]	ΔE [m]	ΔN [m]	ΔHz [m]
1	−0.02	−0.11	0.11	0.03	−0.12	0.13
2	−0.02	−0.10	0.10	−0.01	−0.16	0.16
3	−0.04	−0.13	0.14	−0.01	−0.20	0.20
4	0.00	−0.07	0.07	0.03	−0.18	0.18
5	0.00	−0.02	0.02	0.05	−0.06	0.07
6	−0.01	−0.08	0.08	0.04	−0.20	0.20
7	−0.01	−0.06	0.06	0.02	−0.12	0.12
8	−0.01	−0.13	0.13	0.04	−0.08	0.09
9	0.00	−0.08	0.08	0.07	−0.11	0.13
10	0.00	−0.09	0.09	0.07	−0.12	0.14
11	−0.02	−0.09	0.09	0.00	0.01	0.01
12	–	–	–	–	–	–
13	–	–	–	–	–	–
14	0.00	−0.11	0.11	−0.01	−0.14	0.14
15	−0.02	−0.02	0.03	−0.08	−0.02	0.08
16	0.01	−0.03	0.03	−0.02	0.02	0.03
17	–	–	–	–	–	–
18	0.00	−0.01	0.01	0.05	0.11	0.12
19	−0.01	0.00	0.01	–	–	–
20	0.00	−0.04	0.04	−0.01	0.04	0.05
	Min	0.01			Min	0.01
	Max	0.14			Max	0.20

Figure also shows displacements determined by 15 GNSS points from monitoring system determined in period from April 2017 until May 2019 (red color). Position of GNSS points from monitoring system is different than from CPs. The difference between displacements determined from orthophoto and the displacements determined by the GNSS RTK method and by the monitoring system is within the accuracy of the UAV measurement, and the directions of displacements mostly coincide with those determined by the other two methods.

5 Conclusions

In this paper, results of two UAV surveys of Kostanjek landslide have been shown. Flights were performed in April 2017 and May 2019. SenseFly eBee RTK with integrated GNSS RTK receiver was used for UAV surveys. Since, in many cases there are sites where GCPs cannot be placed or measured easily or safely, the achieved accuracy was analyzed without the use of GCPs.

Kostanjek landslide is slow-moving landslide and within the period of two years, monitoring system of 15 GNSS points detected displacements from 0.5 to 16 cm. Evaluation of the accuracy of UAV surveys was done by comparing detected point movements of 20 CPs from orthophotos created from two UAV surveys with movements of the same points measured by GNSS RTK method. Differences between determined coordinates of CPs from orthophotos and coordinates determined by GNSS RTK were up to max 9 cm with a standard deviation of coordinate differences up to 4 cm for both surveys. Better results were obtained for UAV survey May 2019 due to lower flight altitude and resulting higher resolution orthophoto. Displacements determined from orthophotos are in a range from 1 to 20 cm and directions of displacements mostly coincide with displacements determined by GNSS RTK method and by monitoring system. UAV survey proved to be suitable for of the slow-moving landslides and the results of UAV survey indicate that the Kostanjek landslide is still active with defined movement directions.

Acknowledgements. The authors would like to thank Nenad Smolčak (Geomatika Smolčak Ltd.) for his help and technical support during the field measurements.

References

1. Mihalić Arbanas, S., Krkač, M., Bernat, S.: Application of advanced technologies in landslide research in the area of the City of Zagreb (Croatia, Europe). Geologia Croatica: journal of the Croatian Geological Survey 69(2), 231–243 (2016).
2. Cruden, D.M., Varnes, D.J.: Landslide types and processes. In: Landslides: Investigation and Mitigation, Transportation. Research Board Special Report 247, National Research Council, Washington, D.C., 36–75 (1996).
3. Ortolan, Ž.: Development of 3D engineering geological model of deep landslide with multiple sliding surfaces (Example of the Podsused Landslide). Dissertation, University of Zagreb (in Croatian) (1996).
4. Županović, Lj., Opatić, K., Bernat, S.: Određivanje pomaka klizišta Kostanjek relativnom statičkom metodom. Ekscentar, br. 15, 46–53 (2012).
5. Krkač, M., Špoljarić, D., Bernat, S., Mihalić Arbanas, S.: Method for prediction of landslide movements based on random forests. Landslides 14(3), 947–960 (2017).
6. Niethammer, U., Rothmund, S., Joswig, M.: UAV-based remote sensing of the slow-moving landslide Super-Sauze. In: Proceedings of the International Conference on Landslide Processes: From Geomorphologic Mapping to Dynamic Modelling, Strasbourg, France, 69–74 (2009).
7. Lucieer, A., de Jong, S. M., Turner, D.: Mapping landslide displacements using Structure from Motion (SfM) and image correlation of multi-temporal UAV photography, Progress in Physical Geography 38(1), 97–116 (2014).

8. Turner, D., Lucieer, A., de Jong, S. M.: Time Series Analysis of Landslide Dynamics Using an Unmanned Aerial Vehicle (UAV), Remote Sensing 7, 1736–1757 (2015).
9. Pellicani, R., Argentiero, I., Manzari, P., Spilotro, G., Marzo, C., Ermini, R., Apollonio, C.: UAV and Airborne LiDAR Data for Interpreting Kinematic Evolution of Landslide Movements: The Case Study of the Montescaglioso Landslide (Southern Italy), Geosciences 9(6), 248 (2019).
10. Danzi, M., Di Crescenzo, G., Ramondini, M.; Santo, A.: Use of unmanned aerial vehicles (UAVs) for photogrammetric surveys in rockfall instability studies. Rend. Online Soc. Geol. Ital. 24, 82–85 (2013).
11. Forlani, G., Diotri, F., Morra di Cella, U., Roncella R.: Indirect UAV Strip Georeferencing by On-Board GNSS Data under Poor Satellite Coverage. Remote sensing 11, 1765 (2019).
12. Ortolan, Ž., Pleško, J.: Repeated photogrammetric measurements at shaping geotechnical models of multi-layer. Rudarsko-geološko-naftni zbornik 4, 51–58 (1992).
13. Kordić, B.: Development of three-dimensional terrestrial laser scanning method for determining and analyzing of landslide surface movements. Dissertation, University of Zagreb (in Croatian) (2014).
14. Krkač, M., Mihalić Arbanas, S., Nagai, O., Arbanas, Ž., Špehar, K.: The Kostanjek landslide - Monitoring system development and sensor network. In: Landslide and Flood Hazard Assessment, Proceedings of the 1st Regional Symposium on Landslides in the Adriatic-Balkan Region, Zagreb: Hrvatska grupa za klizišta, 27–32 (2014).
15. Krkač, M.; A phenomenological model of the Kostanjek landslide movement based on the landslide monitoring parameters. Dissertation, University of Zagreb (in Croatian) (2015).
16. eBee Classic by sensefly – The Professional Mapping Drone, https://www.sensefly.com/, last accessed 2020/01/31.
17. Marendić, A., Paar, R., Tomić, H., Roić, M., Krkač, M.: Deformation monitoring of Kostanjek landslide in Croatia using multiple sensor networks and UAV, INGEO 2017, 7 th International Conference on Engineering Surveying, Lisbon, Portugal, 203–210 (2017).

GNSS Time Series as a Tool for Seismic Activity Analysis Related to Infrastructure Utilities

Sanja Tucikešić[1]([⊠]), Ankica Milinković[2], Branko Božić[3], Ivana Vasiljević[4], and Mladen Slijepčević[1]

[1] Faculty of Architecture, Civil Engineering and Geodesy, University of Banja Luka, Banja Luka, Bosnia and Herzegovina
sanja.tucikesic@aggf.unibl.org
[2] VEKOM Geo d.o.o., Belgrade, Serbia
[3] Faculty of Civil Engineering, University of Belgrade, Belgrade, Serbia
[4] Faculty of Mining and Geology, University of Belgrade, Belgrade, Serbia

Abstract. GNSS technology tracks the movements of the Earth's crust and its deformation with high accuracy over shorter spatial and temporal periods. Time series of GNSS coordinates are commonly used for geophysical research and have proven to be useful in the research of the cycles of seismic deformations which relate to the whole seismic cycle. The seismic of the Earth's crust could have a great impact on engineering infrastructure. They can damage the objects or cause disaster. GNSS could provide useful information on the deformation of the Earth's crust, which contributes to a better understanding of the occurrence of surface stresses. Earthquake activity is closely related to the dynamics of large tectonic plates. In the article, GNSS time series analysis was used to estimate coseismic displacements with high accuracy and reliability. The GNSS data analysis was related to earthquakes around Durres from January 20th, 2014 to November 28th, 2019, recorded at four permanent GNSS-stations (Ohrid, Dubrovnik, Lecce and Matera). The studied area covers the territory under the influence of the Adriatic microplate which is one of the most important drivers of tectonic processes in the area. The research of interrelation between seismic activity and continuous GNSS measurement could be very useful for earthquake studies and provide good information for designers of engineering projects and maintenance of it.

Keywords: GNSS · Time series · Seismic deformations · Earthquake · Infrastructure utilities

1 Introduction

Earthquake prediction is a hard scientific mission and the main goal of seismology. Specific methods that can predict the place and time of seismic quaver don't exist. One of the methods is to reason interrelationship between seismic events and movements or deformations of the Earth's surface. High-Precision Positioning, today, is an effective tool

© The Editor(s) (if applicable) and The Author(s), under exclusive license
to Springer Nature Switzerland AG 2021
A. Kopáčik et al. (Eds.): *Contributions to International Conferences on Engineering Surveying*,
SPEES, pp. 246–256, 2021. https://doi.org/10.1007/978-3-030-51953-7_21

for capturing movement on the Earth's surface. The three most commonly used space-geodetic techniques are VLBI (Very Long Baseline Interferometry), SLR (Satellite Laser Ranging), and the GNSS (Global Navigation Satellite System). The most significant is researching with high frequency kinematic GNSS- receivers of permanent stations network situated in Japanese GEONET developed at a time of the disastrous magnitude 9 earthquake in 2011. GNSS Time Series allows us to determine movement magnitude and direction of the measured station relative to other locations or reference frames.

Networks of GNSS sensors can track three-dimensional movements of the ground surface even at rates of less than 1 mm per year. It is evident that the spatial vector reflects the movements of GNSS stations under the influence of the tectonic forces. Time series of GNSS stations can describe other processes such as hydrologic loading, volcanic inflation, etc. Time series of GNSS stations are divided into three individual graphs that describe the East, North, and Up components of station motion.

Detecting crustal deformation on various timescales is essential for understanding the cycle of an earthquake event. Measurement of coseismic deformations is of great importance for earthquake studies. Coseismic deformations occur in the vicinity of earthquake epicenters. The detection of coseismic deformations is important information for engineers to improve earthquake resistance for buildings and other structures.

This study aims to evaluate how different data sources can be used and integrated for seismic hazard assessment of an area.

2   Data Sets

Tectonics of the Mediterranean Sea, between Africa and Eurasia, it is complex, and involve the motions of numerous microplates and regional-scale structures. Simplified tectonic map of the Albania region presented in Fig. 1. Albania is a Balkan country with a high rate of seismicity. The west coast of Albania is a convergent plate boundary and there is a fold and thrust belt. Adria plate is diving beneath the Eurasia plate to form the Adriatic collision zone. The convergence here forms the Hellenides Mountains. Albania is geologically and seismo-tectonically a rather complicated region. It is characterized by obvious microseismicity (a high number of small earthquakes), sparse medium-sized earthquakes (magnitude M 5.5–5.9), strong earthquakes (magnitude M 6.0–6.4) and rare large earthquakes (magnitude M ≥ 6.5).

The 2019 Durres, Albania Earthquake involved a series of inland earthquakes beginning on April 23rd of 2019 along the active fault zone on the eastern shores of the Adriatic. Over the past 100 years, seven earthquake M 6 and larger events have occurred within 150 km.

Recent examples of earthquakes in Albania have shown that resonance of soil and structures can significantly increase earthquake damage, and even buildings designed as seismically resistant have suffered more damage than expected when resonance occurred. Knowledge of the long-term seismic stress velocities is critical to determining the seismic hazard of an area.

We will first describe the different data we have used: geodetic data (Time-series GNSS) and seismic data around Durres (7 earthquakes magnitude higher than 4.5).

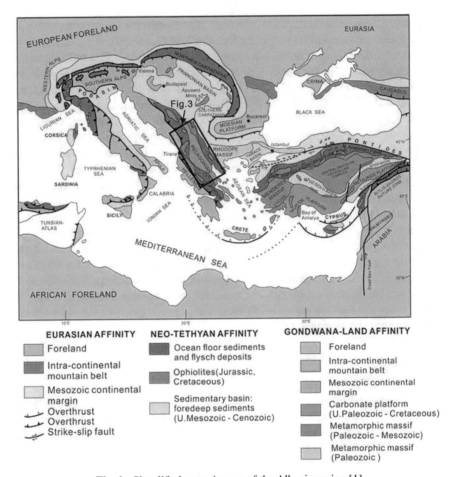

Fig. 1. Simplified tectonic map of the Albania region [1]

2.1 Time Series GNNS—Geodetic Data

For this research, we used time-series data from the station ORID (Ohrid), DUBR (Dubrovnik), MATE (Matera) and USAL (Lecce), Fig. 2.

We processed data from 4 continuously recording GNSS for the period from January 2014 to December 2019. GNSS stations are located at distances of the ~150 km from the epicenter of the Durres earthquake, of which the ORID station is the closest. Daily GNSS time series data from these stations in the IGS14 reference frame are available from the Nevada Geodetic Laboratory at http://geodesy.unr.edu/NGLStationPages/Glo balStationList/. Every week, NGL updates the daily position coordinates of some 10,000 stations. Every day, we update 5-min position coordinates for more than 5000 stations. Every hour, we update 5-min position coordinates for about 2000 stations. These lower-latency products have proved to be useful, for example, for gaining early insight into large earthquakes by measuring permanent coseismic displacements and postseismic

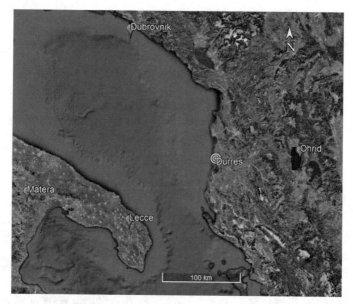

Fig. 2. Spatial layout of the GNSS stations

displacements caused by fault after-slip and upper mantle relaxation [2]. An overview of the downloaded and used time series of GNSS coordinates is shown in Table 1.

Table 1. Time series of GNSS coordinates

Station	N (°)	E (°)	Period	Location
ORID	41.127	20.794	2014-01-01 to 2019-12-07	Ohrid (North Macedonian)
DUBR	42.650	18.110	2014-01-01 to 2019-12-14	Dubrovnik (Croatian)
MATE	40.649	16.704	2014-01-01 to 2019-12-07	Matera (Italian)
USAL	40.335	18.111	2014-01-01 to 2019-12-07	Lecce (Italian)

Stations were installed to meet specific geophysical research criteria based on geographic location and feature standardized instrument configurations (e.g., receiver type and antenna type), monumentation, metadata, and data flow. Used GNSS stations are Class A stations (positions with 1 cm accuracy at all epochs of the period of the used observations). Coseismic deformation is investigated using these stations.

By looking at data from a GNSS stations over a period of time, we can determine whether the ground surface has moved (deformed). By combining the data collected from a GNSS network, it is possible to get a larger view of which areas of the surface are moving as well as the speed and direction of movement.

2.2 Seismic Data

Albania lies across the convergent boundary between the Eurasian Plate and the Adriatic Plate, as a part of the complex collision zone with the African Plate. The region is seismically active, with several M ≥ 6 earthquakes in the last hundred years. In 1979, the largest of these events struck 70 km (43 mi) further north, in Montenegro, killing 136 people (101 in Montenegro and 35 in Albania). The most recent significant earthquake in the area was an M 6.4 event on 26 November 2019 with an epicenter WSW of Mamurras,

Table 2. Earthquakes used for determined coseismic movement

Time	N (°)	E (°)	Depth	Mag	Place
2019-11-28 (23:00:43)	41.37	19.58	10	4.6	2 km NNE of Shijak, Albania
2019-11-28 (10:52:42)	41.56	19.41	10	4.9	23 km W of Mamurras, Albania
2019-11-28 (10:25:05)	41.53	19.43	10	4.5	22 km WSW of Mamurras, Albania
2019-11-27 (20:27:06)	41.54	19.44	10	4.5	21 km WSW of Mamurras, Albania
2019-11-27 (14:45:24)	41.54	19.47	13	5.3	19 km WSW of Mamurras, Albania
2019-11-26 (17:19:13)	41.55	19.48	10	4.7	17 km W of Mamurras, Albania
2019-11-26 (13:05:00)	41.59	19.52	10	4.7	14 km W of Mamurras, Albania
2019-11-26 (07:27:02)	41.54	19.44	10	4.8	21 km WSW of Mamurras, Albania
2019-11-26 (06:08:22)	41.58	19.43	10	5.4	22 km W of Mamurras, Albania
2019-11-26 (03:02:59)	41.42	19.64	10	5.3	3 km NNW of Vore, Albania
2019-11-26 (02:59:23)	41.42	19.53	10	5.1	8 km NNW of Shijak, Albania
2019-11-26 (02:54:12)	**41.51**	**19.52**	**20**	**6.4**	**15 km WSW of Mamurras, Albania**
2019-11-26 (01:47:55)	41.35	19.5	10	4.6	5 km ENE of Durres, Albania
2019-09-21 (22:07:30)	41.32	19.53	10	4.8	3 km SW of Shijak, Albania
2019-09-21 (14:15:52)	41.47	19.54	14	5.1	12 km NW of Vore, Albania
2019-09-21 (14:04:25)	41.34	19.53	20	5.6	3 km WSW of Shijak, Albania
2019-04-23 (08:58:14)	41.55	19.52	10	4.5	14 km WSW of Mamurras, Albania
2018-07-04 (09:01:09)	41.45	19.58	22	5.1	8 km NW of Vore, Albania
2014-01-20 (07:15:07)	41.38	19.58	23	4.5	3 km NNE of Shijak, Albania

Bold values in table signifies earthquake with the largest magnitude which was at the time the most powerful in 30 years and damaged 500 houses. JavaScript Object Notation (JSON) text format was used for processing of all earthquakes on our study. JSON offset file is created based on the data downloaded from USGS (U.S. Geological Survey) at https://earthquake.usgs.gov/earthquakes/search/ for earthquakes with magnitudes higher than 4.5, Table 2 and Fig. 3.

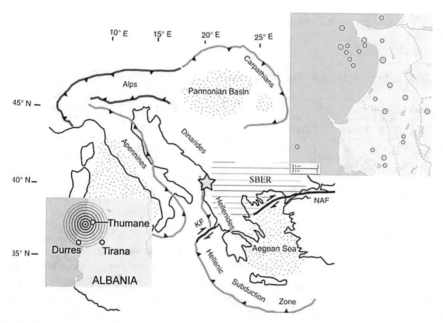

Fig. 3. Map of the region showing the compression, major earthquake (2019-11-26) and earthquakes used for the determined coseismic movement

Earthquake induces coseismic deformations in the Earth's crust and may generate large abrupt changes in the position of geodetic GNSS stations. The stress in the Earth's crust induced by plate tectonics usually releases along faults close to plate boundaries or within the deforming area.

The first earthquake analyzed had a 4.5-moment magnitude scale (M) and occurred on 20 January 2014, at an approximate depth of 23 and 3 km NNE of Shijak, Albania. After this earthquake four years later, an earthquake of 5.1-moment magnitude scale (M) occurred on 4 July 2018, at an approximate depth of 22 and 8 km NW of Vore, Albania. These two earthquakes were followed by a series of earthquakes in 2019, at an approximate depth of 10 km.

The system showed in Fig. 3 today is within the southern Balkan region north of the North Anatolian fault (NAF), shown by the horizontal line pattern. Retreating subduction zones and related backarc extensional areas for the Mediterranean region are shown in blue, and advancing subduction zones a related area of backarc shortening are shown in red). Backarc extensional regions are shown by a dotted pattern. KF = Kefalonia fault zone [3].

3 Results of Numerical Research

The estimation of a GNSS velocity field consists of several phases. First, the accuracy of the GNSS time series is degraded by the presence of offsets. The outliers were eliminated using a robust outlier detection algorithm model (Eq. 1) based on the median and interquartile range (IQR) statistics [4]:

$$\left| \hat{v}_i - median(\hat{v}_{i-w/2}, \hat{v}_{i+w/2}) \right| > 3 \cdot IQR(\hat{v}_{i-w/2}, \hat{v}_{i+w/2}) \tag{1}$$

Table 3. Presented data analysis (epoch and gaps after elimination of outliers using Eq. 1)

Station	Epoch	Gap (%)	Epoch after eliminated outliers	Gap after eliminated outliers (%)
DUBR	2039	6.20	2005	7.70
ORID	2153	0.60	2097	3.20
MATE	2165	0.00	2128	1.80
USAL	2162	0.20	2130	1.70

If the offsets were retained they could dominate in the velocity estimation quality. Presented data analysis, after the elimination of outliers, is shown in Table 3.

After analyzing the time series of individual stations to find out discontinuities and to exclude periods of bad quality data, final velocity estimation can be done by combining all daily solutions using least-squares methods. The mathematical model (Eq. 2) used in analyzing the coordinate components of the GNSS daily time series was given, as [5]:

$$y(t_i) = a + \sum_{i=1}^{n} b_i(t_i - t_0)^i + \sum_{i=1}^{n_p}(c_i \sin(2\pi t_i/p_i) + d_i \cos(2\pi t_i/p_i)) + \sum_{i=1}^{n_e}(t_{e_i}) \tag{2}$$

Our research focuses on the southern part of Europe. We have determined the rate of tectonic displacement of this area, Table 4. For all stations, variations are almost perfectly linear for the horizontal components that are east and north, and this clearly shows the tectonic motion, Fig. 4. As a result of the trend analyses of time series, it was determined that stations were moving in the northeast direction 29.96 mm/year, 26.73 mm/year, 30.85 mm/year and 30.08 mm/year, for DUBR, ORID, MATE and USAL, respectively. This finding is consistent with the region's tectonic plate movements.

Using the latest GNSS measurements, we can obtain geodetic stress velocities and a deformation model for the study area. The geodetic deformation field is representative of the 5-year time series of long-term tectonic deformation.

When processing available earthquakes, the obtained coseismic velocities do not show any significant changes in the time series of GNSS coordinates at the stations DUBR, MATE and USAL. However, GNSS station ORID shows significant changes in the time series of GNSS coordinates by component elevation (Up) to 18.74 mm for earthquakes of 4.6-moment magnitude scale (M) from November 28, 2019, at an

Table 4. Plate motion based on GNSS time series

Station	North linear (mm/year)	East linear (mm/year)	Up linear (mm/year)
DUBR	18.31 ± 0.02	23.72 ± 0.02	−0.94 ± 0.06
ORID	11.90 ± 0.02	23.93 ± 0.02	0.38 ± 0.06
MATE	18.37 ± 0.02	24.79 ± 0.02	−0.49 ± 0.05
USAL	18.69 ± 0.02	23.57 ± 0.02	−1.05 ± 0.05

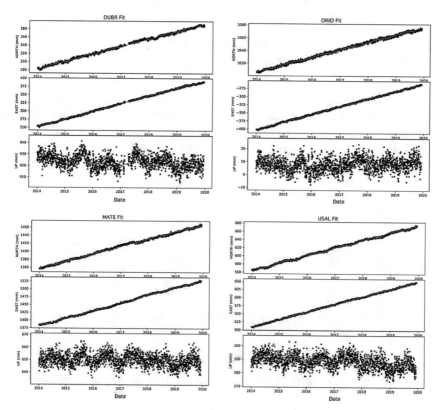

Fig. 4. Detrending daily observations of the sites DUBR, ORID, MATE and USAL including determining semi-annual and annual signals

approximate depth of 10 and 2 km NNE of Shijak, Albania. Time series ORID with all earthquakes from 2014 to 2019 magnitude ~4.5 and earthquake from November 28 2019 shown in Fig. 5. An event was observed that a series of earthquakes in one day causes a greater coseismic velocity than one earthquake. Ohrid station is the closest to the epicenter of the earthquake. So the greatest coseismic velocity is justified. In this study, we see that each earthquake generates coseismic deformations in the region surrounding its epicenter.

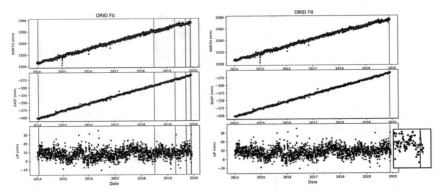

Fig. 5. Time series ORID: with all earthquakes from 2014 to 2019 magnitude ~4.5 (left) earthquake from November 28, 2019 (right)

The first earthquake analyzed on 20 January 2014 obtained coseismic velocities that show significant changes in the time series of GNSS coordinates at the stations ORID by component elevation (Up) to 11.47 mm. Changes in height component of GNSS stations ORID, induced by earthquake occurrence are shown in Fig. 6.

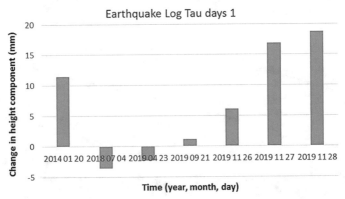

Fig. 6. Change in height component of GNSS stations ORID, induced by earthquake occurrence

Using the methods of spectral analysis in the frequency domain the spectral indices of post fit residuals were estimated. With the Lomb–Scargle algorithm, the periodogram of post fit residuals was also calculated for each position components [6, 7]. The values of spectral indices determined for all stations have a range that corresponds to the fractional Gaussian noise, Fig. 7, which is stationary, uncorrelated with time, and has statistical properties that are invariant over time [8]. These results indicate that used stations Ohrid, Dubrovnik, Lecce, and Matera, respectively are rather stable and reliable.

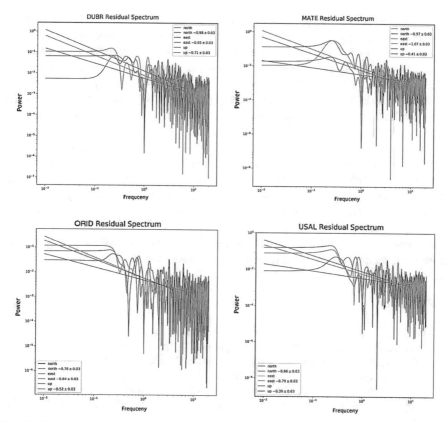

Fig. 7. The Lomb-Scargle periodogram and the estimated spectral indices for GNSS station DUBR, ORID, MATE and USAL

4   Conclusions

The GNSS stations can partially replace the seismograph. The GNSS stations cannot detect high-frequency seismic vibration and they are less sensitive than the seismograph. However, the GNSS station network may have another aspect, which is to look for a link between the slow displacement of the Earth's surface and the generated seismic parameters. Seismology and geodesy in combination provide a new and unique view of the process over the full spectrum of displacements of interest to seismologists and engineers and valuable information that can augment existing EEW systems for earthquakes.

Based on time series for four GNSS points, coseismic velocities for a total of 7 earthquakes in Albania from 2014 to 2019 were determined. In this study, we see that each earthquake generates coseismic deformations in the region surrounding its epicenter. Analyzed real coseismic deformations on the day of the earthquake at station ORID shows significant changes in the time series of GNSS coordinates by a component of elevation. The relationship between vertical movements and seismicity was established long ago and is confirmed by modern technologies, especially InSAR [9]. Much attention is paid to establishing the connection between modern vertical movements and the forces that cause them. One of these forces is the seismic force. Using global satellite geodesy, it is possible not only to obtain high-quality information on modern geodynamics but also to observe all its changes in space and location, which is especially important for the needs of proper seismic zoning and finding relationships in geodynamic processes.

An event was observed that a series of earthquakes in one day causes a greater coseismic velocity than one earthquake, namely, coseismic displacements accumulated during the last earthquake is not only due to large earthquakes but also to the accumulation of many small motions induced by smaller earthquakes.

Earthquake damage cannot be completely avoided. We do not have information about building standards that are prevalent in Albania and to what extent they are respected, but materials for building are important. Materials need to combine stiffness with ductility (bend without breaking). Brittle materials, such as unreinforced masonry or concrete blocks are the most dangerous materials during earthquakes. The priority is to avoid the complete collapse and collapse of the building. Historic buildings could be further strengthened and secured. This is, however, much more expensive than when you build an object from the start so that it is as safe as possible from an earthquake.

References

1. Fahui, X., Jingsui,Y., Paul, T. R., Yildirim, Di., Ibrahim, M., Xiangzhen, X., Yanhong, C., Wenda, Z., Zhongming, Z., Shengming, L., Yazhou, T., Zhu, H..: Petrology and geochemistry of high Cr# podiform chromitites of Bulqiza, Eastern Mirdita Ophiolite (EMO), Albania. Ore Geology Reviews. Vol. 70, pp. 188–207. (2015)
2. Blewitt, G., Hammond, W. C., Kreemer, C.: Harnessing the GPS data explosion for interdisciplinary science, Eos, 99 https://doi.org/10.1029/2018EO104623. Published on 24 September 2018.
3. Burchfiel, B.C., Nakov, R., Dumurdzanov, N., Papanikolaou, D., Tzankov, T. Serafimovski, T., King, R.W., Kotzev, V., Todosov, A., Nurce, B.: Evolution and dynamics of the Cenozoic tectonics of the South Balkan extensional system. Geosphere. Vol. 4, pp. 919–938. (2008)
4. Nikolaidis, M.: Observation of geodetic and seismic deformation with the Global Positioning System (Ph.D. Thesis), University of California, San Diego, San Diego. (2002)
5. Bevis, M., Bedford, J., Caccamise, II D.J. The Art and Science of Trajectory Modelling. In: Montillet JP., Bos M. (eds) Geodetic Time Series Analysis in Earth Sciences. Springer Geophysics. Springer, Cham, pp 1–27 (2020)
6. Scargle, J.D.: Studies in astronomical time series analysis. III-Fourier transforms, autocorrelation functions, and crosscorrelation functions of unevenly spaced data. The Astrophysical Journal. Vol. 343, pp. 874–887. (1989)
7. Schultz, M. Stattegger, K. Spectrum: spectral analysis of unevenly spaced paleoclimatic time series. Computer & Geosciences. Vol. 23 (9), pp. 929–945. (1997)
8. Goudarzi, M. A., Cocard, M., Santerre, R. Noise behavior in CGPS position time series: the eastern North America case study. Journal of Geodetic Science. Vol. 5 (1), pp. 119–147. (2015)
9. Jung, H.-S., Hong, S.-M. Mapping three-dimensional surface deformation caused by the 2010 Haiti earthquake using advanced satellite radar interferometry. PLoS One. Vol. 12 (11). (2017)

Application of Inertial Sensors, GNSS, Navigation

Analysis of the Yaw Observability in GNSS-INS Integration for UAV Pose Estimation

Gilles Teodori[(✉)] and Hans Neuner

TU Wien, Wiedner Hauptstraße 8-10, 1040 Vienna, Austria
{gilles.teodori,hans.neuner}@tuwien.ac.at

Abstract. The quality of pose estimation of unmanned aerial vehicles (UAVs) via the integration of Inertial Navigation Systems (INS) and Global Navigation Satellite Systems (GNSS) is primarily affected by sensor specific errors and flight manoeuvres. Especially the yaw estimation is subject to larger drifts. As part of this paper, a detailed analysis of the latter influence factor is carried out with the aim of improving the yaw estimation within an error state-space Kalman filter. The results indicate that the observability of the yaw error states increases during manoeuvring phases with horizontal acceleration components. As observability refers to the ability of estimating states, such manoeuvring can reduce the drift of the yaw angle significantly.

Keywords: INS/GNSS-Integration · Observability · UAV

1 Introduction

The ability to collect bathymetric data from UAV-based mapping systems leads to significant cost and time savings in the land surveying industry. The possibility to georeference the acquired data via ground control points is not feasible over wider water surfaces. Hence, the attitude, velocity and position determination of the UAV relies primarily on the onboard sensors and their integration. The standard navigation equipment of UAVs consists of Inertial Navigation Systems (INS) and Global Navigation Satellite Systems (GNSS). The loosely coupled integration architecture is commonly used due to its simplicity and its flexible usage in navigation applications. This integration architecture is generally based on an error-state Kalman filter.

The quality of attitude estimation and in particular that of the yaw angle is strongly affected by sensor errors and vehicle dynamics. This is because attitude errors are observed through the impact they have on velocity and position errors [1]. In case of low vehicle dynamics, the yaw error has a small impact on the velocity and the position error in comparison to the roll and the pitch error. Therefore, the yaw error is called a weak observable state and the yaw angle tends to drift from its true counterpart.

The observability of states in an INS/GNSS system can be analysed using an analytic, a data-driven approach or a combination thereof [2–4]. The term observability defines thereby the possibility of estimating states of a system from its external outputs

© The Editor(s) (if applicable) and The Author(s), under exclusive license
to Springer Nature Switzerland AG 2021
A. Kopáčik et al. (Eds.): *Contributions to International Conferences on Engineering Surveying*,
SPEES, pp. 259–269, 2021. https://doi.org/10.1007/978-3-030-51953-7_22

[5]. The analytic approach usually incorporates a theoretical statement of whether a state is observable or not [6]. In this approach it is common to neglect any system and measurement noise [7]. The data-driven approach uses data from computational simulations or from field experiments to verify the theoretical statements under more realistic conditions. The present paper has opted for the data-driven approach in order to outline the effects of noise sources on the attitude estimates. Independent of the chosen approach, there is common agreement that manoeuvring is beneficial regarding the observability of certain states, such as the yaw error state.

Note that the problem of yaw observability becomes less important where additional measurements are taken into account, which directly measure the yaw angle. Eling et al. [8, 9] were able to provide a precise attitude solution from a combined use of inertial sensors, magnetic field sensors and onboard GPS baselines. The magnetic field sensors as well as the GPS baselines provide absolute yaw information, which adds supplementary support to the yaw estimation process. Nonetheless, the presence of magnetic disturbance fields (magnetically active materials, electrical lines) can lead to large systematic errors in the outputs of magnetic field sensors. The yaw accuracy from onboard GPS baselines is strongly dependent on the baseline lengths. Accordingly, the dimension of the host vehicle must accommodate the required baseline length, which limits the downsizing tendencies of mobile mapping systems.

Even though the yaw observability has been addressed in many contributions, no one to the best of our knowledge has given an insight in the attitude estimation process. The present paper aims, in the first place, to validate the observability improvements of yaw error states from translational motion based on simulation studies. Moreover, it is visualised how manoeuvres (e.g. different types of accelerations) are passed through the filter mechanisms and contribute—or not—to the yaw error's observability. The simulation results clearly show a decrease regarding the standard deviation of the yaw error state as acceleration changes are performed.

This paper is divided into five sections. Section 2 describes the used INS/GNSS integration architecture. Section 3 introduces the flight scenario of the UAV. In Sect. 4 the results of the simulation studies are presented and the discussion of these will take place in Sect. 5. The final section draws a conclusion and elaborates on future works.

2   Loosely Coupled INS/GNSS-Integration

The inertial navigation system (INS) produces a navigation solution based on an initial or a previous navigation solution and via the integration of gyro and accelerometer measurements. This navigation solution comprise the system's orientation, velocity and position resolved about the axes of the local navigation frame. Due to initialization errors, inertial measurement errors and processing approximations the forward propagation of the inertial navigation solution undergoes some distinctive errors.

These attitude, velocity and position errors are expressed by the error state vector (Table 1 for symbology):

$$\delta x^n = \left[\delta\phi_{nb}^n,\ \delta\theta_{nb}^n,\ \delta\psi_{nb}^n,\ \delta v_{eb,N}^n,\ \delta v_{eb,E}^n,\ \delta v_{eb,D}^n,\ \delta\varphi_b,\ \delta\lambda_b,\ \delta h_b \right]^T, \tag{1}$$

Table 1 Symbology

	Sub/Superscript	Description
Coordinate frames	i, e	Earth-centered inertial and earth-fixed
	b, n	Body and local (level) navigation
Filter phase	−, +	Predicted and updated quantities
	k	Actual filter epoch
Axes alignment	NED	North, East and down components
Submatrices	a, v, p	Attitude, velocity and position

where $\delta\phi^n_{nb}$, $\delta\theta^n_{nb}$ and $\delta\psi^n_{nb}$ are the roll, pitch and yaw errors, $\delta v^n_{eb,N}$, $\delta v^n_{eb,E}$ and $\delta v^n_{eb,D}$ represent the velocity errors in north, east and down direction, and $\delta\varphi_b$, $\delta\lambda_b$ and δh_b indicate the latitude, longitude and height errors. The dynamic evolution of these errors represents the system model in the opted error state-space Kalman filter (ESS-KF). The presentation of the time derivative of the error states is highly space consuming. Therefore, and in view of simplicity reasons, only the propagation of the velocity error due to an attitude error is specified, as this is required for the discussion section of this work. The complete derivation of the error states dynamics may be found in [10]. The approximated coupling of attitude errors into velocity errors equals [1]

$$\delta\dot{v}^n_{eb} \approx \left[\delta\boldsymbol{\Psi}^n_{nb}\wedge\right]\left(\boldsymbol{C}^n_b f^b_{ib}\right), \tag{2}$$

where $\delta\dot{v}^n_{eb}$ denotes the time derivative of the velocity errors and $\delta\boldsymbol{\Psi}^n_{nb}\wedge$ represents the skew-symmetric matrix of the attitude errors. The last term indicates the true specific-force measurements f^b_{ib} rotated from the b-frame to the n-frame by the true transformation matrix $\boldsymbol{C}^n_b$. The main objective of this Kalman filter is to estimate the navigation errors from GNSS position and velocity solutions. For an ESS-KF, the input metrics (measurement innovation) consist of the difference between INS and GNSS position and velocity solutions. As no lever arm is assumed, the measurement innovation results in [1]

$$\delta z^-_k = \begin{bmatrix} \boldsymbol{p}_{b,GNSS} - \hat{\boldsymbol{p}}^-_{b,INS} \\ \boldsymbol{v}^n_{eb,GNSS} - \hat{\boldsymbol{v}}^{n,-}_{eb,INS} \end{bmatrix}_k, \tag{3}$$

it follows that the measurement matrix is given by

$$\boldsymbol{H} = \begin{bmatrix} \boldsymbol{0}_{3x3} & \boldsymbol{0}_{3x3} & -\boldsymbol{I}_{3x3} \\ \boldsymbol{0}_{3x3} & -\boldsymbol{I}_{3x3} & \boldsymbol{0}_{3x3} \end{bmatrix}. \tag{4}$$

The estimation of the error state vector is performed following [11] by using the gain matrix $\boldsymbol{K}_k$ and the measurement innovation

$$\delta\hat{\boldsymbol{x}}_k = \boldsymbol{K}_k \delta z^-_k \tag{5}$$

with

$$\boldsymbol{K}_k = \boldsymbol{P}^-_k \boldsymbol{H}^T \left(\boldsymbol{H}\boldsymbol{P}^-_k \boldsymbol{H}^T + \boldsymbol{R}\right)^{-1}, \tag{6}$$

where R denotes the measurement noise covariance matrix and is assumed to be constant over time. The predicted error covariance matrix P_k^- results from the time-discrete representation of the system model, the so-called transition matrix T_{k-1} and from the system noise covariance matrix Q_{k-1}

$$P_k^- = T_{k-1} P_{k-1}^+ T_{k-1}^T + Q_{k-1}. \tag{7}$$

Using the symbology from Table 1, the submatrix representation of P_k^- is

$$P_k^- = \begin{bmatrix} P_{a,a}^- & P_{a,v}^- & P_{a,p}^- \\ P_{v,a}^- & P_{v,v}^- & P_{v,p}^- \\ P_{p,a}^- & P_{p,v}^- & P_{p,p}^- \end{bmatrix}_k. \tag{8}$$

Therefore, out multiplying Eq. (6) with the measurement matrix from (4) gives

$$\begin{bmatrix} K_{a,p} & K_{a,v} \\ K_{v,p} & K_{v,v} \\ K_{p,p} & K_{p,v} \end{bmatrix}_k = - \begin{bmatrix} P_{a,p}^- & P_{a,v}^- \\ P_{v,p}^- & P_{v,v}^- \\ P_{p,p}^- & P_{p,v}^- \end{bmatrix}_k \left(\begin{bmatrix} P_{p,p}^- & P_{p,v}^- \\ P_{v,p}^- & P_{v,v}^- \end{bmatrix}_k + \begin{bmatrix} R_{p,p} & 0_{3x3} \\ 0_{3x3} & R_{v,v} \end{bmatrix} \right)^{-1}. \tag{9}$$

The update of the error covariance matrix equals

$$P_k^+ = (I - K_k H) P_k^-. \tag{10}$$

Finally, the integrated navigation solution is formed by adding the estimated navigation errors to the inertial navigation solution

$$\hat{x}_k^+ = \hat{x}_k^- + \delta \hat{x}_k. \tag{11}$$

This integrated navigation solution is then utilized for the upcoming computation of the inertial navigation solution at epoch $k + 1$.

3   Flight Scenario

The propagation of INS errors depends on the vehicle dynamics [10]. Therefore, a wide variety of flight manoeuvres has to be analyse in order to fully understand the behaviour of the attitude estimation in a loosely coupled INS/GNSS-integration. Initial studies were performed using common flight phases, including stationary, constant velocity, constant acceleration and time-varying acceleration phases. Furthermore, different travel directions (north and east) as well as 90° turning and circular motions have been investigated. Their examination revealed similarities in terms of attitude estimation. For the sake of simplicity, the simulation studies are based only on one trajectory, which however summarize the essential features regarding the yaw error observability. The trajectory incorporates planar translational motions in north direction with phases of zero, constant and linear time-varying acceleration. Figure 1 illustrates the time history of these manoeuvres. Note that the vehicle's initial north velocity is 5 ms^{-1} and it remains level during the whole flight time. Furthermore, the UAV travels in north direction with a 0° yaw angle. During the period from 20 to 40 s and the period from 60 to 80 s, the vehicle's acceleration changes at a rate of 0.05 ms^{-3} from 0 to 1 ms^{-2} and vice versa.

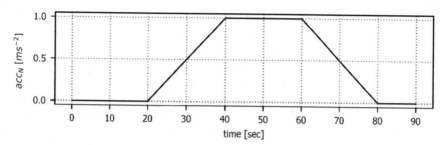

Fig. 1 Time history of the UAVs northern acceleration

4 Simulation Results

In Kalman filtering, it is assumed that the system and measurement noise have a zero-mean symmetric distribution. For that reason, the true IMU (Inertial Measurement Unit) and GNSS measurements are distorted by Gaussian white noise sequences of the following noise densities and standard deviations (STD), respectively:

- gyro: 0.01 $°s^{-1}$ $(\sqrt{Hz})^{-1}$; accelerometer: 60 μg $(\sqrt{Hz})^{-1}$,
- position: 0.05 m; velocity: 0.01 ms^{-1}.

The IMU provides outputs at the rate of 200 Hz. The GNSS output rate is fixed to 1 Hz, which also indicates the measurement-update rate of the Kalman filter. The system noise covariance matrix Q_{k-1} and the measurement noise covariance matrix R are set using the values of the afore-mentioned list. The initial diagonal error covariance matrix P_0 is obtained using an attitude STD of 0.3° and using the velocity and position STD of the GNSS measurements.

As indicated in the introduction, the focus of this work lies on the attitude estimation. Furthermore, the description of the simulation results will be limited about the estimation of the roll and yaw errors. This is because similarities can be noted between the estimation process of roll and pitch errors. To leave complete statements, these similarities will be pointed out in Sect. 5 of this work. The simulation results shown below are organized as follows: Firstly, a closer look is taken at the time histories of correlations between attitude and velocity error states. These correlations play a major role in the computation of the gain matrix (see Eq. (9)). Hence, the weighting of the measured velocity innovation is displayed underneath the correlation figures. Secondly, the outcomes of the error state and error covariance update are displayed for both studied attitude quantities. Finally, the residuals in the integrated roll and yaw estimates and the results from the Monte Carlo (MC) simulation are presented. One possible use of MC simulation is as a quantitative measure for the uncertainty in estimates. The term itself already indicates that this sampling technique involves a computer-based method, where the inputs are distorted repetitively by a chosen distribution and passed through the filter algorithm. The resulting large set of filter outputs, also called samples, allows quantifying the output distribution. In this contribution, a MC simulation was carried out, because the yaw residuals of one simulation run were regarded as not being representative for assessing possible biases of the estimation. Therefore, the simulation of the trajectory was repeated 1000 times and the mean of the sample was compared to the true value. Furthermore, the estimation of epoch-wise standard deviations from the 1000 samples may validate conclusions drawn from the estimated covariance matrix in the filter.

Figure 2a shows that acceleration changes provoke the building of correlations between the yaw and the eastern velocity error states. However, this cannot be observed during phases of zero or constant acceleration. The algebraic sign of these correlations is paired to those of the acceleration changes. No substantial correlation changes are observable between the yaw error and the velocity error states of both other directions. The high-frequency fluctuations of the correlations is due to the prediction and the update steps of the filter mechanism. Comparing Fig. 2a, b, it becomes recognizable that the correlations dominate the weighting of the eastern velocity innovation. The weights of the velocity innovation in north and down direction are de facto zero during the whole flight time.

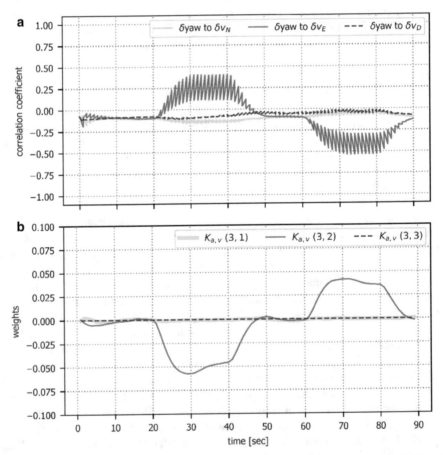

Fig. 2 a Correlations between yaw and velocity error states; **b** weighting of the measured velocity innovation (3rd row of $K_{a,v}$)

The temporal progression of the correlations between the roll and the velocity error states are noticeable different in comparison to those of the yaw error state (see Figs. 3a and 2a). First, a rapid increase of the correlations with the eastern velocity error is perceptible, which then varies within -1 and -0.5 during the manoeuvring. The manoeuvring tends also to correlate the roll to the velocity error in north and down direction. The weighting of the eastern velocity innovation decreases from about 0.9 to reaches equilibrium around 0.2 in the period of the first five seconds (see Fig. 3b). The velocity innovation in north and down direction are weighted with zero over the flight time.

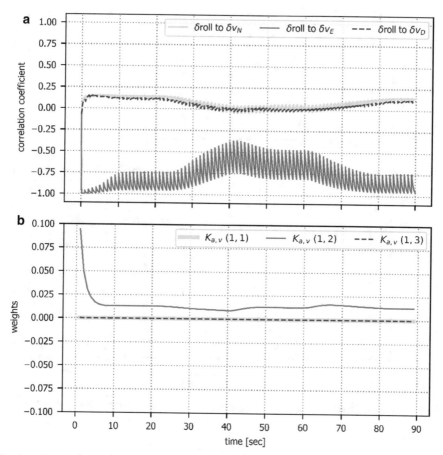

Fig. 3 a Correlations between roll and velocity error states; **b** weighting of the measured velocity innovation (1st row of $K_{a,v}$)

The error state estimation and the error covariance update involve the gain matrix (see Eqs. (5) and (10)). As a result, the estimated yaw and roll errors as well as their respective STD, reflect what the previous figures suggest (see Fig. 4a, b). The yaw error state is only then observable and therefore estimable, when acceleration changes are performed. During such manoeuvres, the STD of the yaw error decreases, whereas in flight phases of zero or constant accelerations, the STD of the yaw error lightly increases. In opposition, the roll error state is estimable from the beginning as demonstrated by the significant size of its first estimate and by the fast convergence of its STD. During the manoeuvring however, the STD of the roll error state increases on a small-scale.

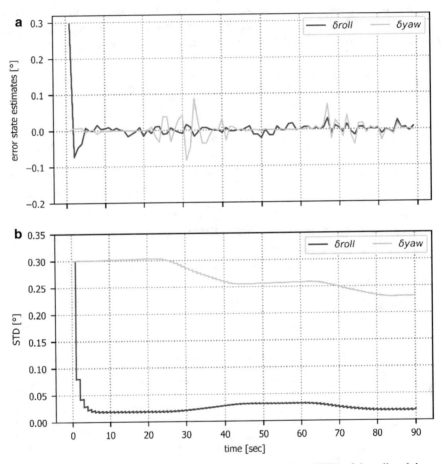

Fig. 4 **a** Estimated roll and yaw error states; **b** standard deviations (STD) of the roll and the yaw error states

In contrast, the outcomes of Fig. 5 slightly differ from the findings of Fig. 4a, b about the intended yaw angle improvements. Even if the yaw residuals tend to decrease over periods of acceleration changes, a clear approval cannot be finalised from Fig. 5. The empirical sample mean and sample STD of the yaw residuals are shown in comparison to the estimated STD (see Fig. 6). Both STD are hardly distinguishable and the sample mean is very close to its expected value of 0°. These results emphasize the validity that acceleration changes increase the accuracy of yaw error estimates and thus minimise the yaw angle drift.

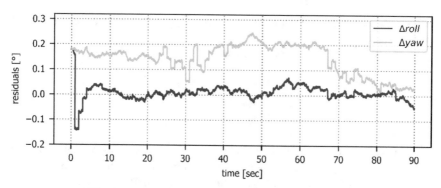

Fig. 5 Residuals in the integrated roll and yaw estimates

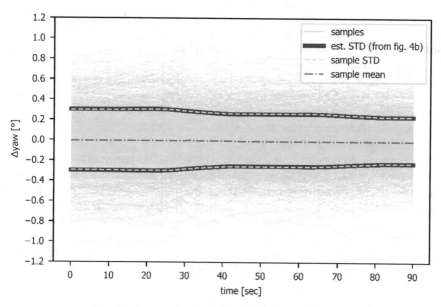

Fig. 6 Monte Carlo simulation (1000 simulation runs)

5   Discussion

The conducted simulation studies have demonstrated the importance of acceleration changes for the observability of the yaw error state. Consequently, the STD of the yaw error states has been significantly reduced in periods of such manoeuvres. The visualisation of this mechanism provides additional support to previous research.

One downside regarding our methodology is that a wide variety of flight manoeuvres has to be studied in order to fully understand the behaviour of the attitude estimation. Rest on our simulation studies, we have been able to identify the acceleration changes as the required condition to estimate attitude error states. However, it has to be pointed out that this is contradictory to explanations given by [1], who argues based on the integrated version of Eq. (2).

In fact, rewriting this equation yields in

$$\delta v_{eb}^n = \begin{bmatrix} \delta\psi_{nb}f_{ib,y}^n - \delta\theta_{nb}f_{ib,z}^n \\ \delta\phi_{nb}f_{ib,z}^n - \delta\psi_{nb}f_{ib,x}^n \\ \delta\theta_{nb}f_{ib,x}^n - \delta\phi_{nb}f_{ib,y}^n \end{bmatrix} t. \tag{12}$$

This equation reflects the so-called coupling of attitude errors into velocity errors and is the reason for the composition of correlations between those error states (see Figs. 2a and 3a). Note that only accelerations in the horizontal plane are able to couple yaw errors into velocity errors. During stationary and constant velocity periods, the yaw error state cannot be observed through this specific coupling. As a result of noisy IMU measurements, the yaw angle performs a random walk and its STD increases during these periods (see Figs. 5 and 4b). The roll and pitch errors contribute to the east and north velocity error, respectively, via the measured reaction to gravity. Due to the fact that gravity is omnipresent and relatively large compared to the vehicle's accelerations, the roll and pitch errors are strong observable states. Regarding the observability of attitude errors, it can be stated that they are observed through the growth in the velocity and position error they produce [1]. It is worth mentioning, without going into excessive detail, that the attitude errors are also coupled with the position errors. This can be seen by the double-time integration of Eq. (2).

Returning to the contradiction, which comprises the fact that our simulation studies claims that horizontal acceleration changes are needed to observe the yaw error state, but the underlying system model requires exclusively horizontal accelerations. It is very likely that this contradictory issue arise from the simulated system and measurement noise. Noisy inputs and/or measurements can mask the effect of weak observable states such as the yaw error state [12]. One possible way to take into account these noise sources is to analyse the stochastic observability of INS/GNSS systems [13]. Further investigations on this subject are currently in progress.

6 Conclusion and Future Works

This study gives an insight into the attitude estimation process in a loosely coupled INS/GNSS integration. The conducted simulation studies support the fact that manoeuvring is beneficial for the observability of the yaw error state. Based on the chosen medium-grade IMU, this manoeuvring presumably consist of horizontal acceleration changes. Clearly, this work has some limitation regarding the investigated manoeuvres and the general statements about the observability of states in a loosely coupled INS/GNSS integration. Despite this, we believe that within this framework we should be able to not only describe the effects of certain manoeuvres, but also allow predictions of the INS/GNSS system performance based on sensor characteristics and planed UAV trajectories. The interpretation of future simulation results would undoubtedly benefit from an accompanying theoretical formulation. In this perspective, a considerable effort is planned in the near future.

References

1. P. Groves, *Principles of GNSS, Inertial, and Multisensor Integrated Navigation Systems*, 2nd ed. Artech House, 2013.
2. Y. Tang, Y. Wu, M. Wu, W. Wu, X. Hu, and L. Shen, "INS/GPS integration: Global observability analysis," *Veh. Technol. IEEE Trans. On*, vol. 58, pp. 1129–1142, 2009, https://doi.org/10.1109/tvt.2008.926213.
3. Y. Wu, H. Zhang, M. Wu, X. Hu, and D. Hu, "Observability of Strapdown INS Alignment: A Global Perspective," *IEEE Trans. Aerosp. Electron. Syst. - IEEE TRANS AEROSP ELECTRON SY*, vol. 48, 2011, https://doi.org/10.1109/taes.2012.6129622.
4. S. Hong, M. Lee, H. Chun, S.-H. Kwon, and J. Speyer, "Observability of error States in GPS/INS integration," *Veh. Technol. IEEE Trans. On*, vol. 54, pp. 731–743, 2005, https://doi.org/10.1109/tvt.2004.841540.
5. Y. Yoo, J. Park, D. Lee, and C. Park, "A Theoretical Approach to Observability Analysis of the SDINS/GPS in Maneuvering with Horizontal Constant Velocity," *Int. J. Control Autom. Syst.*, vol. 10, 2012, https://doi.org/10.1007/s12555-012-0210-2.
6. F. M. Ham and R. Z. Brown, "Observability, Eigenvalues, and Kalman Filtering," *IEEE Trans. Aerosp. Electron. Syst.*, vol. AES-19, pp. 269–273, 1983.
7. Y. Li, X. Niu, Y. Cheng, C. Shi, and N. El-Sheimy, "The Impact of Vehicle Maneuvers on the Attitude Estimation of GNSS/ INS for Mobile Mapping," *J. Appl. Geod.*, vol. 9, pp. 183–197, 2015, https://doi.org/10.1515/jag-2015-0002.
8. C. Eling, P. Zeimetz, and H. Kuhlmann, "Development of an instantaneous GNSS/MEMS attitude determination system," *GPS Solut.*, vol. 17, 2012, https://doi.org/10.1007/s10291-012-0266-8.
9. C. Eling, L. Klingbeil, M. Wieland, and H. Kuhlmann, "A Precise Position and Attitude Determination System for Lightweight Unmanned Aerial Vehicles," *Int Arch Photogramm Remote Sens Spat. Inf Sci*, vol. XL-1/W2, pp. 113–118, 2013, https://doi.org/10.5194/isprsarchives-xl-1-w2-113-2013.
10. M. S. Grewal, A. P. Andrews, and C. G. Bartone, *Global navigation satellite systems, inertial navigation, and integration*, Third Edition. John Wiley & Sons, 2013.
11. D.-J. Jwo and T.-S. Cho, "Critical remarks on the linearised and extended Kalman filters with geodetic navigation examples," *Measurement*, vol. 43, pp. 1077–1089, 2010, https://doi.org/10.1016/j.measurement.2010.05.008.
12. D. Gebre-Egziabher and S. Gleason, *GNSS Applications and Methods*. Artech House, 2009.
13. M. H. Lee, S. Hong, J. H. Moon, and H.-H. Chun, "On the observability and estimability analysis of the Global Positioning System (GPS) and Inertial Navigation System (INS)," pp. 35–38, 2009.

Adaptive Kalman Filter for IMU and Optical Incremental Sensor Fusion

Pavol Kajánek$^{(\boxtimes)}$, Alojz Kopáčik , Ján Erdélyi , and Peter Kyrinovič

Faculty of Civil Engineering, Slovak University of Technology, Radlinského 11, 810 05 Bratislava, Slovakia
pavol.kajanek@stuba.sk

Abstract. The main problem of the use of inertial measurement unit (IMU) lies in the systematic increase of position and orientation errors resulting from their functional principle. Position and orientation calculation based on the integration of inertial measurements (accelerations and angular velocities) leads to a rapid accumulation of errors of the inertial sensor with increasing time of measurement. Existing solutions for the elimination of these errors focus on a combination of IMU with several additional sensors to increase long-term positioning and orientation accuracy. The article deals with the elimination of systematic errors of IMU, based on optical incremental sensor measurements and conditions resulting from the actual kinematic state of IMU. Incremental encoder measurements are used for the correction of speed and travelled distance. The proposed processing model uses the automatic identification of the kinematic state of the IMU, which is realized based on inertial measurements. Based on the current kinematic state of IMU, the process model has been modified, which allows a better way of modelling and eliminating systematic error. The verification of the efficiency of the designed model is realized by experimental measurements. During experimental measurements, the measuring system (IMU and optical encoder) is installed on the trolley. The predefined trajectory of the trolley's movement is given by the known position of reference points, based on which the differences in distance, the orientation differences, the positional differences and their development with increasing time of measurement are analyzed.

Keywords: Inertial measurement unit IMU · Optical incremental encoder · Systematic error · Kinematic state of the system

1 Introduction

Inertial measuring unit (IMU) has a wide range of applications from low-cost devices for ordinary users (application in sports, gaming and healthcare) to complex navigation and mapping systems. Many complex mapping systems used IMU because of their benefit as independence from external signals and the ability to measure 3D motion with high sampling frequency [1]. Determination of position based on inertial measurements is affected by systematic errors of inertial sensors that rapidly accumulate in position

© The Editor(s) (if applicable) and The Author(s), under exclusive license
to Springer Nature Switzerland AG 2021
A. Kopáčik et al. (Eds.): *Contributions to International Conferences on Engineering Surveying*,
SPEES, pp. 270–282, 2021. https://doi.org/10.1007/978-3-030-51953-7_23

and orientation error with increasing time of measurement. Accumulation of systematic errors results from the functional principle of IMU, which is based on the integration of inertial measurements.

The most common ways of eliminating the systematic error of inertial sensors are the fusion of IMU with other sensors, in which the position, speed and orientation of the system is corrected at regular intervals depending on the type of sensor, which is used. In outdoor areas, a combination of IMU and GNSS is most commonly used [2–4]. In indoor areas, there are used additional sensors as cameras (computer vision based on a pair of cameras with a common base) [5, 6], odometer [7], magnetometer [8], lidar [9] and others.

The next important method for the elimination of errors is based on the use of kinematic states of the IMU. In most cases, only the zero velocity update (ZUPT) is mainly used to eliminate the errors of accelerometers at each time when the system is stopped. However, this method has a greater benefit when IMU is placed on the human body. When IMU is placed on a vehicle or trolley, corrections are generated after long time intervals during which the sensor error is difficult to model.

The article deals with the inertial measuring unit IMU CPT, which belongs to the tactical grade and is usually used for the mobile mapping system where the combination of IMU and GNSS is assumed. The article describes the design of the measuring system for indoor areas, which combines IMU with an optical incremental encoder. For data processing, the Kalman filter was designed, which can adapt the process model to the current kinematic state of the system. The algorithm for automatic kinematic state identification was designed too. This algorithm automatically identifies one of the four kinematic models based on inertial measurements.

2 Design of the Measuring System

The proposed measuring system use components of Novatel SPAN CPT system (IMU CPT and datalogger ProPack 6), which is designed for outdoor applications. The developed measuring system is designed for indoor areas, where the GNSS signal is not available and therefore, an optical incremental sensor (optical encoder) was used instead of using GNSS, to producing additional information about the system movement. The optical encoder measures the incremental velocity of the system, which is used to reduce the systematic errors of accelerometers.

The measuring system is placed on a transport trolley, which consists of a steel plate (dimension 910×610 mm) on which four inflatable rubber wheels are placed. The inflatable wheels were chosen because they allow shock absorption and easier to overcome floor irregularities (e.g. door sills). On the rear axle, there are wheels with a movable pulley for maneuvering of the trolley and on the front axle there are wheels with a fixed pulley. The IMU is located centrally above the front wheel axle. Characteristics of used inertial sensors are shown in Table 1. The orientation of the coordinate system of IMU (body frame) relative to the reference coordinate system is shown in Fig. 1. The sampling frequency for inertial measurements is 100 Hz, and data are logging to the receiver (datalogger) ProPack 6. Both the IMU and the receiver ProPack 6 are powered from 12 V battery. The optical encoder LPD3806 is located on the front left-hand wheel.

The sampling frequency for incremental measurements is 20 Hz. The size of one increment corresponds to the shift of the trolley by 0.2 mm. The optical encoder is controlled by the microcontroller Arduino UNO. The power supply of the microcontroller and the incremental sensor is realized from the control PC.

Table 1. Characteristic of inertial sensors from IMU CPT [10]

Sensor	Accelerometer	Gyroscope
Type	Micro electromechanical sensors (MEMS)	Fibre optic gyroscope (FOG)
Bias	± 50 mg (-41 mg, 11 mg, 38 mg)	$20°/h$
Scale factor	1500 ppm (1036 ppm, 535 ppm, 1300 ppm)	4000 ppm
Bias instability	0.25 mg (0.62 mg, 0.63 mg, 0.52 mg)	$1°/h$ ($0.23°/h$, $1.09°/h$, $0.43°/h$)
Noise	55 $\mu g/\sqrt{Hz}$ (47 $\mu g/\sqrt{Hz}$, 48 $\mu g/\sqrt{Hz}$, 20 $\mu g/\sqrt{Hz}$)	$0.0667°/\sqrt{h}$ ($0.0298°/\sqrt{h}$, $0.0307°/\sqrt{h}$, $0.0305°/\sqrt{h}$)

Note parameters in parentheses were determined based on self-testing (multi-position test and Allan variance). The order of the parameters corresponds to the X, Y, Z axes of the coordinate system of IMU

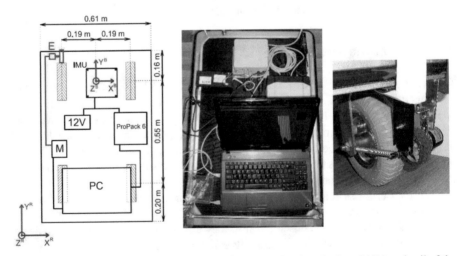

Fig. 1. Measurement system: **a** scheme (on the left), **b** nadir view (in the middle), **c** detail of the optical encoder mounted on the left-hand wheel (on the right)

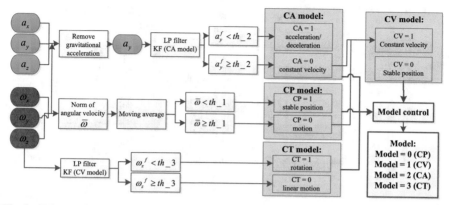

Fig. 2. Scheme of the proposed algorithm for automated identification of kinematic models based on inertial measurements

3 Data Processing

The aim of the data processing is to estimate parameters, which describe the state of the measuring system and its development over time. For this purpose, the Kalman filter was designed. The use of the Kalman filter can be justified by the fact that it is possible to combine individual approaches to eliminate systematic errors of IMU. Based on a combination of multi-sensor measurements an optimal estimator of the system state is determined, where the weight of each measurement is based on the accuracy of measurement relative to the accuracy of the element of the state vector, which is defined by the state covariance matrix. The next advantage is the possibility to combine multiple kinematic models within one processing model. During the motion of the measuring system, its kinematic state changes and the processing model should be adapted to this change.

Four kinematic models (Fig. 3) can be identified within the trolley movement, which can be referred as a constant position model (CP model), constant velocity model (CV model), constant acceleration model (CA model) and constant turn model (CT model) [11]. Before identifying the model, the measured acceleration is transformed from the coordinate system of IMU to a reference coordinate system and after that is the gravitational acceleration eliminated. Finally, accelerations are back-transformed to the coordinate system of IMU. The rotation matrix, which defines the change of orientation between the coordinate systems is determined based on gyroscope measurements. The identification of kinematic models is realized automatically based on inertial measurements and the proposed algorithm (Fig. 2).

The constant position model is identified at the moment when the measuring system is stopped. At this moment, the processing model used ZUPT to correct the calculated velocity and update the bias of inertial sensors. Norm of accelerations or norm of angular velocity can be used to identify this model. In our case, the norm of the angular velocity was used because the bias instability of gyroscopes is smaller than the bias instability of accelerometers.

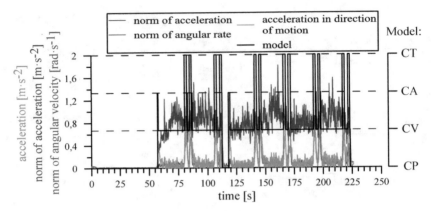

Fig. 3. Identification of kinematic states of the trolley based on inertial measurements

The constant acceleration model (CA model) is identified based on accelerations measured in the direction of the travel of the trolley. The CA model is identified during the acceleration or deceleration of the system.

The constant velocity model is identified when the system is in motion (CP = 1) and at the same time the constant acceleration model is not identified (CA = 0).

The constant turn model is identified based on the angular velocity measured in the direction of the sensitive axis of the gyroscope, which is parallel to the axis of rotation of the trolley. This model is identified when the azimuth of trolley changes.

The thresholds th_1, th_2, th_3 used for automatic identification of kinematic models are determined based on the variance of the corresponding measurements when the system is in a stable position. The threshold th_1 was empirically determined as three times the variance of the norm of angular velocity, which was calculated based on the angular velocity at the start of the measurement when the system is in a stable position. The threshold th_2 was set as 5 times the variance of acceleration measured in the direction of Z axis. A higher coefficient was used because the noise of acceleration is higher than the noise of angular velocity. The threshold th_3 was set as 3 times the variance of angular velocity measured in the direction of Z axis.

In the first step of the Kalman filter (Fig. 4), elements of the state vector $\hat{\boldsymbol{\Theta}}_t^-$ (3D position, velocity, and acceleration) are predicted. The prediction in equations is marked with the symbol "−" in the upper index. For up-date is used the symbol "+". The prediction of the state vector $\hat{\boldsymbol{\Theta}}_t^-$ and its state covariance matrix $\boldsymbol{\Sigma}_{\hat{\boldsymbol{\Theta}},t}^-$ are realized (1, 2) based on an aposterior estimation of the state vector $\hat{\boldsymbol{\Theta}}_{t-1}^+$ and state covariance matrix $\boldsymbol{\Sigma}_{\hat{\boldsymbol{\Theta}},t-1}^+$ from the previous epoch. The transformation matrix $\mathbf{F}_i$ is a Jacobian matrix, which defines the change of elements of the state vector and covariance matrix in time. The vector of control inputs $\mathbf{u}_t$ is calculated based on the measured accelerations and angular velocities, which are used to calculate elements of the state vector. The matrix $\mathbf{B}_i$ is a Jacobian matrix of control inputs. The uncertainty of the control inputs is defined by the matrix $\mathbf{w}_i$. The matrix $\boldsymbol{\Sigma}\varepsilon_i$ defines process noise. The index i represents the identifier of the current kinematic model identified by the above-defined algorithm. The transition between the kinematic models changes the size of the matrices, e.g. transition from the CA model to the CV model, resulting in that the accelerations are predicted as zero and any changes of velocity are modelled as noise. The dimension of matrices can increase too (e.g. transition from the CV model to CT model), where the model estimates the acceleration value.

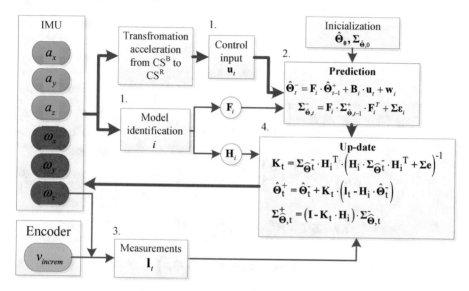

Fig. 4. Scheme of the Kalman filter used

$$\hat{\boldsymbol{\Theta}}_t^- = \mathbf{F}_i \cdot \hat{\boldsymbol{\Theta}}_{t-1}^+ + \mathbf{B}_i \cdot \mathbf{u}_t + \mathbf{w}_i \tag{1}$$

$$\boldsymbol{\Sigma}_{\hat{\boldsymbol{\Theta}},t}^- = \mathbf{F}_i \cdot \boldsymbol{\Sigma}_{\hat{\boldsymbol{\Theta}},t-1}^+ \cdot \mathbf{F}_i^T + \boldsymbol{\Sigma}\varepsilon_i \tag{2}$$

Transformation matrices $\mathbf{F}_i$ describing the development of state vector elements (in a single axis direction). For each model can be defined as follows:

$$\mathbf{F} = \begin{bmatrix} f & 0 & 0 \\ 0 & f & 0 \\ 0 & 0 & f \end{bmatrix}, \ \mathbf{f_{CP}} = [1], \ \mathbf{f_{CV}} = \begin{bmatrix} 1 & dt \\ 0 & 1 \end{bmatrix}, \ \mathbf{f_{CA}} = \begin{bmatrix} 1 & dt & 0.5 \cdot dt^2 \\ 0 & 1 & dt \\ 0 & 0 & 1 \end{bmatrix}$$

$$\mathbf{f_{CT}} = \begin{bmatrix} 1 & \frac{\sin(\omega \cdot dt)}{\omega} & \frac{1-\cos(\omega \cdot dt)}{\omega^2} \\ 0 & \cos(\omega \cdot dt) & \frac{\sin(\omega \cdot dt)}{\omega} \\ 0 & -\omega \cdot \sin(\omega \cdot dt) & \cos(\omega \cdot dt) \end{bmatrix} \tag{3}$$

The initial values of the state vector are defined based on the initial position $[X_0, Y_0, Z_0]$, initial velocity $[0, 0, 0]$, and initial acceleration $[0, 0, 0]$.

The state vector $\hat{\boldsymbol{\Theta}}_{t-1}^+$ is updated based on measurements. In our case, the measurement vector $\mathbf{l}_t$ consists of the velocity of the trolley expressed in the reference coordinate system (5). These elements of the measurement vector were calculated from incremental velocity v_{increm} and angular rate ω_z. Incremental velocity was defined based on measured increments inc and the time interval Δt between measurements (4). The relation between the state vector and the measurement vector is defined by the measurement matrix H_i(6). The weight of innovation is defined by Kalman gain K_t (7), which is calculated from the noise of measurements Σe and the covariance matrix of the predicted state vector $\Sigma_{\hat{\Theta}_t}^-$.

The accuracy of the updated state vector $\hat{\Theta}_t^+$ (8) is defined by the covariance matrix $\Sigma_{\hat{\Theta}_{t-1}}^+$ (9):

$$v_{increm_t} = (inc_t - inc_{t-1}) \cdot \Delta t \tag{4}$$

$$\mathbf{l}_t = \begin{bmatrix} v_{increm_t} \cdot \cos(\omega_{zt} \cdot dt) \\ v_{increm_t} \cdot \sin(\omega_{zt} \cdot dt) \end{bmatrix}. \tag{5}$$

$$\mathbf{H} = \begin{bmatrix} h & 0 \\ 0 & h \end{bmatrix}, \quad \mathbf{h}_{CV} = \begin{bmatrix} 0 \\ 1 \end{bmatrix}^T, \quad \mathbf{h}_{CA} = \begin{bmatrix} 0 \\ 1 \\ 0 \end{bmatrix}^T, \quad \mathbf{h}_{CT} = \begin{bmatrix} 0 \\ 1 \\ 0 \end{bmatrix}^T \tag{6}$$

$$\mathbf{K}_t = \Sigma_{\hat{\Theta}_t}^- \cdot \mathbf{H}_i^T \cdot \left(\mathbf{H}_i \cdot \Sigma_{\hat{\Theta}_t}^- \cdot \mathbf{H}_i^T + \Sigma e \right)^{-1} \tag{7}$$

$$\hat{\Theta}_t^+ = \Theta_t^- + \mathbf{K}_t \cdot \left(\mathbf{l}_t - \mathbf{H}_i \cdot \hat{\Theta}_t^- \right) \tag{8}$$

$$\Sigma_{\hat{\Theta},t}^+ = (\mathbf{I} - \mathbf{K}_t \cdot \mathbf{H}_i) \cdot \Sigma_{\hat{\Theta},t}^- \tag{9}$$

where the index:
T—indicates transposition,
-1—indicates inversion.

The noise of measurements are defined by a matrix Σe, whose diagonal elements are root mean square error in determining the velocity of the trolley based on measurements from an incremental sensor and gyroscope:

$$\Sigma e = \begin{bmatrix} \Sigma e[1, 1] & 0 \\ 0 & \Sigma e[2, 2] \end{bmatrix} \tag{10}$$

$$\Sigma e[1, 1] = \cos^2(\omega_z \cdot dt) \cdot \sigma_{v_{increm}}^2 + v_{increm}^2 \cdot \sin^2(\omega_z \cdot dt) \cdot dt^2 \cdot \sigma_{\omega_z}^2 \tag{11}$$

$$\Sigma e[2, 2] = \sin^2(\omega_z \cdot dt) \cdot \sigma_{v_{increm}}^2 + v_{increm}^2 \cdot \cos^2(\omega_z \cdot dt) \cdot dt^2 \cdot \sigma_{\omega_z}^2. \tag{12}$$

where:
$\sigma_{v_{increm}}$—root mean square error of incremental velocity,
σ_{ω_z}—root mean square error of angular rate in the direction of Z axis.

4   Experimental Measurement

The experiment aimed to verify the model developed for data processing as well as for the evaluation of the uncertainty of the determined position and orientation. During the experimental measurement, the measuring system was moved along a predefined trajectory in the shape of a polygon. Coordinates of reference points (RPs) of the trajectory were determined by independent terrestrial measurement and that way could be used for evaluation of the measuring system accuracy. The predefined trajectory consists of a small polygon (Fig. 5—RPs 1 to 4) and a large polygon (Fig. 5 RPs 5 to 11). The single measurement consists of a small polygon, which was passed 6 times and the large polygon, passed once. A small polygon is more appropriate for evaluation of the orientation quality, due to the greater number of RPs and straight lines between them. The larger polygon represents a more complex movement of the system (passing through the doorstep, maneuvering the trolley when passing through the door). The total trajectory length is approximately 562 m and includes the part of six small polygons (372 m) and the part build by the large polygon (190 m).

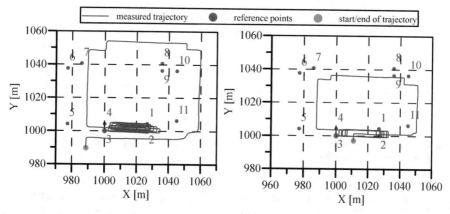

Fig. 5. Trajectory determined by inertial measurement using ZUPT (on the left) and using partial kinematic models (on the right)

RPs are located on the intersection of two straight lines with different orientation of the calculated trajectory. Therefore, are characterized as points, where the orientation of the movement is significantly changed. This fact enables us to identify RPs using the measured angular velocity and could be identified as the midpoints of the trajectory part, where the CT model was identified.

The trajectory measured by the system was calculated using a proposed model of data processing, which combines the inertial measurements with measurements from the optical encoder using the Kalman filter. The whole model includes partial models (procedures), which describe different kinematic states (CP, CV, CA and CT) of the system and enable the effective elimination of systematic errors of inertial sensors. The application of these partial kinematic models enables the usage of conditions that resulting from their characteristics. To evaluate the impact of individual processing procedures, the trajectory was initially calculated based on inertial measurements only (Fig. 5—left), in the second step using partial kinematic models (Fig. 5—right) and in the third phase, inertial measurements were combined with measurements from the optical encoder. The application of external data enables the effective elimination of systematic errors (Fig. 6).

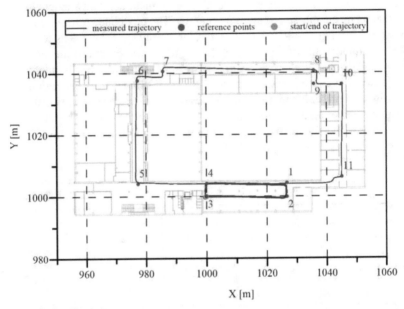

Fig. 6. Trajectory determined by inertial measurement using ZUPT, partial kinematic models and data from the optical encoder

The position uncertainty could be evaluated using position differences Δp calculated at all RPs. The difference Δp in position is calculated as the Euclidean distance between the measured and the given position of each RP determined by independent terrestrial measurement. When inertial measurements with ZUPT are used, the drift in relative position is 0.027 m s^{-1} (Fig. 7—blue line). This is caused by the application of ZUPT in non-regular time intervals (different time intervals between two stops). Using the partial kinematic models in data processing, the drift in relative position decreased to 0.014 m s^{-1} (Fig. 7—green line). The usage of kinematic models allows the better modelling of the sensor's noise. The CT model distinguishes between linear motion and movement with changing orientation, which leads to better elimination of errors of measurements in the direction perpendicular to the main direction of motion. In the third phase of data processing the Kalman filter was applied for the fusion of datasets of the IMU and the optical encoder as well as for the application of partial kinematic models. This solution significantly reduces position differences, which also shows on the drift of relative position, which decreasing to 0.001 m s^{-1} (Fig. 7—red line). A significant increase of positional accuracy is related to the use of the optical encoder, which effectively eliminates the systematic errors of accelerometers without necessary stops. The

Kalman filter optimizes the estimation of the state vector based on measurements and the accuracy of prediction and measurements.

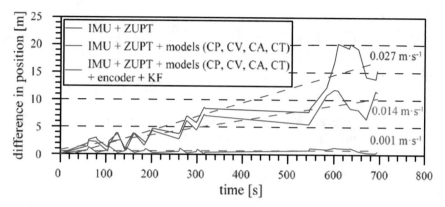

Fig. 7. Comparison of trajectories calculated based on differences and drift in relative position

The relative position difference δp could be calculated, which is defined as the ratio between the position difference Δp and the length of the trajectory d_{TTD} travelled to the given RP (Fig. 8):

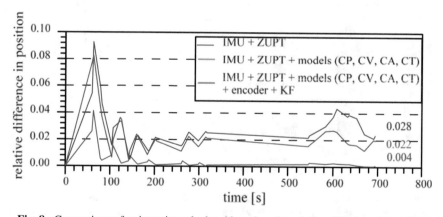

Fig. 8. Comparison of trajectories calculated based on the relative difference in position

$$\delta p = \frac{\Delta p}{d_{TTD}} \tag{13}$$

The uncertainty of the trolley orientation along the trajectory was evaluated using straight sections of the trajectory (between points 1 and 4), which were repeatedly travelled within a small and large polygon. The drift of the trolley orientation calculated based on time of the measurement and the average difference in orientation for a straight

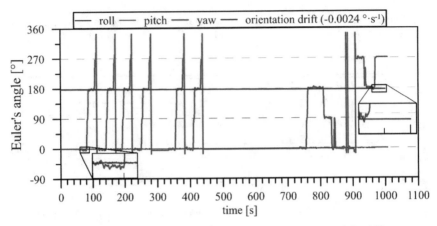

Fig. 9. The orientation of trolley defined by Euler's angles and the drift

part of the trajectory achieved the value of $-0.0024° \text{ s}^{-1}$ (Fig. 9). This represents an increase of the orientation error by $-8.64°/\text{h}$.

The efficiency of elimination of systematic errors using different data processing procedures can be seen in Fig. 10, where the values above the bar represent a decreased error relative to processing using only inertial measurements with ZUPT.

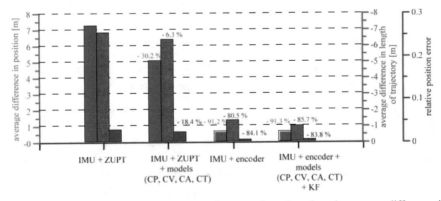

Fig. 10. Comparison of different data processing procedures based on the average difference in position, the average difference in trajectory length and the average relative position error. *Note* the percentages above the bars represent an improvement in the parameter relative to the first method

5   Conclusion

IMUs build a significant part of different navigation and mapping systems, due to their independence from external signals and high sampling rate. The used Novatel system is primarily designed for outdoor mapping and use the combination of IMU and GNSS. The paper aims to adapt the measuring system for indoor use. For this purpose, GNSS has been replaced by an optical encoder.

Measurements from the optical encoder were used for the reduction of systematic errors of accelerometers, which leads to significant improvement in velocity and position accuracy. The main part of the proposed processing model is the Kalman filter, which combines inertial measurements with the measurements from the optical encoder. The state vector was estimated by using a process model that adapts to the actual kinematic state of the system. The current kinematic state of the measuring system is identified based on inertial measurements using the algorithm described in the second part of the paper.

Based on the results of experimental measurements, the effectiveness of different procedures used for the reduction of systematic errors was evaluated. The paper describes the usage of partial kinematic models (CP, CV, CA, CT), which are developed for effective elimination (reduction) of systematic errors of inertial sensors according to the actual state of the system. The application of data generated by the optical encoder using the Kalman filter implemented in the data processing brings a significant increase in the position accuracy. When inertial measurements are used with ZUPT, only the drift in relative position at the level 0.027 m s^{-1} was achieved. By application of partial kinematic models, this value is decreasing to 0.014 m s^{-1}. The combination of the IMU and the optical encoder, as well as the usage of different kinematic models for processing, systematic errors are reduced intensively, which is represented by the reduction of the drift in relative position to 0.001 m s^{-1}.

The usage of the Kalman filter resulting in a 5% improvement in the accuracy of the determination of the trajectory length and a 1% improvement in position accuracy. In addition, the usage of the proposed data processing procedure allows the incorporation of another sensor (lidar, magnetometer, pressure sensor), which measures position and orientation and its functional principle is not based on a recursive algorithm.

Acknowledgements. This publication was created with the support of the Scientific Grant Agency of the Ministry of Education, science, research and sport of the Slovak Republic and the Slovak Academy of Sciences for the project VEGA-1/0506/18.

References

1. El-Sheimy, N.: Am overview of mobile mapping system, In: Fig working week, Cairo, Egypt, April 16–21., 24 p. (2005).
2. Munguía, R.: A GPS-aided inertial navigation system in direct configuration. *Journal of applied research and technology* 12(4), 803–814 (2014).
3. Li, X., Xu, Q., Song, X.: A highly reliable and cost-efficient multi-sensor system for land vehicle positioning. Sensors 16(6), 755 (2016).
4. Liu, Y., Fan, X., Lv, Ch. Li, L.: An innovative information fusion method with adaptive Kalman filter for integrated INS/GPS navigation of autonomous vehicles. In: *Mechanical Systems and Signal Processing*, 18(100), 605–616 (2018).
5. Alatise, M. B., Hancke, G. P.: Pose estimation of a mobile robot based on fusion of IMU data and vision data using an extended Kalman filter. Sensors 17(10), 2164 (2017).
6. Poulose, A., Han, D. S.: Hybrid indoor localization using IMU sensors and smartphone camera. In Sensors 19(23), 5084–5101 (2019).

7. Kopáčik, A., Kajánek, P., Lipták, I.: Systematic error elimination using additive measurements and combination of two low cost IMSs. IEEE Sensors Journal 16(16), 6239–6248 (2016).

8. Won, D., Ahn J., Sung S., Heo, M., Im, S., Lee, Y. j.: Performance improvement of inertial navigation system by using magnetometer with vehicle dynamic constraints. Journal of Sensors, 11 (2015).

9. Zhang, S., Guo, Y., Zhu, Q., Liu Z.: Lidar-IMU and Wheel Odometer Based Autonomous Vehicle Localization System. In: 2019 Chinese Control And Decision Conference (CCDC), p. 4950–4955 IEEE, Nanchang, China (2019).

10. Novatel, SPAN IMU: IMU CPT. https://www.novatel.com/products/span-gnss-inertial-systems/span-imus/imu-cpt/.

11. Li, X. Rong, Jilkov, V. P. Survey of maneuvering target tracking. Part I. Dynamic models. In: IEEE Transactions on aerospace and electronic systems, 2003, vol. 39.4. p. 1333–1364.

CroCoord v1.0—An Android Application for Nearly Real-Time 3D PPP Multi-frequency Multi-GNSS Static Surveying Using Smartphones

Marta Pokupić[1]([⊠]) [ID], Matej Varga[2] [ID], Zvonimir Nevistić[2] [ID], Marijan Grgić[2] [ID], and Tomislav Bašić[2] [ID]

[1] Bilogorska 20, Vukosavljevica, 33404 Špišić Bukovica, Croatia
`marta.pokupic95@gmail.com`
[2] Faculty of Geodesy, University of Zagreb, Zagreb, Croatia
`{matej.varga,zvonimir.nevistic,marijan.grgic,`
`tomislav.basic}@geof.unizg.hr`

Abstract. The hardware of new-generation smartphones enables navigation, positioning and surveying (NPS) using Global Navigation Satellite Systems (GNSS). Moreover, unprecedented progress has been made in the last few years related to the accuracy and reliability of smartphone NPS giving three significant changes: (1) accessing raw code and carrier GNSS measurements; (2) collecting multi-frequency measurements and; (3) collecting measurements from multiple global satellite navigation constellations (GPS, GLONASS, Galileo, BeiDou-3). We developed CroCoord v1.0—an Android application for nearly real time PPP multi-GNSS static positioning and surveying. An application offers increased accuracy compared to other solutions given by smartphone applications, as it uses PPP GNSS post-processing method to correct collected multi-frequency multi-GNSS measurements for several sources of errors, including ephemeris, iono-sphere, troposphere, satellite clocks, etc. The positional accuracy of static survey-ing using smartphones (including also our application) was tested on one control test site with three GNSS processing methods (SPP, PPP, and DGNSS). The posi-tional accuracy of CroCoord in PPP mode is below 1 m and may even reach 0.5 m for an observation period longer than 1 h. The positional accuracy of CroCoord is higher for 65% compared to the accuracy obtained by default processing routines of android smartphone. Furtherly, the combination of smartphone and permanent GNSS station data (a network-DGNSS mode) reached the possibility to obtain the accuracy of 9 cm for 2-h static session. Our results confirm the enormous poten-tial of smartphone multi-frequency multi-GNSS measurements in all civilian and professional applications which have benefits from increased positional accuracy.

Keywords: CroCoord android application · Multi-frequency · Multi-GNSS · Positioning · Post-processing · Raw GNSS measurements

A. Kopáčik et al. (Eds.): *Contributions to International Conferences on Engineering Surveying*, SPEES, pp. 283–296, 2021. https://doi.org/10.1007/978-3-030-51953-7_24

1   Introduction

The Global Navigation Satellite System (GNSS) provides the position, velocity, and time (PVT) for any point in a global reference 3D coordinate system by using radio signals transmitted from satellites orbiting around the Earth [1]. Currently, there are four full constellations of global navigation satellites: GPS (NAVigational Satellite Time and Ranking Global Positioning System), Russian GLONASS (rus. GLObal'naya NAvigatsionnaya Sputnikovaya Sistema), Galileo and BeiDou-3. These four global navigation systems, supplemented by one fully operational Regional Navigation Satellite System (RNSS), the Japanese Quasi-Zenith Satellite System (QZSS), and agumentation systems (such as EGNOS), form the so called multi global navigation satellite system multi-GNSS. In the past few years, thanks to the technological advancement of hardware, newer generations of smartphones have been shown to achieve a positioning accuracy of several decimetres in nearly real time applications. Moreover, a number of studies have been performed to verify the positioning accuracy with these smartphones for different purposes [2–11].

However, until 2017 all Android smartphones have used the signal on one frequency band only (L1, code pseudoranges), and most often have received signals only from GPS satellites. The accuracy of such absolute positioning is in the range of 5–15 m. The first dual-frequency smartphone to receive signals within the L1/E1 and L5/E5 bands is the Xiaomi Mi 8 that was released in May 2018. Since then, new dual-frequency smartphones have been constantly appearing on the market and may now receive L5 band of the GPS system with a median frequency of 1176 MHz, or on the E1 and E5 band of the Galileo system with a median frequency of 1575.42 and 1191.795 MHz. To date, there are 41 dual-frequency smartphone models on the market, from 10 manufacturers [12]. The first commercial chip to have this ability is the BCM47755, which can simultaneously receive the following signals: GPS (L1-C/A and L5), GLONASS (L1), BeiDou-2 (B1), QZSS (L1 and L5) and Galileo (E1 and E5a). The second chip Quectel LC79D integrated in smartphones with the ability to obtain measurements on multiple frequencies was designed in the late 2019 [13]. Chip and smartphone manufacturers are very aware of the capability that dual-frequency offers in terms of positioning accuracy for location based services. The possibility to obtain measurements on second frequency has several advantages, for example: increased reliability to users—if one of the frequency bands fails, the other can be used as backup; reduced signal acquisition time and improved accuracy of positioning and timing; reduced problems caused by obstructions such as buildings and other obstacles; reduced problem to multipath errors; the possibility to eliminate ionospheric refraction [12].

Another improvement in smartphone positioning is related to the open access of users to raw GNSS measurements, instead of getting "black-box" coordinates calculated by internal hardware and software of smartphones. Currently, the user may obtain actual measurements in the form of clock values and can use them for computation of code pseudoranges or carrier-phases by itself. Furthermore, one has access to additional data, such as SNR (Signal to Noise Ratio) estimates, Doppler and navigation messages. By making raw GNSS measurements available, users have full control over processing and obtaining coordinates, and may use advanced processing techniques such as optimizing multiple GNSS solutions, selecting satellites, frequencies, etc. Developers are able to leverage the access to raw GNSS measurements and the enhanced accuracy offered by this dual-frequency capability to create new applications requiring high accuracy robust positioning.

In a broader sense, satellite positioning is performed by using a large number of surveying methods and techniques, which depend on the user's needs for precision and speed. Positioning can be divided into absolute and relative. Absolute positioning is a method of determining the coordinates of a single receiver, while at relative positioning a minimum of two or more receivers, are used at the same time. Furthermore, positioning can be static or kinematic, depending on whether the receiver is moving or not. The collected measurements can be based on the Code modulation or Phase modulation of the carrier wave and processed in real time or in post-processing [1]. In this paper, the emphasis will be on static measurements with three methods: Single Point Positioning (SPP), Precise Point Positioning (PPP) and differential GNSS (DGNSS). SPP uses broadcast satellite orbits and clocks, and no other corrections except for ionosphere in case measurements contain multiple frequencies. PPP is a method of absolute static positioning of a single receiver in real time, where various satellite corrections and ephemeris are applied. It uses precise orbits and clocks instead of the broadcast message. The precondition for PPP method is to have a GNSS receiver (in this case a smartphone) which can collect multi-frequency (MF) measurements. PPP method is used in everyday practice by the number of users, civilians and geoscientists in applications such as smart cities, augmented reality applications, construction, sports, transportation, etc. DGNSS/RTK technique uses a base or reference receiver at a known location (such as from another GNSS device in surrounding or permanent GNSS station) collecting GNSS data at the same time as a receiver at an unknown surveyed location.

2 Raw GNSS Measurements

Raw GNSS measurements which user gets from the chip of Android smartphones are accessed by the Application Program Interface (API) which is a set of routines, protocols, and tools for building software programs and applications. Raw measurements include: (i) clock values of the receiver (smartphone) and their deviation, (ii) clock values of satellites, accumulated delta ranges (number of whole waves of the signal carrier), (iii) bits of navigation messages, and (iv) other metadata [2]. Raw GNSS measurements can be accessed through Android 7.0 and later versions with an API Level 24, which contains *GnssClock*, *GnssNavigationMessage*, and *GnssMeasurement* Java classes. The *GnssClock* and *GnssMeasurements* classes contain methods that provide the data necessary to calculate code pseudoranges and carrier phases from clock values of the smartphone and the satellite. Methods from the *GnssNavigationMessage* class provide a navigation message for all satellite constellations. API also includes Android classes *OnNmeaMessageListener* and *Location*. *OnNmeaMessageListener* class, using *onNmeaMessage* method, retrieves NMEA information about clock values of GNSS satellites. NMEA stands for National Marine Electronics Association, which defines and controls the NMEA 0183 standard for communicating with marine electronic devices such as transmitters, sonar, gyroscope, autopilot, GPS receiver, etc. [14]. The *Location* class contains methods which provide the location and associated metadata.

Java code for all methods is available freely and developers can use them in building Android applications [15]. We used them in developing CroCoord application in order to build the function which performs collection (surveying) of GNSS measurements [2].

2.1 Smartphone Hardware Design

The quality of GNSS data, obtained by a smartphone depends on several hardware components, mainly the GNSS chip. Smartphone hardware is designed in the way that it saves battery power, minimizes Time To First Fix (TTFF), and improves accessibility and position continuity. Smartphone design influences the selection of hardware components and software algorithms. Hardware consists of low-cost components; for positioning and navigation the most important are the antenna and oscillator. An oscillator is located outside GNSS chip and compensates temperature variations which affect the stability of the frequency. An alternative, low-energy crystal oscillator controls the time when GNSS is off with an accuracy degradation of 6 s per week. The antenna used in smartphones is called PIFA (Planar Inverted-F Antenna) which is a laminated copper plate on which all components are soldered, mechanically fixed and electrically connected. Because of its linear (instead of circular) polarization and directionality of the radiation diagram, it results in a decrease of radio signal strength by several dB. The position of the antenna depends more on the design of the smartphone than on the signal constraints, and for this reason, the interaction with the user's hand further weakens the reception of the GNSS signal. The smartphone antenna is subject to multiple reflections and its signal-to-noise ratio is reduced by about 10 dB/Hz compared to the geodetic receiver [4, 5, 16, 17].

3   CroCoord Android Application

Several Android GNSS data processing applications are available for static positioning and geodetic surveying using smartphones. They all may provide coordinates from collecting GNSS measurements using SPP. However, CroCoord Android application was developed with an idea of improving the accuracy by using advanced post-processing of raw multi-frequency multi-GNSS measurements compared to other applications. Positioning with CroCoord is based on the PPP method. The main result of processing collected raw GNSS measurements are coordinates of surveyed points in WGS84 reference coordinate system, and in official reference coordinate systems of the Republic of Croatia (HTRS96, HDKS, HVRS71). Results are available in nearly real time, because collected raw measurements take few seconds to process on the external server. Our application was developed in Android studio, its entire functionality is written in Java programming language, whereas the design is in XML format. Application logo (see Fig. 1) and the name itself, "Cro" from "Croatian" and "Coord" from "Coordinates", refer to the suitability of the application for positioning and surveying in Croatia. The CroCoord methodology opens up the possibility of simply extending the application to other countries, but with the need for additional data such as geoid models (to obtain orthometric height of surveyed point) and transformation parameters for transformation from the epoch of measurement to the epoch of the country-specific coordinate reference frame.

Fig. 1. CroCoord application logo

For full functionality of our application only requirements are internet access and permission to collect satellite positioning signals. Currently, source codes of the application are not available due to development. However, Android application executable files can be obtained at any time after the request of the corresponding author, followed by instructions for use. Also, the server is not active all the time, it must be started by author when planning to use the application.

3.1 Application Flowchart

CroCoord application runs as follows:

1. User starts application and collects single or multi-frequency multi-GNSS measurements via GUI for any selected point and session duration. The application uses all Java classes and methods from Sect. 2.
2. Collected measurements are pushed (uploaded) to the synchronization server via mobile internet.
3. The client downloads collected measurements from the server.
4. RTKLIB GNSS post-processing software processes data (calculates coordinates and other metadata).
5. Client uploads processed results back to the synchronization server.
6. User pulls (downloads) results from the synchronization server.
7. GUI displays downloaded results in the form of point coordinates and position of users on the map.
8. Results and all metadata are saved in the internal memory of a mobile phone as a CSV file.

The CroCoord GUI consists of three tabs: *Points*, *Map*, and *Info* (see Fig. 2). On the *Points* tab, the user enters the name of the point and its description (such as a "polygon point", "tree", "curb", etc.). The user selects the duration of the session using *Timer* button. The surveying of a point starts by pressing the *Start* button and is stopped either after expiration of time set by timer, or after pressing the *Stop* button. After the user stops the collection of measurements, collected data are sent to the synchronization server, as a file under its own ID, using the *sendPostRequest* method.

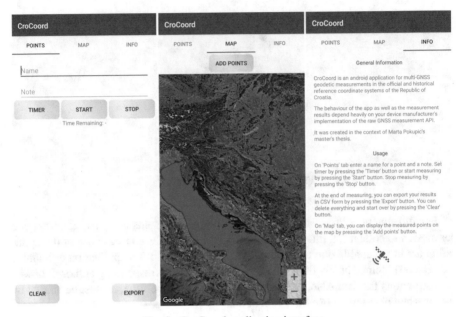

Fig. 2. CroCoord application interface

The server then returns the processing results (response time under 10 s) and the application retrieves them by using the *sendGetRequest* method based on the measurement ID.

Some processing results are displayed on the *Points* tab in the form of coordinates (Easting, Northing) in HTRS96/TM (Transverse Mercator) projection and orthometric height of point in the HVRS71 vertical reference system. On the *Map* tab user can add the position of a measured point to a map view by clicking the *Add points* button, which calls the *addPointsToMap* method (see Fig. 3). All measured points, as well as those that are not displayed in the interface, can be stored on internal memory in CSV format and shared by pressing the *Export* button. The entire process of using a CroCoord application, i.e. how to use it and position with it, is described in the third *Info* tab.

3.2   Raw GNSS Measurements of Android Application

By pressing the *Start* button, the application starts simultaneously collecting raw multi-GNSS measurements by using the *Location, GnssNavigationMessage, OnNmeaMessageListener, GnssClock* and *GnssMeasurement* Java classes methods, and writing raw GNSS measurements into a new "rawData" text string using the *startNewLog* method. When completed, the application sends the file with measurements to the synchronization server in *.txt format. Raw GNSS data are just numeric values that cannot be directly processed in GNSS measurement software packages, but this problem is solved by the client that converts the measurement file to RINEX v2 format.

After CroCoord application sends collected measurements to the *synchronization server*, the measurements are being processed on the desktop computer, so in addition

Fig. 3. Surveying results displayed

to the Android application, there have been created program scripts that serve both as the sync server and program scripts to process the collected raw GNSS measurements. The sync server serves only as the bridge between Android GUI and the measurement processing script called *client*, which is located on a desktop computer. The computer performs post-processing of the collected measurements and calculates coordinates of measured point using RTKLIB [18]. The sync server and the client are collectively referred to as the *server* and the whole concept of application ↔ server work is illustrated in Fig. 4.

3.3 Synchronization Server

The synchronization server is a script written in the Python programming language that establishes a connection between Android application and the client-computer that performs post-processing of collected GNSS measurements. The scripts are located in a cloud through an online Heroku platform that supports several programming languages.

The sync server can be accessed from multiple devices, at any time and from different places with an available internet connection. The sync server and the client communicate through directories. The sync server stores raw GNSS measurements in measurements directory from which the client retrieves measurements, and after processing them, saves their results to the results directory. The client converts the result *.txt file into *.JSON format, adding the finished state to the file for the sync server to recognize the right file and respond to an Android application that then uses the necessary data to display it.

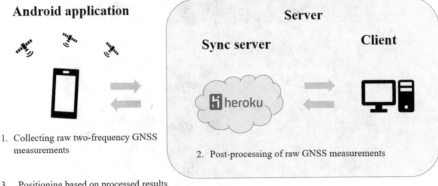

Fig. 4. Concept of application ↔ server

3.4   Client

The client is the central element of the entire process that enables CroCoord application positioning. The part of the client that communicates with the sync server is written in the Python programming language, and the scripts that process the measurement data are written in the MatLab programming language and they are located on the computer. The main MatLab post-processing script defines the names of the input files and result files, as well as directories of their downloads or storage. It also runs other scripts and packages via batch files. The whole work of the client can be shown in the following:

1. Retrieval of the *.txt file of raw GNSS measurements from the measurement directory from the sync server.
2. Conversion of *.txt file of raw GNSS measurements to RINEX v 2.11 format.
3. Registering the time of the last measurement from the RINEX file by which GPS week, GPS day of the year, GPS seconds of the week and Julian Day are calculated.
4. Broadcast ephemeris from NASA FTP are retrieved.
5. Processing RINEX measurement file by using RTKLIB software in PPP batching mode, which as a result provides coordinate series (latitude, longitude, height) for each measurement epoch.
6. Filtering outliers from the coordinate series using reliability factors (99%).
7. Calculation of definite coordinates of points and their standard deviations using the statistical function of the median set.
8. Transformation of coordinates from WGS84 (measurement epoch) to desired coordinate system (HTRS96, HDKS and HVRS71).
9. Saves the results (orthometric heights in HVRS71 and HVRS1875 vertical coordinate systems, terrestrial coordinates, $\bar{E}$ and $\bar{N}$ in HTRS96/TM and $\bar{y}$ i $\bar{x}$ in Gauss-Krüger (HDKS), and their metadata) in a *.json file to the *results* directory from where the results are loaded to the sync server.

Obviously, the central part of computing coordinate of the surveyed point is using RTKLIB in PPP mode. RTKLIB is an open source software for GNSS positioning and analysis, which supports GPS, GLONASS, Galileo, QZSS, BeiDou and SBAS systems [18, 19]. It is known to be used in some other GNSS post-processing services, such as Rokubun Jason [20]. The main MatLab script starts running RTKLIB using a batch file in which it defines all the required parameters and settings, specifying the processing method and type of output results. Some of the settings which are defined prior to processing are: positioning method, used frequencies, elevation mask, SNR mask, Earth tidal wave corrections, dynamic models, ionospheric correction, tropospheric correction, satellite ephemeris, ambiguity solution method for different systems, ambiguity determination settings, maximum age differential, GDOP threshold, result format (coordinates), time format, height type, carrier-based measurement settings, antenna settings, etc. After processing of measurements, RTKLIB gives results in the form of the coordinate series (latitude, longitude and height) in the WGS84 reference coordinate system for each measurement epoch together with the associated standard deviations and other metadata such as a number of satellites, DOP, etc.

4 Accuracy of Static GNSS Surveying Using Smartphones

One of the biggest advantages of CroCoord, compared to the limits in accuracy of smartphone-based GNSS data processing, is its accuracy, since it uses RTKLIB for PPP processing of collected data. In order to verify it, we performed one static session on December 31st, 2019 which lasted 2 h from 09:00 to 11:00 UTC on one selected control point of the University of Zagreb Faculty of Geodesy. Collected data (from smartphone and ZAGR CROPOS (CROatian POsitioning System) permanent station), as well as additional data used in post-processing (clocks, ephemeris, EOP), may be downloaded on following permanent https://doi.org/10.5281/zenodo.3633374.

Reference (true) point is part of the well-stabilized geodetic network of six points used by the Faculty of Geodesy. The network is reobserved and readjusted each year with six GPS + GLONASS L1/L2 professional grade GNSS receivers using method of relative static and 20 measurement sessions lasting minimally 20 min. The reference coordinates of control point were obtained through several measurement campaigns which have lasted for more than 24 h in total. The estimated accuracy of the coordinates of control points is expected to be at few mm level which is well suitable for comparison with smartphone measurements.

The specific goal of this test was not only to verify improvement in accuracy of CroCoord application compared to other Android-based smartphone GNSS processing software, but also to investigate achievable accuracy of smartphone GNSS surveying for three positioning methods: SPP, PPP, and DGNSS. For SPP and PPP testing, only smartphone collected measurements were used, whereas the performance of DGNSS was tested using the additional measurements from permanent station ZAGREB which is a part of CROPOS GNSS network. In DGNSS method, ZAGR station was fixed with coordinates $X = 4,282,037.523$ m, $Y = 1,224,886.395$ m, $Z = 4,550,534.641$ m. Smartphone measurements on control point were collected with dual-frequency Xiaomi Mi Mix-3.

Measurements were processed using several commercial and free software packages including Trimble Business Center 5.00, Leica GeoOffice 8.4, RTKLIB, Google GNSS analysis software, as well as with online post-processing services such as Rokubun Jason, Trimble VRX, and NRCan CSRS-PPP.

Comparison of definite plane coordinates (Easting, Northing) for each solution was performed in conformal Gauss-Krüger projection using the following parameters: $\varphi_0 = 0°$, $\lambda_0 = 16.5°$, linear scale 0.9999, shift in Easting 500,000 m. The accuracy of each solution was analysed using positional (plane) residuals between map projection coordinates (Easting, Northing) of two quantities: reference coordinates E, N of control point and E, N of each solution (coordinates obtained by processing of measured data):

$$\sigma = \sqrt{\Delta E^2 + \Delta N^2} = \sqrt{(E_{control} - E_{solution})^2 + (N_{control} - N_{solution})^2} \quad (1)$$

In Table 1 and on Fig. 5 we present and describe only the best solutions for each GNSS positioning method, so the results obtained with other software and measurements from other constellations are not shown. The first two solutions (S1 and S2) are obtained by using static data from the smartphone and permanent station in network-DGNSS method. The solution S1 represents the optimal solution in which coordinate of the point was obtained by using network-DGNSS method in Leica GeoOffice where both GPS and Galileo dual-frequency were exploited. The position residual between reference coordinate and this solution is 9 cm. Solution S2 is obtained in the same way as S1, but

Table 1. The results of GNSS data processing for measurement session on December 31st, 2019, from 09:00 to 11:00 UTC

Solution #	Software	Description	SPP/PPP/DGNSS	ΔE [m]	ΔN [m]	Error [m]
Reference solution (true)	TBC		Relative static			
S1	Leica Geooffice	GPS + Galileo, L1/E1/L5/E5	Network-DGNSS	−0.08	−0.03	**0.09**
S2	RTKLIB	GPS + Galileo, L1/E1/L5/E5	Network-DGNSS	−0.22	0.27	**0.35**
S3	CSRS-PPP	GPS L1/E1/L5	PPP	−0.31	−0.33	**0.45**
S4	RTKLIB	GPS + Galileo, L1/E1/L5/E5	PPP	−0.01	0.73	**0.73**
S5	Google GNSS analysis tools, Android-default	GPS L1	SPP	−0.15	2.12	**2.12**

Static measurements were performed using smartphone on control point. **S4** is the solution obtained by CroCoord application, whereas **S5** is the solution obtained in smartphone by its default routines from Google GNSS analysis tools

RTKLIB was used. The positional residual is 35 cm, which is almost 4 times worse than solution S1. Leica GeoOffice as commercial software seems to be more accurate than RTKLIB, even if RTKLIB has more customizability than Leica Geoffice.

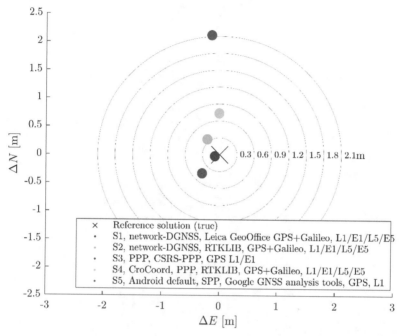

Fig. 5. The results of GNSS data processing for measurement session on December 31st, 2019, from 09:00 to 11:00 UTC

The last three solutions are the analysis of SPP or PPP solutions using three software. S3 solution, obtained from Canadian CSRS-PPP online post-processing service, is the most optimal of these three compared. The solution used only GPS measurements on two frequencies. The positional residual is 45 cm, which is nearly the same as S2 solution in network-DGNSS method. PPP using RTKLIB resulted in positional residual of 73 cm. This result shows that we can expect the positional accuracy of CroCoord application to be around $\pm$ 1m for shorter and medium observation times (<1 h), and sub-meter accuracy for sessions longer than 1 h. Solution S5 is the SPP solution which can be expected from default smartphone GNSS positioning applications which do not take care of any corrections and usually use only GPS on one-frequency. The positional residual is 2.1 m for our two-hour long session. The accuracy of such positioning might be in the range between 2 and 5 m. The comparison between S4 and S5 solutions is important as it shows that the positional accuracy of CroCoord is supposedly to be higher for around 65% compared to other android GNSS applications.

Definite coordinates of all solutions are given in Table 2.

Table 2. Definite coordinates of all solutions

Solution #	Geodetic on GRS80			Cartesian geocentric on GRS80			Plane	
	φ [°]	λ [°]	h [m]	X [m]	Y [m]	Z [m]	E [m]	N [m]
Reference solution (true)	45.8082391	15.9658304	164.835	4,281,943.22	1,225,064.34	4,550,534.07	497,343.982	5,074,772.381
S1	45.8082389	15.9658293	165.038	4,281,943.40	1,225,064.30	4,550,534.20	497,343.899	5,074,772.355
S2	45.8082416	15.9658276	163.927	4,281,942.49	1,225,063.90	4,550,533.61	497,343.766	5,074,772.655
S3	45.8082361	15.9658265	164.794	4,281,943.51	1,225,064.10	4,550,533.81	497,343.677	5,074,772.048
S4	45.8082457	15.9658303	163.500	4,281,941.83	1,225,063.93	4,550,533.63	497,343.972	5,074,773.115
S5	45.8082582	15.9658285	175.920	4,281,949.23	1,225,065.90	4,550,543.50	497,343.834	5,074,774.500

5   Conclusion

In order to explore the potential of smartphones which may collect raw multi-frequency multi-GNSS measurements, we developed CroCoord v1.0 application which provides the position of surveyed points in WGS84, and official geodetic reference coordinate systems of the Republic of Croatia. We debugged and verified application for completeness and consistency. The application can be used to collect static multi-frequency GNSS measurements from all global satellite navigation systems and for any session duration. The speed of the application is nearly real time (user gets results in the smartphone almost immediately after the end of measurements) and is limited solely by the internet connection needed for the transfer of the data to and from the server and time needed for the processing of GNSS measurements. It is possible to replace RTKLIB as GNSS post-processing tool with any other software having the option for simultaneous batch processing.

We verified that CroCoord application has greater accuracy, using 2-h long static session and SPP/PPP processing method, CroCoord yielded 65% more accurate results compared to other android based GNSS processing algorithms and software (0.73 m with CroCoord compared to default smartphone solution with 2.12 m accuracy), as it exploits benefits of more sophisticated software RTKLIB running on an external server. The analysis of our comparison non-related to CroCoord showed that on-line GNSS post-processing service such as CSRS-PPP also has potential for processing of mobile-phone data in PPP mode as it outperformed the accuracy of RTKLIB processing for 38% (from 0.45 m with CSRS-PPP to 0.73 m with RTKLIB). Finally, our test showed high potential for the usage of corrections and GNSS data from surrounding permanent GNSS stations in DGNSS, where even sub-dm positional accuracy for longer observation periods can be expected.

The executables of application can be obtained upon the request to the corresponding author, after which instructions for usage will be given. Authors are still working on developing and further improving the reliability of the post-processing, and to extend the capabilities of the application regarding the choice of parameters and positioning methods.

Note

Measurements data can be downloaded from following link https://doi.org/10.5281/zenodo.3633374. Data include measurements from smartphone (in android-based format, Rinex 2.11 and 3.03), permanent station ZAGR (in Rinex 2.11 and 3.03), along with all supplementary data authors used in processing such as broadcast and precise ephemerides (in Rinex 2 and MGEX).

References

1. Zrinjski, M., Barković, Đ., Matika, K. (2019). Razvoj i modernizacija GNSS-a. Geodetski list, 73(1), 45–65.
2. Pokupić, M. (2019). CroCoord - Android application for multi-GNSS surveying (Master's thesis). (https://www.bib.irb.hr/1037146).

3. Dabove, P., Di Pietra, V. (2019). Towards high accuracy GNSS real-time positioning with smartphones. Advances in Space Research, 63(1), 94–102.
4. Humphreys, T. E., Murrian, M., Van Diggelen, F., Podshivalov, S., Pesyna, K. M. (2016). On the feasibility of cm-accurate positioning via a smartphone's antenna and GNSS chip. In 2016 IEEE/ION Position, Location and Navigation Symposium (PLANS) (pp. 232–242). IEEE.
5. Pesyna Jr, K. M., Heath Jr, R. W., Humphreys, T. E. (2014). Centimeter positioning with a smartphone-quality GNSS antenna. In Radionavigation Laboratory Conference Proceedings.
6. Masiero, A., Guarnieri, A., Pirotti, F., Vettore, A. (2014). A particle filter for smartphone-based indoor pedestrian navigation. Micromachines, 5(4), 1012–1033.
7. Wang, L., Li, Z., Zhao, J., Zhou, K., Wang, Z., Yuan, H. (2016). Smart device-supported BDS/GNSS real-time kinematic positioning for sub-meter-level accuracy in urban location-based services. Sensors, 16(12), 2201.
8. 21. Wang, L., Li, Z., Yuan, H., Zhao, J., Zhou, K., Yuan, C. (2016). Influence of the time-delay of correction for BDS and GPS combined real-time differential positioning. Electronics Letters, 52(12), 1063–1065.
9. Adjrad, M., Groves, P. D. (2018). Intelligent urban positioning: Integration of shadow matching with 3D-mapping-aided GNSS ranging. The Journal of Navigation, 71(1), 1–20.
10. Al-Azizi, J. I., Shafri, H. Z. M. (2017). Performance evaluation of pedestrian locations based on contemporary smartphones. International Journal of Navigation and Observation, 2017.
11. Fissore, F., Masiero, A., Piragnolo, M., Pirotti, F., Guarnieri, A., Vettore, A. (2018). Towards Surveying with a Smartphone. In New Advanced GNSS and 3D Spatial Techniques (pp. 167–176). Springer, Cham.
12. European GSA, dual-frequency. Available online: https://www.gsa.europa.eu/newsroom/news/market-understands-value-dual-frequency (accessed on 30 January 2020).
13. Quectel LC79D chip. Available online: https://www.quectel.com/infocenter/news/563.htm (accessed on 30 January 2020).
14. OnNmeaMessageListener Java class. Available online: https://developer.android.com/reference/android/location/OnNmeaMessageListener.html (accessed on 27 January 2020).
15. Android location API. Available online: https://developer.android.com/reference/android/location/package-summary (accessed on 27 January 2020).
16. Van Diggelen, F. S. T. (2009). A-GPS: Assisted GPS, GNSS, and SBAS. Artech House.
17. European GNSS Agency (GSA). Using GNSS raw measurement on android devices. https://www.gsa.europa.eu/system/files/reports/gnss_raw_measurement_web_0.pdf.
18. Takasu, T. (2013). RTKLIB ver. 2.4. 2 Manual. RTKLIB: An Open Source Program Package for GNSS Positioning, 29–49.
19. RTKLIB software. Available online: http://www.rtklib.com/ (accessed on 28 January 2020).
20. Rokubun Jason software. Available online: https://paas.rokubun.cat/ (accessed on 28 January 2020).

Cultural Heritage and Rural Development

Preliminary Analysis of Criteria for Land Consolidation in Protected Areas

Hrvoje Tomić(✉) ⓘ, Siniša Mastelić Ivić ⓘ, Rinaldo Paar ⓘ, and Ante Marendić ⓘ

Faculty of Geodesy, University of Zagreb, Zagreb, Croatia
{htomic,ivic,rpaar,amarendic}@geof.unizg.hr

Abstract. Land Consolidation is the agrarian and technical operation that aims to group and collect the segmented and fragmented holdings into one or more rounded holdings to achieve a more rational agricultural production. Modern approaches to land consolidation moved its primary purpose from the agricultural sector into the environmental and recreational sectors, with the aim of creating multi-purpose land consolidation. This type of Land Consolidation includes the re-allocation of parcels together with a broad range of other measures that contribute to rural development. To assess different future scenarios considering all the identified problems, a comprehensive spatial database is needed. The paper proposes an approach to achieve sustainable rural development using the case study of UNESCO's World Heritage Site Stari Grad Plain in Croatia. Land use classification using Normalized Difference Vegetation Index (NDVI) calculated from fixed-wing UAV multispectral images was used as an example of quantitative indicator. These indicators can be used in the process of land valuation as an important part of land consolidation procedure, increasing information-based decision making in rural development programme.

Keywords: Land consolidation · Land fragmentation · Sustainable rural development · Multi-criteria analysis

1 Introduction

Agricultural land consolidation is the agrarian and technical operation that aims to group and collect the segmented and fragmented holdings into one or more rounded holdings in order to achieve a more rational agricultural production [1]. As a result of inappropriate land policy on the Croatian territory, agricultural holdings are extremely fragmented [2, 3]. The land fragmentation is manifested by a large number of relatively small and spatially divided land parcels of each owner. The land parcels are often very irregular in shape and there is a lack of appropriate road access, which hinders an effective application of modern agricultural machinery. Additionally, there are ownership fragmentation and outdated land administration system data to be considered. Fragmented agricultural land leads to increased costs of agricultural production and affects land abandonment.

Agricultural land consolidation procedures were implemented on the Croatian territory in the past. Those procedures were focused on agricultural development,

A. Kopáčik et al. (Eds.): *Contributions to International Conferences on Engineering Surveying*,
SPEES, pp. 299–308, 2021. https://doi.org/10.1007/978-3-030-51953-7_25

which is visible from the land fragmentation indicators. Most of the land consolidation works refer to the eastern part of the country (Eastern Slavonia and Baranya). However, the concept of rural development has changed over time and it now includes the environmental awareness and a range of non-agricultural applications [4]. Current Croatian land policy includes a rural development policy as an integral part of the EU's Common Agricultural Policy (CAP). Within its framework, Rural Development Plans (RDP) are guiding policy to support the development of agricultural competitiveness, together with the sustainable management of natural resources and balanced territorial development [5].

In year 2015, a new Act on Land Consolidation was introduced. The Act provides the land consolidation projects to be implemented in accordance with multi-year and annual programme. The multi-year programme (for a period of five years) must be adopted by the Croatian Parliament, and annual programme must be approved by the Croatian Government. Although some preparatory activities have been undertaken, none of the planned programs has been developed to date. The preparatory activities on a national level include the determination of the priority areas, i.e. the areas for which it is assumed that redistribution of land is most needed and have certain prerequisites for successful agricultural production together with the interests of people and state.

This paper considers an approach to achieve sustainable rural development using land consolidation procedure on case study of UNESCO's World Heritage Site Stari Grad Plain, the island of Hvar in Croatia. The main characteristic of today's landscape is low intensity, mixed agriculture (dominated by olive groves and vineyards), high density of stone wall enclosures, terraces, channels, dwellings, shelters etc. The selected area represents the case of overlapping of several different protection regimes, creating problems in administration and management. This leads to abandonment of land, and, in general, abandonment of rural communities. The abandonment and depopulation are two closely connected problems that can lead to devastation of natural resources and potentials. Uncultivated land is significantly changing the landscape and can reduce the biodiversity. The importance of nature and landscape protection must be taken into account since the inappropriate land consolidation projects may have great impact on the rural development.

2   Materials and Methods

The purpose of this paper is to analyse the possibilities of using different spatial datasets as quantitative indicators in a multi-criteria analysis to help making a decision based on size, coverage, comprehensiveness, costs and benefits, as well as other characteristics of selected Land Consolidation procedure.

The study was conducted based on the datasets collected during the implementation of the project Development of Multipurpose Land Administration System—DEMLAS [6] supported by the Croatian Science Foundation. The project pilot area was UNESCO's World Heritage Site Stari Grad Plain, the island of Hvar in Croatia. Various sensors and processing procedures were used to create spatial datasets (GNSS RTK, total stations, UAV or terrestrial use of so called "Structure-From-Motion" photogrammetric method using different imaging sensors: visible spectrum—RGB, Near Infrared and Red Edge).

The results of these measurements are vector or raster spatial datasets of different scales and extents.

All available spatial datasets were used to analyse its applicability on the case study area using the quantitative indicators and measures of land consolidation suitability, some of which are the results of our previous research on land consolidation suitability assessment [7, 8] or agricultural land fragmentation analyses [9, 10]. These indicators assess the agricultural land taking fragmentation of holdings and fragmentation of ownership into consideration. Indicators referring to the fragmentation of agricultural holdings assess the number, size and shape of the land parcels (it is difficult or impossible to use machinery on narrow land parcels or those with sharp corners), their position (distance from village and distances between them) and distribution of land parcels sizes. Ownership fragmentation refers to the number of owners registered in the official land register—Land Book. Unfortunately, some of the Land Book records are missing or outdated.

The research methodology included detailed literature review to assess possible solutions and best practices [4, 11] in the countries facing similar problems [12–17]. Those findings were used to synthesize appropriate indicators [18–23] and measures which might be useful and applicable when selecting an optimal land consolidation solution. Different types of multi-criteria analyses [7, 24–29], based on spatial data characteristics, have proven to be effective in making decisions and are used as a basis of decision-support systems in the processes of land consolidation.

2.1 The Study Area

The study area is located on the Island of Hvar in Croatia. The Stari Grad Plain (Fig. 1) represents a comprehensive system of land use and agricultural colonisation by the Greeks, 4th century BC [30]. The criteria for the inclusion in UNESCO are: the land parcel system bears witness to the dissemination of the Greek geometrical model for the division of agricultural land (Criterion II), the plain has remained in continuous use, with the same initial crops being produced for 2400 years (Criterion III), and the plain is an example of very ancient traditional human settlement which is today endangered by modern economic development, particularly by rural depopulation and the abandonment of traditional farming practices (Criterion V).

The area covers 13,77 km^2, the altitude ranges between 1 and 86 meters above the sea level and it is gently sloped—more than 95% of the area has the slope less than 10%. The main characteristics of today's landscape [31] are low intensity, mixed agriculture (dominated by olive groves and vineyards), high density of stone wall enclosures, terraces, channels, dwellings, shelters etc.

The selected area represents the case of overlapping of several different protection regimes, creating problems in administration and management. This leads to abandonment of land, and, in general, abandonment of rural communities. The abandonment and depopulation are two closely connected problems that can lead to devastation of natural resources and potentials.

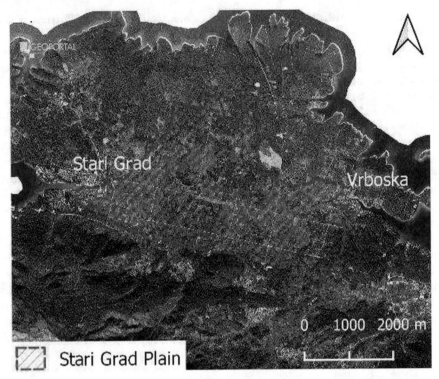

Fig. 1 UNESCO World Heritage Site "Stari Grad Plain" location

3 Analysis and Evaluation

The official cadastral dataset of the study area consists of 9093 cadastral land parcels that are divided through a total of six cadastral municipalities (Stari Grad 47%, Vrbanj 27%, Vrboska 17%, Dol 5%, Svirče 2% and Vrisnik 2%). The administrative boundaries divide the plain across two local self-government municipalities: Stari Grad and Jelsa (Fig. 2).

The Stari Grad Plain area has been overlapped by several regimes of protection (Fig. 3). The protected coastal zone (shown in blue) id efined as the area between 1000 meters inland and 300 meters offshore from the coastline, meaning there are several restrictions on land use regulated by the Spatial Planning Law and Nature Protection Law.

The whole study area is contained within Natura 2000 special protection area (Site: HR100036, Sitename: "Srednjedalmatinski otoci i Pelješac", shown in green line pattern) and southern and a small northern part fall within Natura 2000 Proposed site for community importance (Site: HR2001428, Sitename: "Hvar—from Maslinice to Grebišća", shown in purple).

In order to assess different future scenarios related to all identified problems, a comprehensive spatial database is needed.

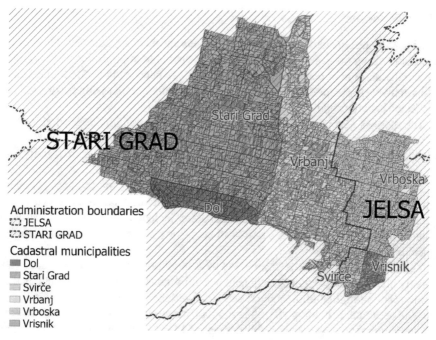

Administration boundaries
JELSA
STARI GRAD
Cadastral municipalities
Dol
Stari Grad
Svirče
Vrbanj
Vrboska
Vrisnik

Fig. 2 Case study administrative division

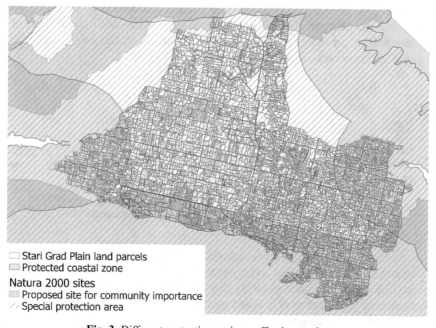

Stari Grad Plain land parcels
Protected coastal zone
Natura 2000 sites
Proposed site for community importance
Special protection area

Fig. 3 Different protection regimes affecting study area

3.1 Indicators and Measures

Based on the data registered in official land registers and other registers and development of remote sensing measurement technologies, it is possible to determine various indicators and measures to be used in a multi-criteria analysis needed to make a decision based on size, coverage, comprehensiveness, costs, and benefits as well as other characteristics of selected land consolidation procedure.

The datasets generated by different sensors (LIDAR, multi/hyperspectral sensors) are vector or raster spatial datasets of different scales and extents. These datasets can cover large areas and they should be processed automatically in order to quantify their results.

With reference to the results of our previous studies [8–10], the cadastral parcel has been selected as the base unit, meaning all the other required data are associated with the cadastral parcel data. The used cadastral dataset included information on registered owner or co-owners, making it possible to perceive the parcel spatial distribution of agricultural holdings.

The land fragmentation is manifested as a large number of relatively small and spatially divided land parcels of each landowner. Fragmented agricultural land leads to increased costs of agricultural production: the utilized agricultural land area of an average Croatian agricultural holding (UUA) is 5,6 ha (European Union UUA is 16,1 ha) and each agricultural holding consists of an average of 15 cadastral parcels [10]. The average total of agricultural land area per landowner at the study area (Fig. 4) is 0,6 ha and more than one third of the landowners own less than 0,2 hectares. Though the distribution of a number of land parcels per landowner may indicate that there is a lot of landowners owning only one land parcel, further analysis [9] showed that this indicator cannot be used separately, without an indication of the size of land parcels. The average agricultural holding at the study area consists of 5,9 land parcels (Fig. 5).

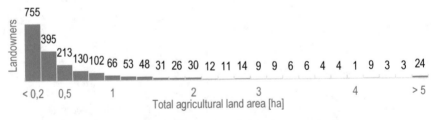

Fig. 4 Distribution of total agricultural land area per landowner

Additional indicators needed to assess the different land use scenarios include various anthropogenic, biophysical and structural factors together with the inclusion of legal regulations and physical planning documentation. These indicators can include already mentioned indicators of land and ownership fragmentation of agricultural holdings (determined from official land and other registry records) as well as pedological data, data on aspect, height above sea level, slope, land use, transportation, landscape and cultural character, protected sites, buildings, dry stone wall structures and so on. In

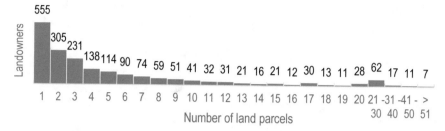

Fig. 5 Distribution of total number of land parcels per landowner

order to use these data in multi-criteria analyses, they need to be joined to each individual land parcel. Joining of the data may involve different methods of data classification (image or point cloud classification and vectorizing) and/or spatial analysis (spatial joins, intersections of buffer analyses).

4 Results and Discussion

Our previous research on land consolidation suitability assessment and agricultural land fragmentation analyses showed that out of all the data registered in the official land and other registers, the land use data records are most outdated records, meaning that the situation in the registers does not correspond to the actual situation in the field. Up-to-date values of this indicator can be determined using remote sensing data. Various data obtained from satellite mission's data are freely available and can be used to determine land use at small scale, but it is only relatively recently that this has been possible using UAV sensors at a large scale, gathering much more detailed land use data.

Land use can be determined using land cover classification based on the Normalized Difference Vegetation Index—NDVI. The NDVI is an indicator which is used to assess whether the target being observed contains live green vegetation or not. The NDVI can be determined based on (visible) red and near-infrared (NIR) spectrum channel of multi-spectral image, using a simple formula:

$$NDVI = \frac{NIR - Red}{NIR + Red} \tag{1}$$

The NDVI determined in Copernicus Sentinel-2A mission (Fig. 6) with a spatial resolution of 20 m can be used for larger scale planning, but the use of UAV sensors allows much smaller ground sample distance (GSD < 10 cm), which allows joining of land use data to each individual land parcel. For this purpose, fixed wing UAV—eBee SQ equipped with Parrot Sequoia multispectral sensor was used. The images were processed using automated "Structure-From-Motion" photogrammetric method, GSD = 8 cm.

The determination of the NDVI-based land use for each parcel involved image classification to identify different land uses, spatial intersection to assign the classification results raster data to land parcels and statistical processing to determine predominant land use.

In a similar way, it is possible to assign other land consolidation indicators, using different sensors and measurements methods. The additional data obtained in the same step by using eBee UAV were digital orthophoto map (Red, Green, Red Edge and Near-Infrared channels) and point cloud data.

Fig. 6 NDVI values, location Mirje-Stari Grad Plain: (left) Copernicus Sentinel-2A MSI, Level-1C, (centre) eBee Sequoia multispectral, (right) eBee Sequoia and cadastral parcels boundaries

5   Conclusion

The results of the paper support the idea of using automatically processed remote sensing datasets generated by different sensors (LIDAR, multi/hyperspectral sensors, UAV) to be used as land consolidation indicators. These datasets are covering relatively large areas and there is a need to process them automatically in order to quantify their results.

The chosen study area represents the case of overlapping of several different protection regimes (Protected coastal zone and Natura 2000 sites) and administrative divisions (physical planning divided into two local self-government units), creating problems in administration and management.

Multi-criteria analyses are commonly used as a stakeholder/decision-maker tool for the selection of optimal solution scenario, using possibly conflicting indicators and measures.

The main benefits of using the quantitative indicators and measures to make a decision based on size, coverage, comprehensiveness, costs, and benefits as well as other characteristics of applicable land consolidation procedure are the effectiveness and the increase of information-based decision making in rural development programme, trying to reconcile interests of farmers, non-farmer landowners, conservators and local government.

Although the paper did not cover the whole range of land consolidation procedures supporting rural development scenarios due to over-extensiveness of the procedure, these findings further emphasize the importance of accurate and up-to-date land administration records from the determined land use indicator, which is given as an example of quantitative measure.

References

1. Medić V. (1992) Devedeset godina komasacija u hrvatskoj. Sociologija sela 1/2:97–106.
2. Hartvigsen M (2014) Land reform and land fragmentation in Central and Eastern Europe: Land use policy 36:330–341. https://doi.org/10.1016/j.landusepol.2013.08.016.

3. Mađer M., Matijević H., Roić M. (2015) Analysis of possibilities for linking land registers and other official registers in the Republic of Croatia based on LADM. Land use policy 49:606–616. https://doi.org/10.1016/j.landusepol.2014.10.025.
4. Food and Agriculture Organization of the United Nations. (2008) Opportunities to mainstream land consolidation in rural development programmes of the European Union. FAO.
5. EC/COM (2010) The CAP towards 2020: Meeting the food, natural resources and territorial challenges of the future. 672:18.
6. Roić M., Vranić S., Kliment T., et al (2017) Development of Multipurpose Land Administration Warehouse. In: Proceedings of FIG Working Week 2017: "Surveying the world".
7. Tomić H., Mastelić Ivić S., Roić M. (2018) Land consolidation suitability ranking of cadastral municipalities: Information-based decision-making using multi-criteria analyses of official registers' data. ISPRS Int J Geo-Information 7:. https://doi.org/10.3390/ijgi7030087.
8. Tomić H., Roić M., Mastelić Ivić S., et al (2016) Use of Multi-Criteria Analysis for the Ranking of Land Consolidation Areas. In: Proceedings of Symposium on Land Consolidation and Land Readjustment for Sustainable Development.
9. Odak I. (2017) Agricultural Land Fragmentation Analysis of Land Holdings and its Application in the Rural Land Management, doctoral thesis. University of Zagreb Faculty of Geodesy.
10. Odak I., Tomić H., Mastelić Ivić S., Zagreb (2017) Vrednovanje fragmentacije poljoprivrednog zemljišta. 021:215–232.
11. UN FAO (2004) Operations manual for land consolidation pilot projects in Central and Eastern Europe. Organization 69.
12. Lisec A., Primožič T., Ferlan M. et al (2014) Land owners' perception of land consolidation and their satisfaction with the results - Slovenian experiences. Land use policy 38:550–563. https://doi.org/10.1016/j.landusepol.2014.01.003.
13. Sklenicka P. (2006) Applying evaluation criteria for the land consolidation effect to three contrasting study areas in the Czech Republic. Land use policy 23:502–510. https://doi.org/10.1016/j.landusepol.2005.03.001.
14. Marinković G., Ninkov T., Trifković M. et al (2016) On the land consolidation projects and cadastral municipalities ranking. Teh Vjesn - Tech Gaz 23:1147–1153. https://doi.org/10.17559/TV-20140316225250.
15. Pašakarnis G., Morley D., Maliene V. (2013) Rural development and challenges establishing sustainable land use in Eastern European countries. Land use policy 30:703–710. https://doi.org/10.1016/j.landusepol.2012.05.011.
16. Miranda D., Crecente R., Alvarez M.F. (2006) Land consolidation in inland rural Galicia, N.W. Spain, since 1950: An example of the formulation and use of questions, criteria and indicators for evaluation of rural development policies. Land use policy 23:511–520. https://doi.org/10.1016/j.landusepol.2005.05.003.
17. Gónzalez X.P., Marey M.F., Álvarez C.J. (2007) Evaluation of productive rural land patterns with joint regard to the size, shape and dispersion of plots. Agric Syst 92:52–62. https://doi.org/10.1016/j.agsy.2006.02.008.
18. Leń P., Mika M. (2016) Determination of the Urgency of Undertaking Land Consolidation Works in the Villages of the Sławno Municipality. J Ecol Eng 17:163–169. https://doi.org/10.12911/22998993/64827.
19. Guanghui J., Xinpan W., Wenju Y., Ruijuan Z. (2015) A new system will lead to an optimal path of land consolidation spatial management in China. Land use policy 42:27–37. https://doi.org/10.1016/j.landusepol.2014.07.005.
20. Janus J., Taszakowski J. (2015) The Idea of Ranking in Setting Priorities for Land Consolidation Works. Geomatics, Landmanagement Landsc 31–43.
21. Hiironen J., Riekkinen K. (2016) Agricultural impacts and profitability of land consolidations. Land use policy 55:309–317. https://doi.org/10.1016/j.landusepol.2016.04.018.

22. Terres J.M., Scacchiafichi LN, Wania A, et al (2015) Farmland abandonment in Europe: Identification of drivers and indicators, and development of a composite indicator of risk. Land use policy 49:20–34. https://doi.org/10.1016/j.landusepol.2015.06.009.

23. Farahani R.Z., SteadieSeifi M., Asgari N. (2010) Multiple criteria facility location problems: A survey. Appl Math Model 34:1689–1709. https://doi.org/10.1016/j.apm.2009.10.005.

24. Bournaris T., Moulogianni C., Manos B. (2014) A multicriteria model for the assessment of rural development plans in Greece. Land use policy 38:1–8. https://doi.org/10.1016/j.landusepol.2013.10.008.

25. Behzadian M., Khanmohammadi Otaghsara S., Yazdani M., Ignatius J. (2012) A state-of-the-art survey of TOPSIS applications. Expert Syst Appl 39:13051–13069. https://doi.org/10.1016/j.eswa.2012.05.056.

26. Cai F., Pu L., Zhu M. (2014) Assessment framework and decision-support system for consolidating urban-rural construction land in coastal China. Sustain 6:7689–7709. https://doi.org/10.3390/su6117689.

27. van Niekerk A., du Plessis D., Boonzaaier I. et al (2016) Development of a multi-criteria spatial planning support system for growth potential modelling in the Western Cape, South Africa. Land use policy 50:179–193. https://doi.org/10.1016/j.landusepol.2015.09.014.

28. Comino E., Bottero M., Pomarico S., Rosso M. (2016) The combined use of Spatial Multicriteria Evaluation and stakeholders analysis for supporting the ecological planning of a river basin. Land use policy 58:183–195. https://doi.org/10.1016/j.landusepol.2016.07.026.

29. Muchová Z. (2019) Assessment of land ownership fragmentation by multiple criteria. Surv Rev 51:265–272. https://doi.org/10.1080/00396265.2017.1415663.

30. Stari Grad Plain - UNESCO World Heritage Centre. https://whc.unesco.org/en/list/1240/. Accessed 20 Jan 2020.

31. Andlar G., Aničić B., Šteko V. (2014) Landscape plan as an instrument for revitalisation of the cultural landscape- case of the UNESCO's Stari Grad Plain landscape plan.

The Role of 3D Cadastre in the Preservation of Historical Cultural Heritage

Nikola Vučić[(✉)] [iD]

State Geodetic Administration, Zagreb, Croatia
nikola.vucic@dgu.hr

Abstract. In this paper is investigated the roles and possibilities of 3D cadastre in the preservation of cultural heritage. Two-dimensional preview of immovable cultural heritage is sometimes not sufficient to respond to all demands imposed by modern society. Thus, it is recommended to register such objects into 3D cadastre. Registering the complexity of property relations and the overlapping of property rights by layers (floors) that may be present in immovable cultural properties is not the primary objective of the cadastral service. However, the introduction of the 3D cadastre can contribute significantly to the quality management of historical cultural assets. The aim of the research is to provide examples of cultural assets and define possibilities for registering cultural assets in 3D cadastre. Legal and technical aspects of registering cultural assets are also defined in this paper. Investigation of current registration and future perspective for registration of cultural heritage in Croatia is based on three locations: Dubrovnik-town walls and fortifications, Split—Diocletian's Palace and medieval Split, and Stari Grad—cultural landscape of Stari Grad Plain. All three sites are under UNESCO protection.

Keywords: Cultural heritage · Historical cultural heritage · 3D cadastre · Land registry · Preservation

1 Introduction

The protection of cultural heritage is currently often discussed topic. United Nations Educational, Scientific and Cultural Organization (UNESCO) play an important role for the promotion of conventions and directions concerning the preservation of the archaeological heritage, the architectural monuments, the historic places etc. Experts in the field have recognized the importance of Information Technologies (IT) in cataloguing, archiving and conserving antiquities, monuments and generally cultural assets [1].

The protection of cultural heritage is a field of public law that applies to three dimensional space, as it is related to the protection of underground and maritime antiquities. It is strongly related to 3D cadastre, as stratification of real property or the application of 3D restrictions on real property facilitate archaeological research [2].

Registering the complexity of property relations and the overlapping of property rights by layers (floors) that may be present in immovable cultural properties is not

© The Editor(s) (if applicable) and The Author(s), under exclusive license
to Springer Nature Switzerland AG 2021
A. Kopáčik et al. (Eds.): *Contributions to International Conferences on Engineering Surveying*,
SPEES, pp. 309–316, 2021. https://doi.org/10.1007/978-3-030-51953-7_26

the primary objective of the cadastral service. However, the introduction of the 3D cadastre can make a significant contribution to the quality management of historical cultural assets. The aim of the research is to give examples of cultural assets and define possibilities for registering cultural assets in 3D cadastre. Legal and technical aspects of registering cultural assets is also define in this paper.

The first section of the paper is introduction, the second section deals with legal aspects of Croatian cultural heritage in cadastre. Current registration and future perspective for registration of cultural heritage in Croatia on three location in Croatia will provide in third section (Dubrovnik-town walls and fortifications, Split—Diocletian's Palace and medieval Split, and Stari Grad—cultural landscape of Stari Grad Plain). The paper then ends with a conclusion on the findings.

2   Cultural Heritage in Cadastre

In the Republic of Croatia there are two sets of records regarding real estate—technical cadastre (the Cadastre) and legal cadastre (the Land Registry). Real property in Croatian real property law is based on the *superficies solo cedit* principle, where a land surface parcel includes everything relatively permanently associated with the parcel on or below the land surface (primarily buildings, houses, etc.) Cadastre in the territory of the Republic of Croatia lacks a 3D data management related to the cultural property, thus in 2018 Law on the Protection of Cultural Property was changed. According to this law, the decision reached by the Ministry of culture for immovable cultural property defines the boundaries of the protected cultural property. Decision on cultural heritage has to be registered in the cadastre and the municipal court/land registry. In the case of an underwater or undersea archaeological site decision makes Ministry of culture in collaboration with harbour master's office.

Exceptionally, the boundaries of the cultural landscape can be defined descriptively (watercourses, roads, railways, etc.) when it is possible to include it the spatial plan. The boundaries of underwater archeological sites are marked with the geo-referenced points/coordinates, while the boundaries of under the sea archeological sites are georeferenced with points in the WGS 84 coordinate system, whereas the decision has to be submitted to the cadastre.

An archaeological building in interest of the Republic of Croatia located below the surface, forming one functional whole, related to the cadastral parcels above the surface protected by the Law on Protection and Preservation of Cultural Property, is considered as a public good and as inalienable property of the Republic of Croatia. When such building extends under one or more cadastral parcels, the registration of the public good will be conducted for the whole building. Per request of local and/or regional self-government, such buildings may be ceded to the management and use of local and/or regional self-government units if the building is located on their territory [3].

2.1   LADM

The LADM is an international standard developed by the Technical Committee ISO/TC 211 (geographic information and geomatics). The standard has the designation ISO 19152:2012 (Geographic information—Land Administration Domain Model) and has been available since 1 December 2012. To the experts from the field of land and land rights registration, the LADM gives a basis for the development of an effective land administration system, but it also enables expansions and adaptation to the local needs. The LADM is an object model described by classes. The LADM core consists of four basic classes that describe the relations between persons and the objects of registration. Those classes are: LA_Party (Person), LA_BAUnit (Basic administrative unit) and LA_RRR (Right, Restriction and Responsibility). The objects of registration are spatial units determined by their spatial position (LA_SpatialUnit).

The first version of the Croatian LADM profile was developed in 2012. (Figure 1) New classes, attributes, and types were added in the code list. For the attributes added in classes HR_SpatialUnit: HR_UsageTypeLand and HR_UsageTypeBuilding, the corresponding code list was created according to the Regulation on Land Cadastre [4] and according to the Regulation on the Content and Form of Real Property Cadastre Documentation [5]. The code list was also created for HR_OwnerType, HR_MonumentMaterial, HR_BoundaryType attributes, in accordance with the current Law on State Survey and Real Property Cadastre [6].

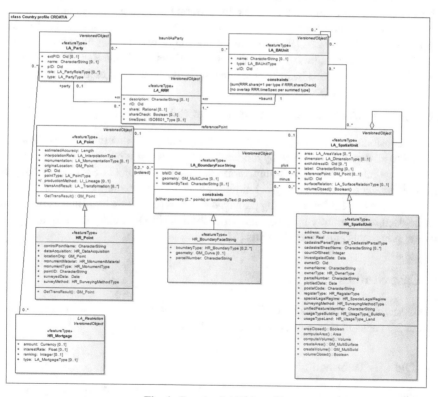

Fig. 1 Croatian LADM profile

Another important contribution to the development of the 3D Cadastre in the Republic of Croatia is the introduction of the unique identifier of special parts of real property, proposed to be implemented in the State Survey and Real Property Cadastre Act and the Land Registration Act [7].

The unique identifier of special parts of real property is a solution for all objects that are needed for the development of 3D Cadastre (buildings of various purposes, underpasses, overpasses, tunnels, bridges, viaducts, underground buildings, etc.).

The unique identifier could be used for denotation of separate parts of buildings, such as flat, apartment, business space, where each separate part gets a unique identifier in the Croatian land management system.

The unique identifier of a special part includes:

- identification number of the cadastral municipality
- number of land registry file
- number of land registry sub-file.

In [1] authors proposed Arheological cadastre based on LADM. Adjustment in the archaeological data introduced changes and external classes in the system in order to serve the construction of a registry for the cultural heritage. In some cases the classes keep the codification of LADM, nevertheless, the majority of classes have been changed so as to represent the entities of the Archaeological Cadastre. In the Republic of Croatia, LADM can also be a good basis for establishing a 3D cadastre of cultural heritage.

2.2 The Role of 3D Cadastre in Protection of Cultural Heritage

Two-dimensional preview of immovable cultural heritage is sometimes not sufficient to respond to all demands imposed by modern society. Thus, it is recommended to register such objects into 3D cadastre. Today, terrestrial laser scanning (TLS) is one of the best and most advanced measurement techniques for the 3D documentation of historical sites. Also the value of TLS for capturing damaged or decayed historic buildings is that scans can be used to accurately measure and virtually reposition partially or fully destroyed statues, ornaments and other elements [8].

3   Case Studies

In this section case studies from Republic of Croatia are provided as an example of current 2D cadastre with some improvements in 3D direction.

3.1   Old City Dubrovnik

The Old Town of Dubrovnik (Fig. 2) is a historical monument with significant fortification architecture (fortresses with city walls, towers and bastions built between 12th–17th century), gothic-renaissance churches, monasteries, palaces and fountains. The settlement was built on the rock at the beginning of the 7th century by the refugees from nearby Epidaurus.

Fig. 2 Old city Dubrovnik on web service www.katastar.hr

Over the years settlement expanded and by the 13th century evolved into the city with the strong defence system. During the next centuries (until 1814), due to the advanced maritime affairs, trading and skilful diplomacy, city managed to retain independency and political freedom paving the way to become one of the most powerful maritime force in the Mediterranean. In addition to material wealth, city manage to bring out the importance of science and art to enhance culture of life. The historic centre of Dubrovnik, with its surroundings belongs to the national and UNESCO cultural heritage, and presents a masterpiece of urban, fortification, cultural, historical and civilization values [9].

The 3D image with the link to the Google street view can be obtained through the service www.katastar.hr (Fig. 3).

Fig. 3 Old city Dubrovnik view from cadastral parcel number 4624 Dubrovnik (link to Google street view service from service www.katastar.hr)

3.2 The Basement Halls of Diocletian's Palace

Diocletian's Palace in Split is one of the best preserved architectural achievements of Late Antiquity. The emperor Diocletian gathered the finest architects of the time to undertake the construction of this palace, which took place during a tumultuous period (295–305

AD). The Palace represented a new type of fortified and luxurious imperial residence, whose northern quarters were designed to accommodate the military and servants, while the southern part served for the Emperor's residential and religious purposes. Subsequent centuries of architectural activities permanently altered the original structure of the Palace's upper floor, but the basement halls in the southern part of the Palace reveal the original floor plan and layout of the Emperor's residential quarters. Construction of the basement halls was determined by the geological substratum consisting of steep sea cliffs stretched from the north to the south [10].

The Law on Cultural Property changed during 2018 specifically due to the 3D open issue with the basement of the Diocletian's Palace (Fig. 4). Basements are located below the earth surface level (Fig. 5), while above surface numerous private houses are built, along with the public buildings, streets and squares.

Ministarstvo kulture, na temelju članka 19.a i 19.b Zakona o zaštiti i očuvanju kulturnih dobara („Narodne novine", broj 69/99, 151/03, 157/03, 87/09, 88/10, 61/11, 25/12, 136/12, 157/13, 152/14, 44/17 i 90/18) donosi

R J E Š E N J E

1.Utvrđuje se da su **Dioklecijanovi podrumi u Splitu** građevina arheološkog značenja od interesa za Republiku Hrvatsku i javno dobro u općoj uporabi u neotuđivom vlasništvu Republike Hrvatske, sukladno članku 19.a stavku 1. Zakona o zaštiti i očuvanju kulturnih dobara.

2. **Građevina iz točke 1. ovog rješenja obuhvaća** punu visinu substrukcija (podrumskih prostorija) Dioklecijanove palače, a nalazi se ispod slijedećih zemljišnoknjižnih čestica: zk. čest. zgr. 1654, zk. čest. zgr. 1669/2, zk. čest. zgr. 1671/1, zk. čest. zgr. 1671/2, zk. čest. zgr. 1675/1, zk. čest. zgr. 1675/2, zk. čest. zgr. 1675/3, zk. čest. zgr. 1676/1, zk. čest. zgr. 1676/2, zk. čest. zgr. 1676/3, zk. čest. zgr. 1677/2, zk. čest. zgr. 1677/3, zk. čest. zgr. 1677/1, zk. čest. zgr. 1678, zk. čest. zgr. 1681, zk. čest. zgr. 1683, zk. čest. zgr. 1684, zk. čest. zgr. 1687, zk. čest. zgr. 1689, zk. čest. zgr. 1690, zk. čest. zgr. 1691, zk. čest. zgr. 1692, zk. čest. zgr. 1693/1, zk. čest. zgr. 1693/2, zk. čest. zgr. 1696/1, zk. čest. zgr. 1697/1, zk. čest. zgr. 1697/2, zk. čest. zgr. 1697/3, zk. čest. zgr. 1698, zk. čest. zgr. 1699/2, zk. čest. zgr. 1701, zk. čest. zgr. 1702, zk. čest. zgr. 1704, zk. čest. zgr. 1705, zk. čest. zgr. 1706, zk. čest. zgr. 1707, zk. čest. zgr. 1717, zk. čest. zgr. 1735/1, zk. čest. zgr. 1735/2, zk. čest. zgr. 1735/3, zk. čest. zgr. 1766, zk. čest. zgr. 1776, zk. čest. zgr. 1777, zk. čest. zgr. 1778, zk. čest. zgr. 1779, zk. čest. zgr.

Fig. 4 Part of Decision of Ministry of culture for underground cultural heritage (basement halls of Diocletian's Palace). *Source* Ministry of culture of the Republic of Croatia [12]

Consequently, regulation, for the buildings of archeological significance of interest to the Republic of Croatia, was proposed. Namely, buiilidings located below the surface forms one functional part, which is not related to the cadastral parcels above it within the meaning of the Law on Protection and Preservation of Cultural Property [3], public property in general use in the inalienable property of the Republic of Croatia.

3.3 Stari Grad Plain (Island of Hvar)

Stari Grad Plain (Fig. 6) represents a comprehensive system of land use and agricultural colonisation by the Greeks, in the 4th century BC. Its land organisation system, based on geometrical parcels with dry stone wall boundaries (chora), is exemplary. This system was completed from the very first by a rainwater recovery system involving the use of tanks and gutters. The land parcel system set up by the Greek colonisers has been respected over later periods. Agricultural activity in the chora has been uninterrupted

Fig. 5 Part of the basement halls of Diocletian's Palace. (*Source* City of Split Museum) [7]

for 24 centuries up to the present day, and is mainly based on grapes and olives. The ensemble today constitutes the cultural landscape of a fertile cultivated plain whose territorial organisation is that of the Greek colonisation [11].

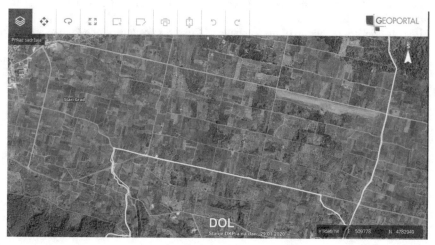

Fig. 6 Stari Grad plain on Geoportal of State Geodetic Administration. (*Source* dgu.geoportal.hr)

4 Conclusion

With a two-dimensional cadastre it is not possible to resolve the entry of cultural heritage below the surface. The good example of 2D insufficiency is underground cultural heritage buildings which could not be registered before law changes were introduced.

Hence, the 3D cadastre is essential for quality cadastral recording of cultural heritage. Using Terrestrial Laser Scanning (TLS) or Building Information Management (BIM) technology, current cadastral system can be enhanced to correspond to the requirements needed for the cultural heritage evidence.

References

1. Dimopoulou, E., Gogolou, C.: LADM as a Basis for the Hellenic Archaeological Cadastre, (2013) 5th Land Administration Domain Model Workshop, 24–25 September 2013, Kuala Lumpur, Malaysia.
2. Dimopoulou, E., Kitsakis, D.: Addressing Public Law Restrictions within a 3D Cadastral Context, (2017) ISPRS Int. J. Geo-Inf., (7), 182; https://doi.org/10.3390/ijgi6070182.
3. Law on Protection and Preservation of Cultural Heritage, Zakon o zaštiti i očuvanju kulturnih dobara, Official Gazzete No 90/2018,, (2018) https://narodne-novine.nn.hr/clanci/sluzbeni/2018_10_90_1756.html.
4. Regulation on Land Cadastre, Official Gazzete of the Republic of Croatia No. 4/2007 (2007).
5. Regulation on the Content and Form of Real Property Cadastre Documentation, Official Gazzete of the Republic of Croatia No. 142/2008 (2008).
6. Law on State Survey and Real Property Cadastre, Official Gazzete of the Republic of Croatia No. 112/2018, (2018).
7. Vučić, N.: Support the transition from 2D to 3D cadastre in the Republic of Croatia, (2015), University of Zagreb, Faculty of Geodesy, PhD Thesis.
8. Lepère, G, Lemmens, M.: (2019) Laser Scanning of Damaged Historical Icons, GIM International, The Netherlands.
9. URL 1: Ministry of culture of the Republic of Croatia, https://www.min-kulture.hr/default.aspx?id=7250, page access 23. January 2020.
10. URL 2: City of Split Museum, http://www.mgst.net/dioklecijanovi-podrumi/, page access 20. January 2020.
11. URL 3, UNESCO, https://whc.unesco.org/en/list/1240/, page access 15 January 2020.
12. URL 4, Ministry of culture of the Republic of Croatia, https://www.min-kulture.hr/userdocsimages/2005/rje%C5%A1enje%20Dioklecijanovi%20podrumi.pdf, page access 29 January 2020.

Author Index

A. Kopáčik et al. (Eds.): *Contributions to International Conferences on Engineering Surveying*,
SPEES, pp. 317–318. https://doi.org/10.1007/978-3-030-51953-7

Printed in the United States
by Baker & Taylor Publisher Services